T0139904

Advances in Intelligent Systems and Computing

Volume 861

Series editor

Janusz Kacprzyk, Polish Academy of Sciences, Warsaw, Poland
e-mail: kacprzyk@ibspan.waw.pl

The series "Advances in Intelligent Systems and Computing" contains publications on theory, applications, and design methods of Intelligent Systems and Intelligent Computing. Virtually all disciplines such as engineering, natural sciences, computer and information science, ICT, economics, business, e-commerce, environment, healthcare, life science are covered. The list of topics spans all the areas of modern intelligent systems and computing such as: computational intelligence, soft computing including neural networks, fuzzy systems, evolutionary computing and the fusion of these paradigms, social intelligence, ambient intelligence, computational neuroscience, artificial life, virtual worlds and society, cognitive science and systems, Perception and Vision, DNA and immune based systems, self-organizing and adaptive systems, e-Learning and teaching, human-centered and human-centric computing, recommender systems, intelligent control, robotics and mechatronics including human-machine teaming, knowledge-based paradigms, learning paradigms, machine ethics, intelligent data analysis, knowledge management, intelligent agents, intelligent decision making and support, intelligent network security, trust management, interactive entertainment, Web intelligence and multimedia.

The publications within "Advances in Intelligent Systems and Computing" are primarily proceedings of important conferences, symposia and congresses. They cover significant recent developments in the field, both of a foundational and applicable character. An important characteristic feature of the series is the short publication time and world-wide distribution. This permits a rapid and broad dissemination of research results.

More information about this series at http://www.springer.com/series/11156

António Pedro Costa · Luís Paulo Reis
António Moreira
Editors

Computer Supported Qualitative Research

New Trends on Qualitative Research

Springer

Editors
António Pedro Costa
Department of Education and Psychology, Didactics and Technology in Teacher Education (CIDTFF) Research Centre
University of Aveiro
Aveiro, Portugal

António Moreira
Department of Education and Psychology, Didactics and Technology in Teacher Education (CIDTFF) Research Centre
University of Aveiro
Aveiro, Portugal

Luís Paulo Reis
Informatics Engineering Department, Faculty of Engineering, LIACC Research Center
University of Porto
Porto, Portugal

ISSN 2194-5357 ISSN 2194-5365 (electronic)
Advances in Intelligent Systems and Computing
ISBN 978-3-030-01405-6 ISBN 978-3-030-01406-3 (eBook)
https://doi.org/10.1007/978-3-030-01406-3

Library of Congress Control Number: 2018956172

This Springer imprint is published by the registered company Springer Nature Switzerland AG
The registered company address is: Gewerbestrasse 11, 6330 Cham, Switzerland

Preface

This book contains a selection of the articles accepted for presentation and discussion at the 3rd World Conference on Qualitative Research 2018 (WCQR 2018), held in Lisbon, Portugal, October 17–19, 2018. WCQR 2018 was organized by Lisbon School of Nursing (ESEL) and Ludomedia. The conference organization also had the collaboration and/or sponsoring of several universities, research institutes and companies, including University of Aveiro/CIDTFF, University of Porto/LIACC, Tallinn University, National Centre for Research Methods (NCRM), MicroIO, webQDA, atlas.ti, DiscoverText and TLAB.

WCQR 2018 builds upon several successful events, including CIAIQ/ISQR 2017 held at Salamanca, Spain, and CIAIQ/ISQR 2016 held at Porto, Portugal. The conference focus was on Qualitative Research with emphasis on methodological aspects and their relationship with research questions, theories and results. This book is mainly focused on the use of Computer-Assisted Qualitative Data Analysis Software (CAQDAS) for assisting researchers in using correct methodological approaches for Qualitative Research projects.

WCQR 2018 featured four main application fields: Education, Health, Social Sciences, and Engineering and Technology. The conference included seven main subjects: rationale and paradigms of Qualitative Research (theoretical studies, critical reflection about epistemological, ontological and axiological dimensions); systematization of approaches with qualitative studies (literature review, integrating results, aggregation studies, meta-analysis, meta-analysis of qualitative meta-synthesis, meta-ethnography); qualitative and mixed methods research (emphasis on research processes that build on mixed methodologies but with priority to qualitative approaches); data analysis types (content analysis, discourse analysis, thematic analysis, narrative analysis, etc.); innovative processes of qualitative data analysis (design analysis, articulation and triangulation of different sources of data —images, audio, video); Qualitative Research in Web context (eResearch, virtual ethnography, interaction analysis, internet latent corpora, etc.); qualitative analysis with the support of specific software (usability studies, user experience, the impact of software on the quality of research and analysis).

In total, after a careful review process, with at least three independent reviews for each paper, a total of 20 high-quality papers from WCQR 2018 were selected for publication in this volume, with a number of authors totalling over 50, from 14 countries, including Brazil, Canada, Colombia, Costa Rica, Ecuador, Finland, France, Israel, Poland, Portugal, Russia, Spain, Sweden and USA. The volume also features six invited papers from distinguished researchers in the field of computer supported qualitative data analysis.

We would also like to take this opportunity to thank the WCQR 2018 organization members not mentioned so far (Conceição Ferreira, Cristina Baixinho, Eunice Sá, Helena Bronze, Helena Presado, João Santos, Manuela Paias, Mário Cardoso, Óscar Ferreira and Sónia Mendes) for their hard and exceptional work on the scientific management, local logistic arrangements, publicity, publication and financial issues. We also express our gratitude to all the members of the WCQR Program Committees and to the additional reviewers, as they were crucial for ensuring the high scientific quality of the event. We would also like to acknowledge all the authors and delegates whose research work and participation made this event a success. Last but definitely not least, we acknowledge and thank all Springer staff involved for their help during the production of this volume.

August 2018

António Pedro Costa
António Moreira
Luís Paulo Reis

Organization

Program Committee

Alberto Mesaque	Federal University of Minas Gerais, Brazil
Ana Amélia Carvalho	University of Coimbra, Portugal
Ana António	CeiED - Interdisciplinary Research Centre for Education and Development, Portugal
Ana Isabel Rodrigues	Polytechnic Institute of Beja, Portugal
Ana María Pinto Llorente	University of Salamanca, Spain
Ana Paula Sousa Santos	University of the Azores, Portugal
Ana Vanessa Antunes	Atlântica University, Portugal
Andrius Puksas	Mykolas Romeris University, Lithuania
António Carrizo Moreira	University of Aveiro, Portugal
António Dias de Figueiredo	University of Coimbra, Portugal
Beverly Palmer	California State University, Dominguez Hills, USA
Branislav Radeljic	University of East London, UK
Brigitte Smit	University of South Africa, South Africa
Bruno Braga	Federal Institute of Education, Science and Technology of Brasília, Brazil
Carl Vogel	Trinity College Dublin, Ireland
Carla Galego	CeiED - Interdisciplinary Research Centre for Education and Development, Portugal
Carolina Afonso	ISEG- Lisbon School of Economics & Management, Portugal
Catarina Brandão	Faculty of Psychology and Educational Sciences, University of Porto, Portugal
Cecília Guerra	University of Aveiro, Portugal
Celina Pinto Leão	University of Minho, Portugal
Daniela Schiek	Universität Bielefeld, Germany
Effie Kritikos	Northeastern Illinois University, USA

Elisabete Pinto da Costa	Lusófona University of Porto, Portugal
Elizabeth Pope	University of West Georgia, USA
Fátima Marques	ESEL - Lisbon School of Nursing, Portugal
Fernando Albuquerque Costa	University of Lisbon, Portugal
Francisco Carreiro da Costa	Lusófona University, Portugal
Gilberto Tadeu Reis da Silva	Federal University of Bahia, Brazil
Graeme Smith	Journal of Clinical Nursing, USA
Hélder José Alves Rocha Pereira	University of the Azores, Portugal
Helen Noble	School of Nursing and Midwifery, UK
Helena Presado	ESEL - Lisbon School of Nursing, Portugal
Helga Rafael	ESEL - Lisbon School of Nursing, Portugal
Henrika Jormfeldt	Halmstad University, Sweden
Hong-Chi Shiaù	Shih-Hsin University, Taiwan
Inês Amaral	Miguel Torga Institute of Higher Education, Portugal
Isabel Alarcão	University of Aveiro, Portugal
Isabel Cabrita	University of Aveiro, Portugal
Isabel Huet	University of West London, UK
Isabel Pinho	Aveiro University, Portugal
Jaime Ribeiro	Polytechnic of Leiria, Portugal
João Amado	University of Coimbra, Portugal
Jorge Remondes	ISVOUGA, Portugal
Kathleen Gilbert	Indiana University, USA
Katrin Niglas	Tallinn University, Estonia
Lakshmi B. Nair	Utrecht University, Netherlands
Lia Oliveira	University of Minho, Portugal
Lídia Oliveira	University of Aveiro, Portugal
Lubomir Popov	Bowling Green State University, USA
Lucila Castanheira Nascimento	University of São Paulo, Brazil
Marcel Ausloos	University of Liege, Belgium
Marco Lamas	Polytechnic Institute of Porto, Portugal
Marcos Teixeira de Abreu Soares Onofre	University of Lisbon, Portugal
Mari Carmen Portillo	University of Southampton, UK
Maria Filomena Gaspar	ESEL - Lisbon School of Nursing, Portugal
Maria José Brites	Lusófona University of Porto, Portugal
Maria Suzana Pacheco	University of Azores, Portugal
Martina Gallarza	University of Valencia, Spain
Miguel Serra	ESEL - Lisbon School of Nursing, Portugal
Mirliana Pereira	University of Chile, Chile
Miroslav Rajter	University of Zagreb, Croatia
Neringa Kalpokaite	Atlas.ti, Germany
Óscar Ferreira	ESEL - Lisbon School of Nursing, Portugal

Contents

Describing the Experience of Young Researchers in Interdisciplinary Qualitative Research Based on Critical Discourse Analysis (CDA) Using NVivo®

Carolina Lopes Araujo[1](✉), Eliane Almeida do Carmo[2], and Raiza Gomes Fraga[3]

[1] Faculdade UnB Planaltina, Universidade de Brasília, Brasília, Brazil
carolinalopesaraujo@yahoo.com.br
[2] Fundação Osvaldo Cruz (FIOCRUZ – Brasília), Brasília, Brazil
eliane.adm@gmail.com
[3] Centro de Desenvolvimento Sustentável, Universidade de Brasília, Brasília, Brazil
raiza.fraga@yahoo.com.br

Abstract. This paper describes a study pathway carried out by young researchers to develop interdisciplinary research on international agreements related to sustainable development. The study is based on critical discourse analysis (CDA) and uses NVivo®, a software program for qualitative analysis. The study pathway aimed to help overcome initial challenges related to the complexity of the theoretical-methodological approach in CDA and to users' lack of proficiency in operating the software. With the goal of aiding young researchers in developing qualitative research, this paper also presents a table that lists the most common sociolinguistic categories from CDA that can support analytical procedures.

Keywords: Critical discourse analysis (CDA) · NVivo® · Interdisciplinarity Undergraduate researchers · Sustainable development

1 Introduction

Working with qualitative research in the social sciences poses significant methodological challenges to researchers, especially to those with no experience with this kind of research. The use of software programs to assist in qualitative research may offer substantial advantages, as argued by Mozzato, Grzybovski, and Teixeira [1], but it may also increase challenges. This happens because such programs require specific expertise to operate their tools, and this kind of expertise can be achieved only through ongoing use.

Difficulties often arise in researchers' initial forays into qualitative interdisciplinary research. They are quite common in studies supported by critical discourse analysis (CDA) and pose an obstacle particularly for researchers with backgrounds in fields other than linguistics.

A. P. Costa et al. (Eds.): WCQR 2018, AISC 861, pp. 1–11, 2019.
https://doi.org/10.1007/978-3-030-01406-3_1

This paper describes a research project conducted by a multidisciplinary group of young researchers, based on the critical discourse analysis of the 2030 Agenda for Sustainable Development text [2]. The 2030 Agenda was adopted by the United Nations (UN) in 2015 and established the Sustainable Development Goals (SDG). The research project is ongoing and is expected to finish in August 2018. Thanks to the research team's multidisciplinary background, the researchers are able to gather contributions from different perspectives. Furthermore, that varied background facilitates the implementation of the transdisciplinary approach in studies such as this one that (1) employ the texts of UN environmental agreements as objects of study, (2) use CDA as their methodological approach, and (3) use the NVivo® software to assist in their analyses.

2 Critical Discourse Analysis as a Theoretical-Methodological Approach

Critical discourse analysis is both a theory and an analytical method that focuses on discourse representation and power relations among social actors [3]. An undergirding assumption of the CDA approach is that a text's aesthetics reveals the ethics and motivations of those who produced it [4]. Although textual analysis is the basic procedure in research endeavors supported by CDA, texts are not their object of study per se. The investigative focus is instead on the causal and ideological effects of texts, on the textual aspect of the social practices, processes, and relations and their effects on power struggles [5, p. 16, 4, p. 441].

Critical discourse analysis differs from simply interpreting a text in that it is based on sociological theory and its main concern is to establish connections between discourse elements and generative mechanisms that unlock social change [3, 6, p. 99]. According to Resende [7, p. 2012], interdisciplinarity is at the heart of critical discourse analysis. The author explains that analyzing the social issues reflected in discourse calls for the deployment of concepts and analytical categories from the social sciences, which cut across disciplinary boundaries. According to Resende and Acosta [8, p. 429], CDA studies require deep knowledge of the social framework to accomplish an effective investigation of social practices through the discourse apparatus. Another characteristic of CDA is that research projects may abandon scientific neutrality and examine social life in moral and political terms [5, p. 15, 9, p. 157]. With its origins in sociolinguistics and systemic functional linguistics approaches, critical discourse analysis proposes discursive analyses based on texts, rather than textual analysis. To that end, CDA requires a profound and detailed investigation of the historical social context, in an interdisciplinary and critic approach that combines linguistics with other social sciences [9, p. 155].

CDA methodological procedures consist of the systematic application of sociolinguistic categories, which reveals meaning-making processes linked to different social practices and structured according to private interests [10]. The analytical categories for social discursive analysis help map the connections between the discursive and the nondiscursive [8, p. 248]. According to Resende and Acosta, however, the analytical categories to be used in a research project should not be defined beforehand, for it is necessary to access data, which will be provided by the texts, in order to identify the most productive analytical categories for the research [8, p. 432]. In this sense, Fairclough [5, p. 16] says that "textual description and analysis should not be seen as prior to and independent from a social analysis and critique—it should be seen as an open process which can be enhanced through dialogue across disciplines and theories".

The complexity of CDA analyses, the interdisciplinarity of research, and the inability to define analytical categories in advance are factors that can act as stumbling blocks to inexperienced researchers working with this qualitative approach. Nonetheless, because the methodological-theoretical apparatus for qualitative interdisciplinary studies is a robust one, we proposed (and experimented with) a research guide laying out procedures to help inexperienced researchers overcome the initial difficulties presented by this kind of research. We will describe this investigative experience in the next sections.

3 A Proposal for an Interdisciplinary Study Based on CDA: The Description of an Experience

This paper draws on the collaborative experience of advising a multidisciplinary group of young undergraduate researchers in analyzing international sustainable development agreements adopted by the UN. The group was inspired by a research project that had analyzed the document "The Future We Want" [11] to assess how social demands have been captured in the outcomes of Rio+20 [12].

Adopting CDA as a theoretical and methodological basis was a bold option that demanded extra effort from the young researchers to get acquainted with such a complex and unusual approach. The variety of discourses present in the debate on sustainability required a solid critical-theoretical background that would allow the researchers to understand the social-historical context of the term "sustainable development",[1] identify the interests involved in the world's developmental geopolitics, and then analyze the hegemonic struggles in the specific topic chosen by each researcher.

The starting point for choosing the subject for each individual project was Araújo's [12, p. 73] mapping of 17 topics of interest raised by the Major Groups (representatives of social groups identified in the Agenda 21) in their speeches at the Rio+20 High Level Panel. Based on that list, the researchers selected the topics for their individual research projects, as summarized in Table 1.

After the topics had been chosen, the young researchers were advised to study the literature on sustainable development and on the decolonization approach, which helped them develop a critical framework. In addition, they were encouraged to look for references that dealt with their specific research issues.

[1] Some authors, however, do not consider "sustainable development" a concept per se. Veiga [23, p. 4] judges "sustainable development" much too imprecise a term to be considered a concept. Furthermore, Nascimento [24] says that many players use the term "sustainability" in their discourses with different senses, according to their interests and objectives: "Sustainability is no longer a concept, notion, or value. It has become an arena for struggles" [24, p. 416].

Table 1. Distribution of topics on the 2030 agenda among young researchers

Researcher	Main research topic	Major (first-cycle program)
Student 1	Ecological aspects of sustainable development	Environmental management
Student 2	Gender issues	International relations
Student 3	Children and youth	International relations
Student 4	Education	Environmental management
Student 5	Peace and conflict resolution	Environmental management

Information provided by the authors.

3.1 Getting Acquainted with CDA

While studying the theoretical background, the researchers took part in activities that introduced them to CDA's sociolinguistic categories and its ontological basis. In advising sessions, the research team discussed results from other CDA studies. The first exercises to introduce the analytical categories involved discussing the practice of analysis presented in Resende and Ramalho [13, pp. 91–144] and studying Chap. 4 from Ramalho and Resende [14, pp. 111–156]. After completing these tasks, the researchers were invited to choose one of the five analytical chapters from Araujo's thesis [12] and explain it to their peers.

To facilitate the researchers' comprehension of the analytical categories and assist them in their analytical efforts, we developed a summary table of the most common sociolinguistic categories. We are aware that such a summarization cannot be exhaustive, but it serves as a supplementary tool for textual analysis based on sociolinguistic categories. Table 2 presents the summary table. The first three columns list the analytical categories that the group had studied previously. The last two columns aim to help the researcher identify passages in the text that are relevant to their analyses.

Table 2. Summary of the main sociolinguistic categories in CDA

Social functions of language	Analytical categories	Analytical subcategories	What to analyze?	How to analyze it?
Pre-genre				
Ways of acting or (inter)acting in discourse (Actional meaning)	Genre structure (Genre)	Rhetorical movements	Things that the text does (contextualizing the topic, introducing the participants. ..)	Verify how the text is structured in terms of the actions that it takes
		Language	Language styles that "situate" a text in a context	Assess characteristic elements of a given document (manner of address; type of sentence construction, etc.)
		Form and formatting	The wording style that characterizes the text	Examine the line and paragraph divisions, section divisions, pre-textual elements, post-textual elements, and graphic elements that accompany the text
	Intertextuality	Articulation of different voices	Direct and indirect "references" that are present in the text and the relationship between voices—distance/difference from what has been said	Identify the use of quotation marks, *verba dicendi*

(*continued*)

Table 2. (*continued*)

Social functions of language	Analytical categories	Analytical subcategories	What to analyze?	How to analyze it?
	Functions of speech	Demanding, offering, asking, affirming	The text's objectives (generic speech functions may turn into more specific types)	Map verbs and the actions that they connote
	Textual cohesion		The logical nexus established between clauses (and between paragraphs, if possible): causality, conditionality, correction* or contradiction, emphasis or mitigation, distancing...)	Map connectors (conjunctions, etc.)
Ways of representing in discourse (Representational meaning)	Interdiscursivity (identifying the discourses and their modes of articulation)	Enumeration	The order of the enumerations to identify the priority level and/or the distancing of some terms	Observe the order of terms' appearance and the distance from the central idea of the sentence
		Repetition (and synonyms or near-synonyms)	Terms that have similar meaning (or refer to terms present in the text) that appear more than once in a text	Observe repeated words, synonyms, near-synonyms, and derived words
		Activated semantic field	The associations between a word or expression and the meaning fields	Examine the most frequent or most important words in the selected excerpts
	The representation of social actors	Mode of reference	The words employed to represent the social actors referred to in the text and their connotations	Map nouns used to refer to the social actors
		Mode of representation	Strategies for representing actors (individual/group; opposition between us and them; personalization, assimilation, functionalization, aggregation, etc.)	Map nouns used to represent the social actors
	The words' meanings	Textual patterns of co-occurrence or collocation (analysis of the context)	How ideas (represented in words that precede or succeed the key term) are associated with the term or issue being analyzed	Examine the text adjacent to the keywords. It is important to identify the order of the enumeration and repetitions of association/proximity of terms
		Lexical choices (semantic field)	The ways a given topic or actor is represented	Examine the most frequent words
	Processes of transitivity	Relational, verbal, mental, behavioral, material, and existential	The type of experience or event represented in the text and the connection between "the one who does something, to whom, and in what circumstances"	Examine verbal elements, their subjects and predicates
Ways of being in discourse (Identificational meaning)	Metaphors (make it possible to know something in terms of something else; highlight some aspects and mask others)	Conceptual metaphors	Whether concepts are structured in terms of others	Identify relational verbs or words that cause strangeness when immediately associated. For example: Time is money. You are wasting my time
		Orientational metaphors	Spatial-orientation representations and how they reflect physical and cultural experiences	Identify words with spatial references (up, down, etc.) For example: I am feeling down
		Ontological metaphors	Strategies for the "materialization" of experiences and abstract phenomena	For example: Nipping evil in the bud. (Evil is abstract, so it does not have a bud.)

(*continued*)

Table 2. (*continued*)

Social functions of language	Analytical categories	Analytical subcategories	What to analyze?	How to analyze it?
	Evaluation (reveals judgments present in the text)	Evaluation affirmations	What elements the text represents as being positive or negative; necessary or disposable	Examine adjectives, adverbs, and exclamation point usage
		Connotation of preference	The subjective marks that express affinity or approval (or the lack thereof)	Identify mental verbs related to affection (to like, to admire, to love, to detest, etc.)
		Evaluative assumptions (linked to value presuppositions)	Passages where values are inserted via implicit content, which is revealed by tacit significations bearing value judgment	Identify implicit, assumed, and unexpressed judgment manifested in the text
	Modality (relativizes the representation of discourses via emphasis or attenuation)	Epistemic modality (it reveals truths; linked to the existential assumptions)	The level of confidence expressed concerning the genuineness of an affirmation	Examine verbs (may [be]), modal adjuncts (certainly, possibly, rarely), adverbial groups (without doubt, with frequency), and expressions that denote confidence (it is certain, it is possible)
		Deontic modalities (they reveal obligation) (linked to the propositional assumptions)	The level of obligation or permission expressed in the text	Examine verbs (necessity) [may/should], modal adjuncts (necessarily, mandatorily, indispensably) and expressions that denote obligation (it is necessary; it is urgent)
	Presuppositions (implied text related propositions, related to shared meanings, which ward off questioning)	Existential assumptions (they assume something is true)	Elements in the text that represent something as undoubtedly true	Analyze affirmative sentences, adjective clauses
		Propositional assumptions (they assume how things are or how they may be)	Elements that represent how a phenomenon is presented or realized, assuming that as the best or even only way to understand reality	Analyze affirmative sentences, adjective clauses
		Value presuppositions (they differentiate good from bad)	Elements that denote judgment (positive or negative)	Analyze affirmative sentences, adjective clauses

Source: Produced by the authors based on Resende and Ramalho, 2009 [13]; Ramalho and Resende, 2011 [14]; Halliday and Matthiessen, 2004 [15]; Van Leeuwen, 2008 [16].

Once they'd become acquainted with the theoretical framework for their research topics, with CDA's ontological concepts, and with the analytical categories, the researchers started working with the 2030 Agenda with the help of the NVivo® software program.

3.2 Using NVivo®

The text "2030 Agenda for Sustainable Development" is a lengthy one, containing 91 paragraphs over 41 pages. According to Fairclough [5, p. 6], textually oriented discourse analysis "is rather 'labour-intensive' and can be productively applied to samples of research material rather than large bodies of text". NVivo® Pro edition 11 coding tools contributed to the analysis procedures on the 2030 Agenda text.

The basic functionality of this type of program is coding data (in our case, selecting excerpts from the 2030 Agenda) into categories, which the program refers to as "nodes". The nodes work as receptacles for the storage of coded information [1, p. 583]. NVivo® allows for the creation of nodes in accordance with the selected theoretical approach and research questions [17, p. 373]. The nodes can be further divided into subcategories, called "sub-nodes", which can organize a node's hierarchical structure in an NVivo® file (dubbed a "project"). The nodes may be grouped into specific folders. In our NVivo project, we created two coding folders: a subject category folder, gathering together the nodes related to the specific issues of each researcher's investigation (see Table 1), and another folder gathering together the analytical category nodes, in which we replicated the hierarchical organization of the sociolinguistic categories presented in Table 2. We were aware, as Resende and Acosta [8, p. 432] had already warned us, that not all of these sociolinguistic categories would end up being relevant to our analyses. Furthermore, we knew that it might be necessary to create new categories or subcategories, which could be done easily in NVivo®. Regardless, the hierarchical structure of sociolinguistic categories, presented in Table 2, would indeed help young researchers identify into which analytical categories they should code excerpts from the corpus.

This analysis corpus was composed of a single "source" (as research materials are known in NVivo®): the original 2030 Agenda text, imported to the project in PDF format. Other accessory sources—the Portuguese version of 2030 Agenda and the Rio +20 outcome document in its English and Portuguese versions—were also imported to the project to help with the analyses. By creating two folders, we were able to differentiate the functions of these sources in the project. Those folders were labeled "Auxiliary sources" and "Analysis corpus".

The researchers' first analytical task was thematic coding through a systematic and detailed reading of the 2030 Agenda text. In this process, the researchers identified all the excerpts related to their respective research issues. Each set of these excerpts made up the specific corpus of analysis for an individual project. The minimum unit of coding was supposed to be one paragraph. As a paragraph may refer to more than one topic, paragraphs were categorized into as many thematic nodes as seemed fit to the researchers. The NVivo® tool "Show coding stripes" helped to visualize all the coding attributed to a paragraph.

It is important to mention that all the researchers operationalized their coding using the same NVivo® file, which was shared among the users through an online storage tool (Dropbox®). When accessing the file, individual users had to identify themselves by their name and initials. NVivo® can require the users' identification if it is set to do so in the program options. This functionality was of great importance since it made it possible for more than one user to code the same excerpt in different nodes. Through the user identification, the coding could be filtered by user.

For the coding of the excerpts into the analytical categories' nodes, there was no minimum unit of coding, so it could be a word or a sentence, or even a paragraph. Researchers worked on their particular selections of paragraphs, analyzing only those passages that they had, in the previous stage, coded into the thematic nodes related to their specific topics. We started by coding the text into analytical nodes based on the sociolinguistic categories that seemed most evident in the text, such as intertextuality, evaluation, and modality. Using NVivo®'s query tools, specifically "Word frequency", the researchers created word clouds that displayed the most frequently occurring words in their paragraph selections. These word clouds were useful for showing the most common lexical choices in the representation of each topic. The list of the most frequent terms was also useful for revealing the activated semantic field in the representation of a specific topic. From this list of words, the researchers associated the terms with the three pillars of sustainable development (social, economic, and environmental). This procedure helped reveal the interdiscursivity of the 2030 Agenda text.

Textual patterns of cooccurrence or collocation were presented in "word trees". Word trees include the search results from NVivo®'s text search query tool and help reveal the association of key terms with other terms in the text.

Aiming to avoid repeated work or loss of coding, the researchers agreed that a codification could be undone only by the user who had coded it—which was possible to know by using the users' identification tool. Anyone who detected coding mistakes or divergences would add a "memo" note to the project and the corrective measure would be discussed in the next advising meeting. This procedure proved productive in terms of learning, since the researchers were able to discuss and revise their coding decisions together. Additionally, having the coding be viewed by more than one researcher minimized inaccuracies resulting from lack of attention or misinterpretation.

As the researchers progressed in their analyses, they became more confident in their comprehension of the categories and more proficient in their use of the program; this allowed them to go further in their queries and propose more profound analyses. For example, one researcher suggested using a matrix coding query to reveal the juxtapositions of topics represented in the text.

4 Preliminary Results

The initial results of the research group were made public in three presentations delivered at the XII Colóquio Internacional da Rede Latino-americana de Análise do Discurso da Pobreza Extrema (REDLAD) (XII REDLAD), which took place in Santiago, Chile, in October 2017. The subject matter of the event motivated us to map the representation of the issue of poverty in the text of the 2030 Agenda. This analysis resulted in the paper "Para além da Economia: representação discursiva da pobreza na Agenda 2030"[2], which was presented by Carolina Lopes Araújo at the event. Using NVivo®'s matrix coding query tool, we found thematic juxtapositions, which were analyzed in two papers also presented at XII REDLAD. "Meio Ambiente para Quem? Representação discursiva da questão ambiental e da pobreza na Agenda 2030"[3] was presented by Matheus Batista da Silva, who discussed the discursive representation of poverty and economic aspects of sustainable development; Yara Resende Marangoni Martinelli presented "O Gênero dos Objetivos do Desenvolvimento Sustentável:

[2] "Beyond Economics: Discursive Representation of Poverty in the 2030 Agenda" – author's translation.

[3] "Environment for Whom?: Discursive Representation of the Issues of the Environment and Poverty in the 2030 Agenda" – author's translation.

representação discursiva da questão de gênero na Agenda 2030”[4]. These three papers were presented on a panel at XII REDLAD, which included a discussion with the audience of the interdisciplinary (CDA-based) research procedures employed by the young researchers.

This collaborative research project also resulted in a scientific paper titled “From ‘The Future We Want’ to the ‘Sustainable Development Goals’: Discursive Changes Concerning the Environmental Sustainability of Development” [18], presented at the XX Encontro da Rede de Estudos Ambientais de Países de Língua Portuguesa (REALP), in Aveiro, Portugal, in May 2018. In addition, the original Portuguese text of the present article, “Percurso de jovens pesquisadores em investigação qualitativa interdisciplinar embasada na Análise de Discurso Crítica (ACD) com o auxílio do software NVivo®”, was presented at the VII Congresso Ibero-Americano em Investigação Qualitativa (CIAIQ 2018) in Fortaleza, Brazil, in July 2018 and published in the conference proceedings [19].

The research results will be presented at the 24th Congress of Scientific Initiation at the Universidade de Brasília in September 2018, when the young researchers will deliver their final research reports.

It is worth mentioning that the development of the research approach described here began in 2016, during an experience of introducing CDA to researchers from other scientific fields who analyzed the text “The Future We Want”, the outcome document produced by Rio+20. That investigative effort resulted in three papers that have already been published: Carmo and Araujo [20]; Silva and Araujo [21]; and Fraga and Araujo [22].

5 Final Considerations

This experience of guiding an interdisciplinary group of young researchers in conducting a discursive analysis of the text of the 2030 Agenda has shown positive learning results and even led to scientific publications. The investigation experience described here aims to help researchers overcome the primary difficulties associated with the complexity of the theoretical-methodological CDA approach and the operationalization of NVivo®’s tools. By describing our research experience here, we hope to encourage initiatives to introduce young researchers to CDA and increase the use of the NVivo® program, contributing to the development of qualitative research. We also hope that the summary of sociolinguistic categories presented in Table 2 may assist inexperienced researchers in developing analytical procedures for CDA-based studies.

We believe that critical socio-discursive studies on international agreements related to sustainable development can contribute to civil society and academia in overseeing and taking part in measures to increase sustainability while transforming the current model of development.

[4] “Engendering the Goals of Sustainable Development: Discursive Representation of the Issue of Gender and Poverty in the 2030 Agenda” – author’s translation.

References

1. Mozzato, A.R., Grzybovski, D., Teixeira, A.N.: Análises qualitativas nos estudos organizacionais: as vantagens no uso do software Nvivo. Rev. Alcance **24**(4), 578–587 (2016)
2. United Nations General Assembly: Transforming our world: The 2030 agenda for sustainable development. In: Resolution adopted by the General Assembly on 25 September 2015. United Nations, New York (2015)
3. Fairclough, N.: Análise Crítica do Discurso como método de pesquisa social científica. Linhas D'Água **2**(25), 307–329 (2012)
4. Resende, V.d.M.: Representação discursiva de pessoas em situação de rua no 'Caderno Brasília': naturalização e expurgo do outro. Linguagem em (Dis)curso **2**(12), 439–465 (2012)
5. Fairclough, N.: Analyzing Discourse: Textual Analysis for Social Research. Routledge, London (2003)
6. Resende, V.d.M.: Análise de Discurso Crítica e Etnografia: o Movimento Nacional de Meninos e Meninas de Rua, sua crise e o protagonismo juvenil. Thesis, Universidade de Brasília, Brasília (2008)
7. Resende, V.d.M.: Abordagem teórico-metodológica para análise interdiscursiva de políticas públicas. In: Atas CIAIQ 2017, pp. 2012–2020. Salamanca (Spain) (2017)
8. Resende, V.d.M., Acosta, M.d.P.T.: Apropriação da análise de discurso crítica em uma discussão sobre comunicação social. Revista de Estudos da Linguagem **26**(1), 421–454 (2018). https://doi.org/10.17851/2237-2083.26.1.421-454
9. Resende, V.d.M., Marchese, M.C.: 'São as pessoas pobrezitas de espírito que agudizam a pobreza dos pobres': análise discursiva crítica de testemunho publicado na Revista Cais—o método sincônico-diacrônico. Cadernos de Linguagem e Sociedade **12**(2), 150–178 (2011)
10. Fairclough, N.: Critical Discourse Analysis: The Critical Study of Language. Longman, Harlow (2010)
11. United Nations General Assembly: The future we want. In: Outcome of the United Nations Conference on Sustainable Development (Rio+20), 20–22 June 2012. Rio de Janeiro (2012)
12. Araujo, C.L.: As vozes da Rio+20: inserção dos interesses dos grupos sociais nos resultados da Conferência das Nações Unidas para o desenvolvimento sustentável. Thesis, Universidade de Brasília, Brasília (2014)
13. Resende, V.d.M., Ramalho, V.: Análise do Discurso Crítica, 2nd edn. Contexto, São Paulo (2009)
14. Ramalho, V., Resende, V.: Análise de discurso (para a) crítica: o texto como material de pesquisa. Pontes Editores, Campinas (Brazil) (2011)
15. Halliday, M.A., Matthiessen, C.: An Introduction to Functional Grammar, 3rd edn. Edward Arnold, London (2004)
16. Van Leeuwen, T.: The representing of social actors. In: Discourse and Practice: New Tools for Critical Discourse Analysis, pp. 23–54. Oxford University Press, New York (2008)
17. Botelho, E.d.A., Freitag, M.S.B., Borges, C., Teixeira, R.M.: Relato de uma Experiência de Utilização do NVivo® em Pesquisa sobre Desaprendizagem Organizacional. In: Atas CIAIQ 2017, pp. 371–380. Salamanca (Spain) (2017)
18. Araújo, C.L., Silva, M.B.d., Fraga, R.G.: O que avançou do 'Futuro que Queremos' aos 'Objetivos do Desenvolvimento Sustentável' concernente à sustentabilidade ambiental do desenvolvimento? In: Anais da Conferência Internacional de Ambiente em Língua Portuguesa/XX Encontro da Rede de Estudos Ambientais de Países de Língua Portuguesa/XI Conferência Nacional do Ambiente, vol. 3, pp. 176–196. Aveiro (Portugal) (2018)

19. Araújo, C.L, Carmo E.A.d., Fraga, G.R.: Percurso de jovens pesquisadores em investigação qualitativa interdisciplinar embasada na Análise de Discurso Crítica (ACD) com o auxílio do software NVivo®. In: Atas do VII Congresso Ibero-Americano em Investigação Qualitativa (CIAIQ 2018), pp. 154–163. Fortaleza (Brazil) (2018)
20. Carmo E.A.d., Araujo, C.L.: Pobreza, equidade entre gêneros e o futuro do planeta. In: Discurso e pobreza em aproximações diversas: classe, raça, gênero, geração e território, pp. 45–68. Pontes Editores, Brasília (2018)
21. Silva, V.X.S., Araujo, C.L. Quem somos nós em 'O Futuro que Queremos'?: análise do discurso sobre acesso a terra e meios produtivos no documento final da Rio+20. In: Anais International Congress of Critical Applied Linguistics (ICCAL), pp. 55–56. Londrina (Brazil) (2015)
22. Fraga, G.R., Araujo, C.L.: As vozes dos agricultores nos resultados da Rio+20: retórica ou realidade? Acervo On-line de Mídia Regional **11**, 49–67 (2016)
23. Veiga, J.E.d.: O principal desafio do Século XXI. Ciência e Cultura **57**(2), 4–5 (2005)
24. Nascimento, E.P.: Sustentabilidade: o campo de disputa de nosso futuro civilizacional. In: Enfrentando os limites do crescimento: sustentabilidade, prosperidade e decrescimento, pp. 415–433. Garamond, Rio de Janeiro (2012)

Contributions and Limits to the Use of Softwares to Support Content Analysis

Clara Suzana Cardoso Braga[1], Diego de Queiroz Machado[1](✉), Márcia Zabdiele Moreira[1], Rafael Fernandes de Mesquita[2], and Fátima Regina Ney Matos[3]

[1] Programa de Pós-Graduação em Administração e Controladoria, Universidade Federal do Ceará, Fortaleza, Brazil
clarabraga930@hotmail.com, {diegomachado,marciazabdiele}@ufc.br

[2] Universidade Potiguar and Instituto Federal do Piauí, Teresina, Brazil
rafael.fernandes@ifpi.edu.br

[3] Universidade Potiguar and Instituto Superior Miguel Torga, Coimbra, Portugal
fneymatos@globo.com

Abstract. The use of software to support qualitative analysis has been growing since the 1980s when the first software began to be developed through collaborative research by researchers. There are still recurrent discussions about the contributions and limits of the use of these resources in qualitative analyzes. Considering this perspective, this work aims to analyze the contributions and limitations of the use of content analysis software, in order to support the researcher in his decision making when opting for the use of software resources. Several contributions are identified, such as the possibility of treatment and retrieval of a large volume of qualitative data, allowing a greater time of interpretative analysis to the researcher. As a more frequent limitation, we present the absence of contextual analysis of the units of study: words. Thus, the active role of the researcher remains indispensable in view of the fact that the software proposes a support for the mechanical activity of content analysis, but the creation of valid scientific knowledge requires the rationality of the researcher from the interpretation of the data against pertinent theorization.

Keywords: Content analysis · Lexical analysis · Support software

1 Introduction

The qualitative data are extremely valid for scientific studies, defined by the authors as data that are not objectively measurable and which meaning can only gain signification through the interpretative perspective of the researcher [1]. In this sense, due to the increase in the opportunity of text data made available via the web, a renewed interest in content analysis and its techniques, in particular, computer aid [2].

Content analysis is a technique that aims at the objective, systematic and quantitative description of the manifest content of the communication [3]. According to Justo and Camargo [1]: “The classic technique for analyzing textual materials dates back to

A. P. Costa et al. (Eds.): WCQR 2018, AISC 861, pp. 12–21, 2019.
https://doi.org/10.1007/978-3-030-01406-3_2

the beginning of the 20th century and was marked by analyzes of journalistic materials and later transposed to the most diverse areas of knowledge" (p. 3).

Regarding the software to support analysis in qualitative research, its use is been increased [1, 4–8]. The acronym CAQDAS (Computer Aided Qualitative Data Analysis Software) designates these software designed to aid in the analysis of qualitative data. However, its use by researchers has been controversial given the positioning of researchers who highlight the benefits of efficacy and ease of use and by others that question their effective benefits [5]. In fact, the use of computer and statistical resources in the analysis of texts is not intended to give qualitative research a purely quantitative approach, through lexical analysis, but aims at enhancing its reach [1].

Given this context of discussion, it is defined as the objective of this article to analyze the contributions and limits of the use of content analysis software. Thus, it is expected that this essay can contribute as a support to the researcher in their decision-making process about whether or not to use these resources.

2 Content Analysis

Content analysis is a hybrid technique that bridges the statistical formalism and the qualitative analysis of the corpus data of a text under analysis [2]. It is a way to systematically convert text into numerical variables for quantitative data analysis [9, 10].

Content analysis aims to decompose a text into lexical or thematic units, coded as categories and, from this, to establish generalizing inferences [11]. Thus, content analysis allows reducing the complexity of a set of texts, making it possible to produce inferences for their social context in an objective way [2]. It is an analysis, which starts from the generation of quantifiable results, established by statistical frequency of the units of meaning, to guarantee an objective impartiality [11]. It begins, therefore, from the principle that the message or the textual corpus can be understood from the decomposition of its content into fragments, such as words, terms or phrases, that allow to reveal the subtleties contained in the text.

The procedures of content analysis, reconstruct representations in two main dimensions: the syntactic, which focuses on the signals and their interrelations; and semantics, which analyzes what is said in the text, about values, intentions, connotative and denotative senses, and so on [2]. The content analysis can be organized in three approaches: lexical analysis (nature and wealth of vocabulary); syntactic (times and verbal modes); and thematic (themes and frequency) [12]. Such approaches can be used together or separately, depending on the purpose of the research.

Content analysis assumes that the vocabulary used in a text and the frequency with which words are used may reveal the sender's conceptions of their values, choices, and references [11]. As for semantics, the reading of the context, of what can be marked beyond the manifesto in words, can also be analyzed through omissions, ignorance, word preferences, ambiguous terms, among others.

The classic materials of content analysis are written texts, which were previously used for another purpose. However, the procedure can also be applied to images, sounds and transcriptions [2, 11].

To perform a content analysis, representation, sample size, and unit division depend on the research problem and are conducted in an interactive manner from theorization to sampling [2]. The definition of the categories is fundamental to reach the goals established in the research, which should be pertinent to these objectives, allowing to generate meaning from the analysis of the vocabulary units [11]. The coding and classification of the materials collected in the sample are research tasks supported by the theory and it is from this codification that the content analysis allows interpreting the text [2].

It is worth noting what Justo and Camargo [1] alert about content analysis, regarding: "(…) the need for training of the researcher, since he is the one who interprets the data from a reading grid that he needs of theoretical and methodological rigor when passing through the sieve of the classification and interpretation of the data " (p. 3).

Good practice in content analysis is based on reliability, validity, consistency and transparency in the use of the method [2]. The authors present as main forces of content analysis: method of systematic and public data analysis; make use of pure naturally occurring data; the ability to handle large amounts of data; and the fact that the technique offers a set of mature and well documented procedures.

Collis and Hussey add as advantages of using content analysis: relatively inexpensive method; after the sample has been formed, can be subsequently reviewed or reexamined; possibility of access to the database at any time; the objects of study are not affected by the researcher's action; and the very clear systems and procedures [9].

Silva and Fossá emphasize that qualitative studies have been gaining, over time, notoriety in the researches of the Business Administration area [13]. Thus, the technique of content analysis is being widely used in these studies. However, many researchers misappropriate the technique of content analysis, failing to properly follow the steps for constructing the analysis. Thus, some authors present criticisms or common errors of the researchers when carrying out the content analysis. The main errors of the researchers are: lack of methodological rigor, propensity to influence the subjectivity of the researcher; and use in small cases or limited evidence [14].

The failures related to the lack of mastery of the assumptions, methods and techniques of data collection and analysis, which has the consequence of performing a content analysis without methodological clarity [15]. He adds that, in some works, the content analysis consists only of presenting clippings of some passages of the participants' speeches, without specifying the criteria for their selection. In the meantime, Justo and Camargo also highlight: the difficulty of working with extensive databases, which make researchers unable to analyze all content in a cohesive and reliable way; the time required for proper analysis of a large volume of material is incompatible with current deadlines for master's, doctoral, or research edicts [1].

In addition, Bauer and Gaskell list problems regarding sample determination, representativity failures, sample size, sampling unit and coding [2]. Collis and Hussey emphasize as technical weaknesses: an indistinct theoretical basis; may yield trivial conclusions; error in analyzing only words or expressions that the researcher finds important, discarding large amount of data; possibility of document omissions by providing an incomplete sample (accessibility); analysis is time consuming and tedious; choice of written documents (sample definition) for reasons other than those the researcher wishes [9]. Finally, Hair Jr. et al. and Creswell reinforce as a weak point the process of manual codification of qualitative information, considering a labor-intensive and time-consuming approach [10, 16].

Table 1 summarizes the aspects considered as advantages and as errors found in works of the technique of content analysis.

Table 1. Advantages of the content analysis and main errors of the researchers in the use of the technique.

	Description	Reference
Advantages	Method of systematic and public data analysis;	Bauer and Gaskell (2011)
	Makes use of pure data that "occurs" naturally;	Bauer and Gaskell (2011)
	Can handle large amount of data;	Bauer and Gaskell (2011)
	Offers a set of mature and well documented procedures;	Collis and Hussey (2005) Bauer and Gaskell (2011)
	Relatively cheap method;	Collis and Hussey (2005)
	After the sample has been formed, the data can be subsequently reviewed or re-examined;	Collis and Hussey (2005)
	Possibility of access to the database at any time;	Collis and Hussey (2005)
	The objects of study are not affected by the researcher's action;	Collis and Hussey (2005)
Major Research Errors	Lack of methodological rigor, by the researchers, has the consequence of performing a content analysis without methodological clarity;	André (2001) Collis and Hussey (2005) Bauer and Gaskell (2011) Reis *et al.* (2013)
	Propensity to influence the subjectivity of the researcher;	Reis *et al.* (2013)
	Use in small cases or limited evidence;	Reis *et al.* (2013)
	Difficulty working with large databases, which researchers have difficulty analyzing in a cohesive and trustworthy way;	Justo and Camargo (2014)
	Time required for proper analysis of a large volume of material is incompatible with the current deadlines related to masters, doctoral or research edicts;	Hair Jr. *et al.* (2005) Collis and Hussey (2005) Creswell (2010) Justo and Camargo (2014)
	Manual coding of qualitative information, considering a labor-intensive and time-consuming approach;	Hair Jr. *et al.* (2005) Creswell (2010)
	Failure of sample determination;	Collis and Hussey (2005) Bauer and Gaskell (2011)
	No data file to store and make raw data accessible for secondary analysis	Bauer and Gaskell (2011)

The advantages highlight the contributions of content analysis in qualitative research, while errors represent the challenges that can be overcome as this method of analysis is improved. The content analysis, over the years, has undergone reformulations since the first precepts, with a more contemporary analysis, influenced by the use of the computer, given the importance of discussing the use of software in your application [13].

3 Content Analysis Softwares

With the advent of Information and Communication Technologies (ICT), the way of doing research has changed considerably. The computational resources currently available for qualitative data analysis become a reality that is coupled up with several qualitative research practices [17]. The use of software for qualitative analysis of texts (official documents or transcriptions) is based on the technique of lexical analysis, which aims to compare or analyze a textual corpus according to the characteristics of its descriptions, in the form of combinations of words, which is done with the aid of descriptive and relational statistics [1].

According to Prediger and Allebrandt, software to support qualitative data analysis can be grouped into two categories, considering the methodology used and the functionalities. The authors present the categorization proposed by Lewins and Silver [8]:

- "code based theory building software": allow thematic coding of qualitative data, keeping contact with the original data from hyperlinks. They also allow the creation of annotations, association of codes, creation of models and conceptual maps, search and test relations, creation of categories;
- "text retrievers, textbase managers": they enable a complex search of a text, allowing to identify words with similar meanings, frequency tables, graphs.

Most softwares are generic tools, which allow for several methodological approaches; however, these may have some limitations. It is up to the researcher to prioritize the aspects of the theoretical, epistemological and methodological base of his research to make sure that the software he intends to use can support the specificities of his research [17].

As for the researchers' posture in the use of software, Lage and Godoy [5] present the categorization of Puebla (2003), in three groups: those who prefer handmade work techniques without computer support; those working with computer programs not designed for use in research (such as spreadsheets); and those using program packages developed specifically for this type of analysis. Regardless these positions, their contributions and limits in this use must be considered.

3.1 Contributions Associated with the Use of Content Analysis Software

Computers can, in a systematic way, process large volumes of text in a short time, providing the researcher a variety of information that can be interpreted [1].

Analyzing countless units of research materials, such as interviews, open questionnaires, newspaper texts, magazines, blogs and websites, among others, is a very expensive and complex process, often even impossible to do with the proper quality without this not consuming a much longer time than usually the researcher and/or the team have. In addition to this, the activities of stratifying data for analysis, making correlations, reassembling data in various rearrangements to try and extract the necessary information using different sources of information together: text, audio, video, images, photos [17].

This is where Computer Assisted Qualitative Data Analyzes (CAQDAS) comes in, which is also considered an eclectic field of research that works in the context of qualitative research methodology and analysis techniques in general, with different philosophical, theoretical and methodological traditions. The main gains with the use of software to support the qualitative analysis are: possibility of verifying the interpretations of the researcher quickly, generating consistency gains; the possibility of checking results by other researchers, being a way to improve the validity of the results; with the automation of mechanical tasks, the researcher is given time to dedicate himself to the interpretative process; speed gains; the project gains flexibility, since the gain of time provided by the use of the software, related to mechanical tasks, allows the researcher to explore new interpretive possibilities; providing graphical interfaces which allows a representation of the findings appropriate to the publishing process [4].

The use of software makes it possible to encode and analyze texts faster and less laboriously [10]. The possibility of storing as codable data the theories that support the research, as well as new results found, facilitates the elaboration of complex correlations between data-results-theoretical aspects, allowing to test ideas and hypotheses [5].

The great amount and diversity of information generated in qualitative research and the advantage of using qualitative data analysis software that facilitates the researcher's action, allowing greater agility of analysis and less manual effort [19].

It is noted, however, that the software provides gains to the researcher regarding the application of the technique, but errors of application related to the lack of control of the methodology are not compensated by the use of the technology [20]. The researcher must master the technique to produce valid scientific knowledge, with the use of tool or not.

All this discussion goes in the direction to an open space and to filling gaps in the training of new researchers and in the research done by those who have little reluctance to qualitative methodologies, as new paradigms become viable, new organizational contexts and subjects that can contribute to science emerge if they are investigated in depth. The search for aspects not accessible to quantitative analysis brings to light the complexity and opportunity of developing organizational studies in the light of an approach that allows this result [7].

Considering Bauer and Gaskell's statement that content analysis reconstructs representations in two dimensions, syntactic and semantic, it is possible to highlight the possibility of a strong contribution of software of analysis in this reconstruction of syntactic representations, that is, the frequency and ordering of words, vocabularies, grammatical characteristics of a textual corpus [2]. But, it is essential the researcher's interpretation analysis, aiming to reconstruct the semantic procedures, considering the associated meanings, thus making possible to present inferences about the textual productions. This argument is reinforced by Lage and Godoy [5]: The software does not forego the intelligence, intuition and creativity of the researcher, because it is them who will guide the directions of the research, choose the best resources, decide when to stop coding data and deepen the analytical process, as well as decide when the analysis is complete (p. 95).

From this perspective, software should not be understood as qualitative analysis software, since the understanding of meaning cannot (yet) be executed by the computer. Thus they should be termed qualitative analysis support programs [4].

3.2 Critical Aspects of the Use of Software and Its Limitations

Considering the numerous positive aspects observed by the researchers regarding the capacity of the CAQDAS to allow a great interaction of the researcher with the research data, there are negative aspects that have to be pointed out and analyzed.

The main limitations of the use of software to support content analysis refer to the fact that, often expressive language is not able to evidence some ideas or opinions about an object, or difficulties or because it is politically incorrect, out of standards [1]. However, these absences are part of the data, and only the external knowledge of the researcher about the subject is that it allows the interpretation of the data.

Point out that the use of software, by facilitating the processing of large volumes or number of texts, ends up allowing the researcher to neglect his role in the analysis of textual data. In these cases, there is a certain emptying of the relations of the textual material with the context, besides mechanical descriptions of the studied content. It is also observed that there are studies that restrict the analysis of the data to the information present in the software outputs, which is short of the exercise required by the researcher [1].

Despite processing statistical calculations behind the analysis interfaces, the software needs to be handled by an analyst who masters both the research in progress, in terms of its objectives and data characteristics, as well as the analysis techniques to be employed [1].

Lahlou cites as a common error in the use of lexical analysis software, described in the research as a research method, which can be observed in numerous publications, the fact of citing the software itself as if it were the data analysis technique, and sometimes as if it were the research method itself [18]. In addition, although lexicographic analyzes provide numerical evidence supporting the interpretation of data, the researcher interferes in several stages of the analytical process and care must be taken, in order that the representations of the researcher do not overlap with the reading that the same makes the data.

In addition, Creswell explains that there are errors related to shallow knowledge of the software and stresses the importance of the researcher to master the software used to use it effectively, which requires dedication and time [16].

There are disadvantages: the programs encourage complex and detailed coding structures, sometimes generating too much coding, which can cause the researcher to be literally stuck in his data and distant of the original context; the impossibility of communication between systems, since a project being analyzed by one tool cannot be exported to another; the main softwares are deprived of capacity what they have are advanced resources for treatment and consultation of the data, lacking the ability of the researcher who must know "what" to consult and "how" to interpret the results; all CAQDAS-type tools contain an implicit theory of qualitative analysis, forcing researchers to use a particular method of analysis, according to the characteristics of the tool; Finally, the process of learning the tools, which usually occurs through the tutorials distributed with the software packages, focused on the teaching of software tools, but not in the process of qualitative analysis [5].

Bandeira-de-Mello highlights limitations of the use of software to support qualitative analysis: removal of the researcher from the reality of the data, due to automation and focus on software programming commands; indication of the researcher to a superficial analysis, due to the distance of the reality of the researched subjects; methodological bias of the software, resulting from the methodological orientation of the group that developed it; costs of preparing the database and learning how to use the software; use of softwares by multiple authors may potentiate possible tensions among those involved in the peer-checking of interpretations; the software does not solve problems related to failure to collect data [4].

Summing up, content analysis, whether performed manually or with software support, requires the researcher's mastery of the technical details, alignment with the research question and theoretical support related to the object of study. With regard to knowledge of the technique, it is important that the researcher is aware of the main flaws in its use, avoiding also incurring errors that will compromise the results of his research.

4 Conclusions

This work aimed to analyze the contributions and limits of the use of content analysis software, in order to support the researcher in their decision when choosing to whether or not to use software resources.

It has been discussed that the main contributions of the use of software to support the qualitative analysis are related to the possibility of analyzing a more extensive database than those analyzed with manual technique, besides the faster data analysis and the possibility of access to the data from the elaborated codifications. The most recurrent constraints are related to the difficulty of access by several researchers in co-authorship, the need for time for maturation and mastery of the tool by the researcher, possibility of software bias, considering the assumptions of the developers, and the propensity to simplify the conclusions from only the encodings generated by the software. This last limitation is understood to be related to the researcher's posture and mastery of the method and less about the software.

It is concluded that, in order to perform a research using content analysis, the researcher must have mastery of the technique and, if they choose to use software, they must really know the computerized tool, analysis, database expansion, and storage capabilities of the encoded data. For the effective quality of the knowledge produced through this technique, however, it is essential the researcher's stance in search of methodological rigor, of deepening in relation to theorization, in the choice of the sample, and in the coding and interpretative and contextual analysis of the data.

In addition, prior to this process, the researcher needs to really define the need for a software to carry out the analysis of the data of his research and, given the large quantity and possibilities of software currently available to the researchers, to know how to evaluate which one offers the best conditions and specificities to properly align with their research, and then follow the entire process of content analysis of qualitative data.

Acknowledgements. This work was conducted during a period in which one of the authors was financially supported with a scholarship from the Fundação Coordenação de Aperfeiçoamento de Pessoal de Nível Superior - CAPES.

References

1. Justo, A.M., Camargo, B.V.: Estudos qualitativos e o uso de softwares para análises lexicais. In: X SIAT & II Serpro. Universidade do Grande Rio, Duque de Caxias (2014)
2. Bauer, M., Gaskell, G.: Pesquisa Qualitativa com texto, imagem e som: um manual prático. Vozes, Petrópolis (2011)
3. Bardin, L.: Análise de Conteúdo. Edições 70, Lisboa (1977)
4. Bandeira-de-Mello, R.: Softwares em pesquisa qualitativa. In: Silva, A.B., Godoi, C.K., Bandeira-de-Mello, R. (orgs.). Pesquisa qualitativa em estudos organizacionais: paradigmas, estratégias e métodos. Saraiva, São Paulo (2010)
5. Lage, M.C., Godoy, A.S.: O uso do computador na análise de dados qualitativos: questões emergentes. Revista de Administração Mackenzie **9**(4), 75–98 (2008)
6. Mozzato, A.R., Grzybovski, D.: Análise de conteúdo como técnica de análise de dados qualitativos no campo da administração: potencial e desafios. Revista de Administração Contemporânea **15**(4), 731–747 (2011)
7. Mesquita, R.F., Matos, F.R.N.: A abordagem qualitativa nas ciências administrativas: aspectos históricos, tipologias e perspectivas futuras. Revista Brasileira de Administração Científica **5**(1), 7–22 (2014)
8. Prediger, R.P., Allebrandt, S.L.: Uso de softwares em pesquisa qualitativa. Salão do Conhecimento **2**(2), (2016)
9. Collis, J., Hussey, R.: Pesquisa em administração: um guia prático para alunos de graduação e pós graduação, 2nd edn. Bookman, Porto Alegre (2005)
10. Hair Jr., J.F., Babin, B., Money, A.H., Samouel, P.: Fundamentos de métodos de pesquisa em administração. Bookman, Porto Alegre (2005)
11. Chizzotti, A.: Pesquisa qualitativa em ciências humanas e sociais. 4 edn. Vozes, Petrópolis (2011)
12. Oliveira, M., Bitencourt, C.C., Santos, A.C.M.Z., Teixeira, E.K.: Thematic content analysis: is there a difference between the support provided by the maxqda® and nvivo® software packages? Rev. Adm. UFSM **9**(1), 72–82 (2016)
13. Silva, A.H., Fossá, M.I.T.: Análise de conteúdo: exemplo de aplicação da técnica para análise de dados qualitativos. Qualit@s Revista Eletrônica **17**(1) (2015)
14. Reis, A.O.A., Sarubbi Junior, V., Berolino Neto, M.M., Neto, M.L.R.: Tecnologias computacionais para auxílio em pesquisa qualitativa: software evoc. Schoba, São Paulo (2013)
15. André, M.: Pesquisa em educação: buscando rigor e qualidade. Cadernos de Pesquisa **113**, 51–64 (2001)
16. Creswell, J.: Projeto de Pesquisa: métodos qualitativo e quantitativo, 3rd edn. Artmed, Porto Alegre (2010)

17. Lima, J.L.O., Manini, M.P.: Metodologia para análise de conteúdo qualitativa integrada à técnica de Mapas mentais com o uso dos softwares nvivo e freemind. Inf. Inf. **21**(3), 6 (2016)
18. Lahlou, S.: Text mining methods: an answer to Chartier and Meunier. Pap. Soc. Represent. **20**(38), 1–7 (2012)
19. Guizzo, B.S., Krziminski, C.O., Oliveira, D.L.C.: O software QSR Nvivo 2.0 na análise qualitativa de dados: ferramenta para a pesquisa em ciências humanas e da saúde. Revista Gaúcha de Enfermagem **24**(1), 53–60 (2003)
20. Mesquita, R.F., Sousa, M.B., Martins, T.B., Matos, F.R.N.: Óbices metodológicos da prática de pesquisa nas ciências administrativas. Revista Pensamento Contemporâneo em Administração **8**(1), 50 (2014)

Development of Basic Spatial Notions Through Work with Educational Robotics in the Early Childhood Education Classroom and Analysis of Qualitative Data with WebQDA Software

Noelia Bizarro Torres(✉), Ricardo Luengo González(✉), and José Luís Carvalho

Investigation Group CiberDidact, Department of Didactics of Experimental Sciences and Mathematics, Faculty of Education, University of Extremadura, Badajoz, Spain
noebizarro87@gmail.com, {rluengo,jltc}@unex.es

Abstract. With the current research work, it is intended to develop basic spatial concepts (front, back, up, down, right and left) through a didactic proposal based on the use of educational robotics in the classroom of Early Childhood Education. The Roamer Robot is specifically used for the potential it offers to adapt to pupils' needs and the characteristics of all activities. For the work in the classroom, a series of contextualised activities are presented in different types of content, which is why Roamer's versatility as a learning tool is enhanced.

With this research project we intend to know the differences on the acquisition of these concepts before and after the intervention in the classroom with the Roamer robot. For this, a qualitative approach research has been designed using as investigation instruments an experimental test and a drawing test that have been adapted to the evolutive characteristics of children in Early Childhood Education. The WebQDA software has been used to support the qualitative data analysis.

In general, it is concluded that the work with robotics in the Early Childhood Education classroom improves the acquisition of basic spatial concepts in these pupils. However, the right and left concepts are still the most complicated to acquire and assimilate.

Keywords: Spatial notions · Educational robotics · Childhood education Qualitative analysis · WebQDA

1 Introduction

This research work results from the completion of a pilot study prior to the doctoral thesis project "Educational robotics as a support to the development of basic spatial notions in Early Childhood Education" that is being carried out in the Department of Didactics of Experimental Sciences and Mathematics of the University of Extremadura (Spain). In this pilot study, the application and treatment of qualitative data is carried out through the WebQDA software [1].

A. P. Costa et al. (Eds.): WCQR 2018, AISC 861, pp. 22–33, 2019.
https://doi.org/10.1007/978-3-030-01406-3_3

A proposal of educational intervention is presented for the development of basic spatial notions in Early Childhood pupils through the robotic use and programming, specifically the use of the Roamer Robot in the classroom. For this, all the aspects that influence the teaching-learning process have been taken into account and that is reflected through the classroom methodological work based on the EMO (experimentation, manipulation and observation).

The process of development of basic notions implies the work through a series of parallel phases which are related to the psychoevolutive needs of pupils of these ages. Thus, the development of the body scheme itself has been worked on as well as the projection on objects external to our body: Roamer.

In order to obtain information, a data collection is carried out before and after the practical intervention with Roamer through a graphic test and an experiential test that is analysed through the WebQDA software. Through this process of collection, treatment and analysis of data, it is possible to analyse if there are significant differences when working in the classroom with an active methodology based on Robotic programming. The results obtained allow teachers to make decisions about the methodology and tools used in the daily work in the classroom, in order to carry out an educational intervention according to the pupils' needs, their context and the social demands in which we find ourselves immersed.

2 Statement of the Problem and Context

The child in pre-school education is in the preoperational stage (2–7 years) according to Piaget's development theory, in Feldman [2]. In this stage the child develops the symbolic function to act and understand the world around him/her; however, according to this author, can not develop organised formal and logical mental processes. Therefore, it is believed that in the early childhood education classroom it is fundamental to work with robotic programming, since it will help the child to structure sequences and mental processes with a functional and significant logic. This idea is developed as Computational Thinking which "involves solving problems, designing systems and understanding human behaviour, making use of the fundamental concepts of computer science" [3]. In order to work on computational thinking in the children's classroom, it is necessary to carry out a problem solving process that helps the child to structure his/her thinking and develop a logical reasoning. This same competence can be applied to any educational area and stage, although in the stage of Early Childhood Education it becomes more relevant, since through the development of these strategies, cognitive aspects that can help overcome the limitations of preoperational thinking are worked on: focus, egocentrism, irreversibility, incomplete understanding of the transformation [4].

The work with robotics in Early Childhood Education can help the development of spatial basic notions related to the child's own body and to objects around him/her. The structuring of the notion of space gains momentum, as the child develops the dynamic control and coordination of his/her own body, as well as the awareness of objects external to him/her.

Manuel Valencia, director of Argan Bot in an interview to the newspaper "El español" [5] defends that "with programming we manage to structure the head so that problems can be solved in a more logical way". This way, when the child faces a problem he has to look at his cognitive structure to analyse the situation and solve the problem, "milestones" that the child has around to which he/she will establish "routes" to create the work procedure, and then choose the most suitable one for solving the problem, as stated in the Theory of Nuclear Concepts (TCN) [6].

Taking into account the above, some aspects that justify our research Project can be named.

- It is very important to analyse the development of knowledge and control of one's own body regarding the spatial orientation of children in Early Childhood Education using robotic programming through Roamer.
- Analyse and observe if there are differences in the results extracted before and after the educational intervention to check if there are significant differences that allow us to make decisions to improve the teaching-learning process.
- It is worth highlighting the legislative contextualisation of the use of robotics in the classroom. For this, Roamer will be integrated as a tool that will help to enhance the pupil's spatial development as a main aspect and other secondary content related to all areas of the official curriculum of Early Childhood Education in our Autonomous Community of Extremadura.
- Through the work in the classroom, pupils are provided with a globalised approach [7] as an essential principle for educational practice in Early Childhood Education. Thus, through educational robotics, contents of various kinds are worked on: mathematical, language, observation of the environment, socialisation, etc.
- In addition, it is worth highlighting the principle of playing and motivation provided by the work with educational robotics in the classroom. It is an aspect closely linked to the learning needs of children of these ages. This way, pupils' main interests are always highly considered.

Taking into account all these aspects, the main objective of the research is to analyse whether the work in the classroom with educational robotics can improve the spatial orientation of the Early Childhood Education pupil, in terms of his/her own body and external objects. Through the work of this objective it is intended to answer the following research question: do pupils improve the acquisition of basic spatial concepts with the use of educational robotics in the classroom?

The evaluation of the learning process is fundamental, since it offers data about development and acquisition of new knowledge. The following definition of evaluation is highlighted [8] "it is that learning activity that is evaluated to check the knowledge, skills and competences that are being acquired, as well as the requirements and conditions that must be met". Using suitable instruments and tools for data analysis is very important. In the stage in which the present study, Early Childhood Education, is developed, pupils' characteristics must be taken into account in order to develop the evaluation activity.

The WebQDA software, chosen to perform data processing and analysis, allows the completion of all the tasks involved in data processing in our research: organisation, categorisation and analysis of those data [1].

WebQDA is software that has been created by the company Micro IO and Ludomedia in collaboration with the University of Aveiro [9]. The Spanish version was developed with the support of the CiberDidact research group of the University of Extremadura (Spain). It is specific software aimed at qualitative research in general, it allows the analysis of graphic data: images—videos and text. It is also a tool that allows you to edit, store and organise documents, where you can create categories, code, control, filter, investigate and consult research data.

The main reason for choosing WebQDA software for data processing is that it allows the incorporation of different types of sources: textual, graphic, sound, etc. When working with Early Childhood Education pupils, the tests designed are very graphic and this software allows to work with them in a very easy and organised way. In addition, it is important to highlight the ease of organisation of research data through the category system. The fact of being able to share the project with other users is very useful to enhance the collaborative work.

3 Methodology

With regard the proposed objective and the research question, this study uses the qualitative method based on data collection instruments by the teacher's direct observation and by analysis of WebQDA software in its latest version 3.0. This tool was necessary to analyse if the work in the classroom with educational robotics through a work methodology based on exploration, manipulation and observation (EMO).

3.1 Sample

The selection of the sample was obtained by a non-probabilistic convenience sampling. The selection criteria that were taken into account for the research were the following:

- 1 group of pupils enrolled in 5 years of Early Childhood Education.
- The selected pupils must have developed a series of basic contents: know numbers, some letters, colours, etc.
- Pupils must attend school continuously so that an ongoing investigation may be carried out.

An attempt was made to find a sample that, in keeping with the above criteria, was accessible to the researcher, since she is the group's tutor. The sample chosen was determined from the 5-year-old children of CEIP Ntra. Sra. of Chandavila from the town of La Codosera (community of Extremadura—Spain). It is a group of 7 pupils with different characteristics and motivations but with a good level of cognitive development; it is worth mentioning the participation in the group of a child with Special Educational Needs.

Of the 7 participants who regularly attend school, all of them have actively participated in the research, since their parents have expressed their prior written authorisation. It is a very small group which allows a very individualised treatment.

3.2 Phases and Timing

As far as the phases of the research are concerned, the study began with an initial phase of data collection and research design. The practical intervention lasted 3 months, from the end of November to the beginning of February. After this practical intervention, the final phase of data processing was carried out. Due to the characteristics of the work methodology employed in Early Childhood Education, delimiting the timing in the number of specific sessions is complicated, since the work is carried out from a globalising perspective, where contents are constantly interrelated. Due to this characteristic, the following work plan is defined (Table 1):

Table 1. Timing of the research.

Phases	Number of sessions	Description
Initial evaluation	2	Drawing[1] and Experiential Evaluation[1]
Practical activities with the Roamer Robot and psychomotor skills	34 Specific sessions and sporadic sessions contextualised in the contents and needs of the group	Combined activities of educational robotics and psychomotricity and contextualised in the classroom, for example. In the daily assembly the protagonist of the day will solve a proposed challenge: directing the Roamer to the letter "L"
Final Evaluation	2	Drawing[2] and Experiential Evaluation[2]

3.3 Instruments Used

The instruments for data collection that have been used have been selected and designed taking into account the characteristics of the participants and the research itself. Below each selected instrument is described, as well as its justification and application process in our research.

Experiential Evaluation

This activity consists in the following: the pupil identifies a certain part of his or her partner's body related to the concepts under study: front-back. Up-down, right left. This identification was made keeping in mind the most common bodily positions: front, back and lying down. One of the pupils played the role of mannequin (he could not move or say anything to his partner; this figure will change so that everyone can participate). Another colleague placed stickers of a certain colour in the body part of the manikin indicated by the teacher.

This experiential evaluation was carried out to know if the pupil is able to differentiate the concepts that are going to be worked on in a body other than his/her own. In addition, the position of the manikin body is changed to check this same acquisition in different axes of reference: front, back, horizontal. The assessment was recorded through photographs of the mannequins after placing the stickers to check the positive and negative results.

The distribution of coloured stickers for each identified concept is the following (Table 2):

Table 2. Distribution of colours for each concept to be identified in the experiential test.

Front	Back	Up	Down	Right	Left

Evaluation by Drawing

The pupils were presented with a drawing already presented that they had to colour according to instructions. It is a drawing in which we can observe different children in different body positions in which their body axis is not equal to the normal axis of reference (standing). These children are playing different games and one can observe objects that the participants must identify and solve in the best way. The indications are the following:

- María is playing hopscotch: color in dark green the bush that is on the RIGHT of María and of light green the bush that is on the LEFT of María.
- Paula does not play hopscotch, draw a ball in FRONT of Paula.
- Draw a rainbow in the BACK of the bushes.
- Pedro does the somersault, surround the UPPER part of his body with a red colour and the LOWER part of his body with a blue colour (Fig. 1).

Fig. 1. Drawing evaluation test.

3.4 Analysis Procedure

After the data collection with the two instruments used, the data analysis was carried out with the support of the WebQDA software.

Taking into account the whole process described above and the evaluation instruments, for the analysis procedure, a qualitative analysis of the tests to identify previous knowledge in the initial evaluation and of the evaluation tests at the end of the process was carried out.

4 Activities Performed

In accordance with the proposed methodology, the following activities were developed:

1. **First phase:** Research design. In this phase, the problem that was intended to be solved was raised. In addition, the spatial concepts that were intended to be worked on and evaluated in the pupils were chosen, the research was designed and the intervention project in the classroom was also programmed, based on the application of activities with educational robotics and bodily activities.
2. **Second phase:** Initial evaluation. The experiential test and the drawing test were carried out just before starting the work in the classroom. This evaluation allowed to know the previous knowledge that our pupils have regarding the contents that are going to be worked on.
3. **Third phase:** Intervention in the classroom. The intervention in the classroom was carried out in the number of sessions initially scheduled. On the one hand, contents from different areas were worked on in the classroom with educational robotics (Fig. 2), thus, the content was presented in a globalised way, taking advantage of all the potentialities of this tool. And on the other hand, basic psychomotor activities have been developed (Fig. 3) that complement the activities with robotics and that are related to the concepts to be worked on.

Fig. 2. Activities in the classroom with the Roamer Robot.

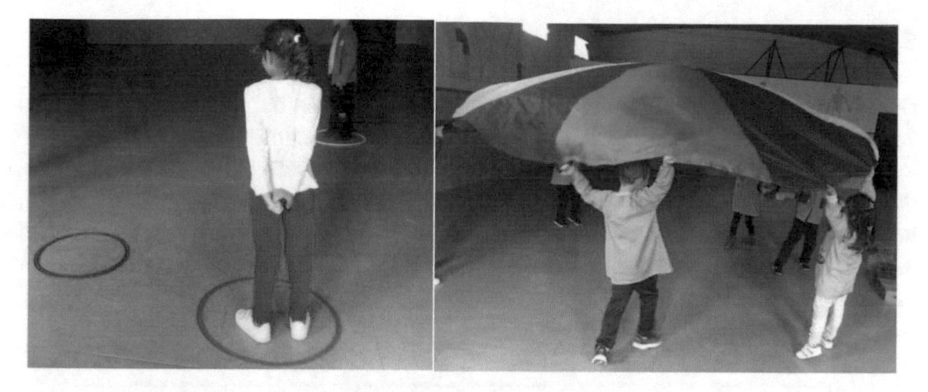

Fig. 3. Activities in the psychomotricity classroom.

4. **Fourth phase:** Final evaluation[2]. Just after the intervention in the classroom with educational robotics and psychomotricity activities, the experiential test and the drawing test were repeated. The aim is to observe if there are any significant differences with regard to the initial evaluation.
5. **Fifth phase:** Data processing through WebQDA software. In this phase several sequenced tasks were carried out:

   – Data transfer to WebQDA: digitised copies of the two tests were incorporated. In the case of the experiential test, photographs were used and in the case of the drawing test, the drawing itself was incorporated into the internal sources of the software. Through WebQDA, those points referring to the contents studied have been indicated and their resolution has been described.
   – Creation of the category system: Subsequently, a system of categories was created to categorise the pupils' answers. The categories are related to the 6 concepts that are being evaluated and the process of acquiring them.
   – Data query: The pertinent consultation was carried out to obtain results that are presented in the following section (Fig. 4).

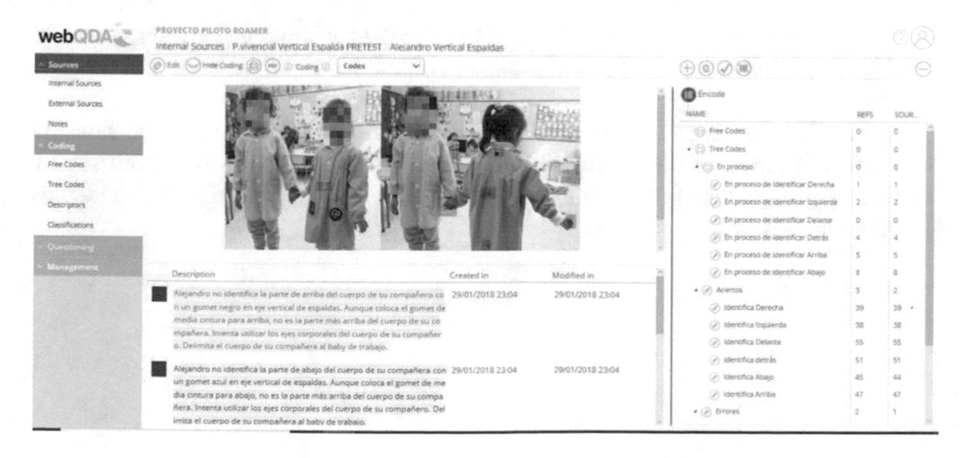

Fig. 4. Data processing with WebQDA Software.

5 Results

Below are the results of the different tests we have carried out taking into account the comparison of the initial and final moments of its application.

In Fig. 5 you can see the results of the experiential test—front. In the initial test, pupils showed greater difficulties in identifying left and right, but in the final test it can be seen how most of them have overcome these difficulties. It is worth noting the difficulty of this test because they are identifying left and right in a mirror position, i.e., in a position contrary to that of their own body.

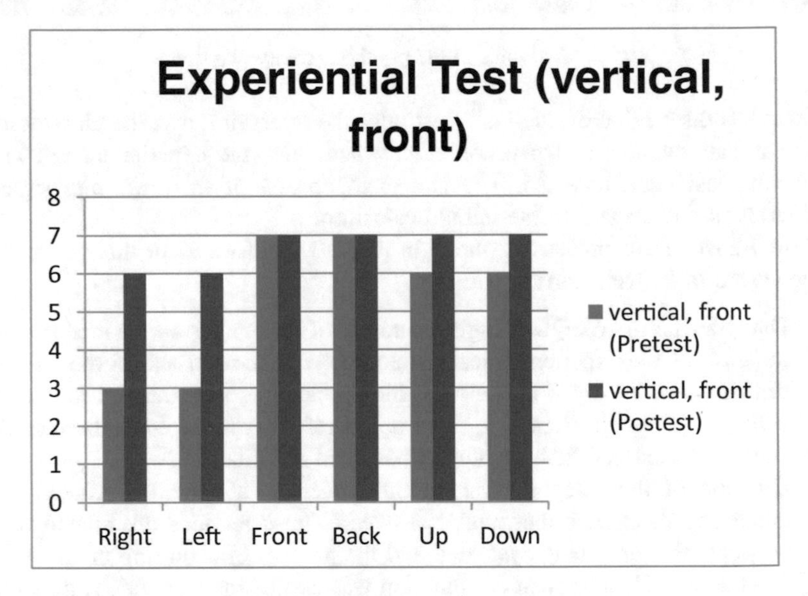

Fig. 5. Experiential test results (front) pretest and postest.

In Fig. 6, referring to the vertical experiential test—back, pupils have improved the identification of all concepts except the concept of left and right, which is still the most complicated.

In Fig. 7, regarding the horizontal experiential test, it can be observed how the correct identification of the concepts which are asked to pupils has improved significantly. It should be noted that in the horizontal position they have been able to identify the right and left of their partner.

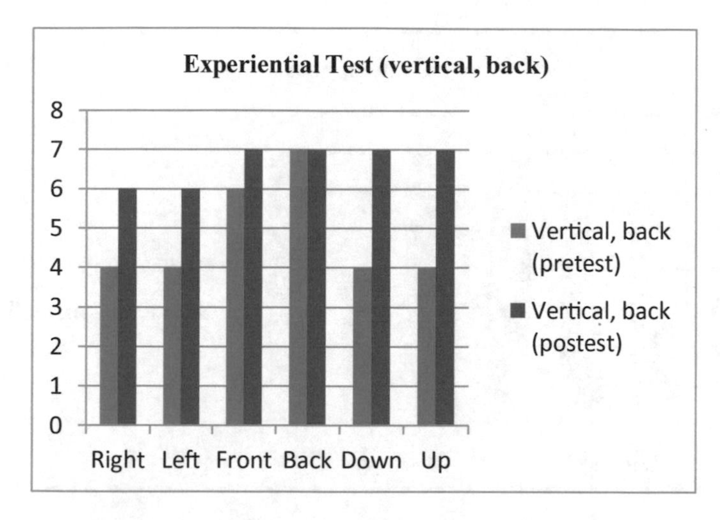

Fig. 6. Experiential test results (back) pretest and postest.

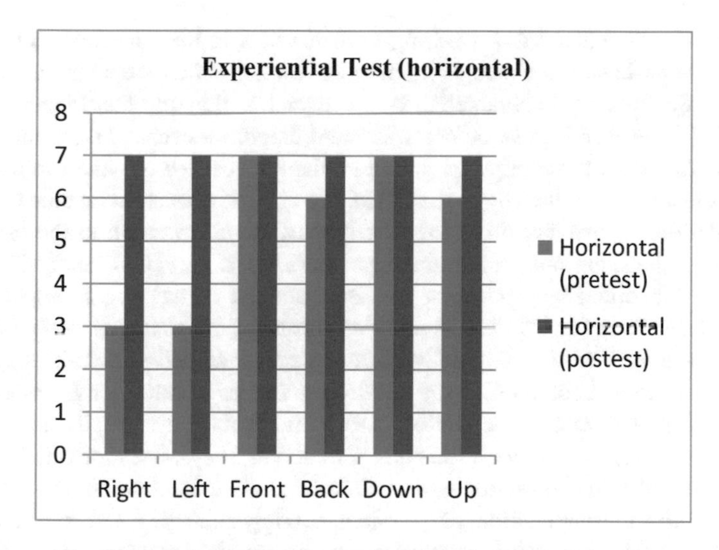

Fig. 7. Experiential test results (horizontal) pretest and postest.

Figure 8 shows the results obtained from the drawing test. Pupils initially presented problems for the identification of the right and the left and also up and down for the position of the figure in which they had to identify these last concepts (Pedro in the drawing test). However, in the final evaluation it has been possible to observe how the pupils improve the identification of all these concepts, although they still have difficulties to identify the concept *down*.

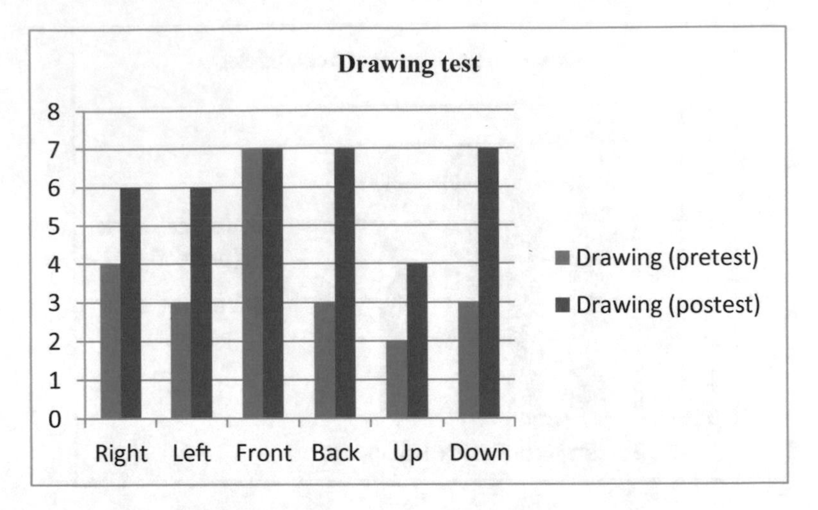

Fig. 8. Initial and final drawing test results.

6 Conclusions and Suggestions

In general, it is concluded that working with robotics in the classroom improves the acquisition of basic spatial concepts in children in early childhood education and helps the child to flexibilise the characteristics of irreversible thinking that Piaget defends in Feldman [2]. However, it must be borne in mind that the concept of right and left is the most complicated for these pupils, since in all the tests carried out one can observe the difficulties that the pupil has for its identification. For this reason, these concepts should be more thoroughly worked on in the activities of the intervention in the classroom.

As far as the questions raised are concerned, the observation, analysis and conclusions on the differences between the tests applied in the two moments of data collection have been carried out. In addition, an educational intervention project has also been designed with manipulative materials and attractive activities for pupils regarding educational robotics in the classroom and psychomotricity, respecting the principle of globalisation, motivation and play. This way, pupils have worked favourably in the classroom with the chosen material. They were motivated and active participants in all activities, so the work with robotics in the classroom is very positive, since it favours the application of problem solving strategies and keeps the pupil motivated. Through this process we have worked on the development of "Computational Thinking" [3].

Using the WebQDA computer software, it has been possible to organise, structure the sources extracted from the evaluation tests; it has also allowed to categorise each concept to be valued and thus the results have been obtained in a very practical and useful way. However, in the thesis project, the comparison of these two evaluation tests is considered with other very useful ones which will allow extracting and comparing results. Thus, we intend to replicate this study using also the extraction of Pathfinder Associative Networks [6], to analyse the cognitive structure of the pupil and the application of a standard test that evaluates the concepts worked on.

Taking into account the conclusions drawn and the proposed limitations, we suggest the replication of the study with a larger sample and observing differences compared to the use of a traditional methodology in the classroom, that is, without the use of robotics in a control group.

Acknowlegement. This work was supported by the European Union and Gobierno de Extremadura under Grant FEDER: Programa Operativo FEDER de Extremadura.

References

1. Ferrão, S., Marques, R., Moreia, A.: WebQDA na análise qualitativa de interações no contexto de uma oficina de formação de professores. Indagatio Didactica **5**(2), 110–121 (2013)
2. Feldman, R.S.: Desarrollo psicológico a través de la vida. Pearson Educación, México (2007)
3. Wing, J.M.: Computational thinking. Commun. ACM **49**(3), 33–35 (2006)
4. Valverde, J., Fernández, M.R., Garrido, M.C.: El pensamiento computacional y las nuevas ecologías del aprendizaje. Revista de Educación a Distancia **46**(3) (2015)
5. Valencia, M.: Cuando los niños aprenden robótica y programación. El Español. Recuperado de (2016). https://www.elespanol.com/ciencia/tecnologia/20160301/106239504_0.html
6. Casas, L., Luengo, R., Godinho, V.: Software MICROGOLUCA: Knowledge representatión in mental calculatión. US-China Education Reviex **1**(4), 592–600 (2011)
7. Alsina, A.: Hacia un enfoque globalizado de la educación matemáticas en las primeras edades. Números: Revista de Didáctica de las Matemáticas **80**, 7–24 (2012)
8. López-Pastor, V.M., Pérez-Pueyo, A., et al.: Evaluación formativa y compartida en Educación: Experiencias de éxito en todas las etapas educativas. Universidad de León, Servicio de publicaciones (2017)
9. Neri de Souza, F., Costa, A.P., Moreira, A. y Neri de Souza, D.: WebQDA—Manual do Utilizador. Universidad de Aveiro, Portugal: Esfera Crítica (2013)

What Is Better to Study: The Printed Book or the Digital Book?: An Exploratory Study of Qualitative Nature

José Luís Carvalho[1](✉), Ricardo Luengo González[1], Luis M. Casas García[1], and Javier Cubero Juarez[2]

[1] Investigation Group CiberDidact, Department of Didactics of Experimental Sciences and Mathematics, Faculty of Education, University of Extremadura, Badajoz, Spain
{jltc,rluengo,luisma}@unex.es

[2] Investigation Group Deprofe, Department of Didactics of Experimental Sciences and Mathematics, Faculty of Education, University of Extremadura, Badajoz, Spain
jcubero@unex.es

Abstract. Several years after the incorporation of digital in educational contexts, there is little systematic knowledge about the attitudes and practices of students with regard to reading in printed and digital books. This study aims to know what the students think of the Master "Research in Teaching and Learning of Experimental, Social and Mathematical Sciences", of the University of Extremadura - Spain (2017–2018 academic year), about the study through printed books or digital books. The design of the research is descriptive and based on the qualitative and quantitative analysis (mixed-method) of the messages of an electronic Forum in which 31 students will participate. The content analysis of the messages of the Forum has been carried out following a process of reduction, organization, coding, obtaining results and determination of conclusions, using webQDA resource, software to support the qualitative analysis of data. The quantitative findings reveal that the largest of university students continues to prefer to study from printed books. The main justifications for this option refer to the importance of making annotations on an object with a life of its own and that reading on paper allows reaching higher levels of concentration and memory through the sensory experience that its use provides.

Keywords: Qualitative research · webQDA · Printed book · Digital book
To study

1 Introduction

The main objective of this study is to know the opinion of the students of the Master "Research in Teaching and Learning of Experimental, Social and Mathematical Sciences" of the University of Extremadura - Spain (academic year 2017–2018), regarding the study from printed books and from digital books.

A. P. Costa et al. (Eds.): WCQR 2018, AISC 861, pp. 34–45, 2019.
https://doi.org/10.1007/978-3-030-01406-3_4

This problem was proposed as a subject for the research work to be carried out by the students in the field of continuous assessment of the subject of the Master "Qualitative Research in Teaching and Learning of Experimental, Social and Mathematical Sciences" of the University of Extremadura - Spain (academic year 2017–2018)

Until a few years ago, the main educational resources for studying were books, encyclopaedias, magazines and paper notes. Currently information and communication technologies provide other resources that can contribute to modifying the way students study and learn.

Electronic books (ebooks) are one of those means that can facilitate the work of teachers and students. In this study it is also important to examine if the students of a Master (post-graduation course), many of them being teachers in practice, with teaching functions in schools and institutes, are adapting to new forms of reading and are taking advantage of the advances that are taking place, or if, on the contrary, they remain rooted to the primitive materials of study and what are the reasons for doing it in one case or the other.

2 Theoretical Foundation

From the end of the Renaissance and the beginning of the 21st century (period named as "Gutenberg era"), the printed text - and the book, specifically - predominated in cultural production and academic production [1]. Nowadays it is not difficult to establish that the printed culture continues to be predominant in the current educational system. However, educational materials on paper, considered as the only reference of knowledge, are beginning to be affected by digital transformation. According to the ISBN Agency [2], in 2015 digital books accounted for 28% of the total production in Spain, due in particular to the adaptation of textbooks to the digital format in Education.

Unlike the book, a tangible, complete, stable object that offers packaged enclosed and structured knowledge on paper, the new electronic media are characterised by spreading intangible books similar to the traditional ones but also other types of books, open, fractioned, flexible, interconnected, multimedia, hypertext and in constant adaptation to the characteristics of the teacher and his group-class. It is the metaphor of the solid culture of books against the liquid culture of the digital ecosystem [3].

A recent study by the Autonomous University of Barcelona and Aula Planeta [4] reveals that the Spanish teacher recognises many potentialities in the digital textbook versus the conventional one. The main advantages of the digital book positively valued by teachers are the following: it helps to motivate students; they can adapt it easily to the development of contents; it allows the immediate search of contents; it presents examples and interactive activities; it favours the development of digital skills; it presents health benefits since they weigh less in students' backpacks; families value the savings they allow and consider that it is a way for their children to stay technologically up-to-date.

Another study carried out in Spain, which gathers the expectations and opinions of compulsory secondary education teachers, highlights that digital books "require greater dedication to elaborate complementary resources since, in general, they present limited and incomplete contents, they must adapt to the different learning levels of the students and, although they consider that they integrate training exercises, they do not have guidelines and tools that allow them to assess the competences acquired by students, nor do they offer them appropriate evaluation formulas with these training practices mediated by ICT." [5]. They also point out that the incorporation of simulations and animations, concept maps, animations, examples, self-correcting and self-assessment tests,…, in digital books can favour the understanding of contents and help students to strengthen their learning.

One of the current ideas of innovation and research in this field is dedicated to the issue of the adoption and self-production of textbooks and digital educational content. For example, in the Italian primary and secondary schools the Ministry of Education promotes, since 2013, the use of digital or mixed texts (printed-digital) in the classroom and, since then, the adoption of the textbook has not been mandatory. Using their autonomy, schools can use teaching materials from publishers, open educational resources or texts produced by themselves. The practices of the Italian educational centres that are innovating range from the creation of content that integrates the school textbook to the "alternative adoption", that is, the creation of independent textbooks which replace completely the editorial texts [6]. The aim is to ensure that innovation is not only technological (from the mere conversion of printed books into digital ones without including any advantage related to audiovisual or interactivity) but that it helps to change the standardised and massified curricula, and to transform the old-fashioned pedagogical model of rote and decontextualized learning.

However, although technology is becoming increasingly popular and easier to use, most studies published since the early 1990s confirm that paper still has advantages over screens as a means of reading [7]. The work published by the prestigious Scientific American magazine [7], which summarises an emerging collection of studies, indicates that people often tend to understand and remember the text on paper better than on a screen. Screens prevent people from navigating efficiently in long texts, which can subtly inhibit reading comprehension.

In a more recent study led by Naomi Baron from the American University [8], which gathered data from quantitative and qualitative surveys of 429 university students from the USA, Japan, Germany, Slovakia and India, the results revealed many opinions on the advantages of printed reading. Almost 92% said they concentrated better when they read from printed text, and more than 80% reported that if the cost were the same, they would prefer to print digital books for school work as well as for reading pleasure.

Turning the pages of a paper book is a similar activity to leaving one footprint after another on a path, there is a rhythm and a visible record of the passing of the printed sheets. Beyond treating individual letters as physical objects, the human brain can also perceive a text in its entirety as a kind of physical landscape. When we read, we build a mental representation of the text. The exact nature of such representations is not clear, but some researchers believe that they are similar to the mental maps we create of the ground, such as mountains and trails, and of interior physical spaces, such as apartments and offices. A paper book makes it easier to form a coherent mental map of the text [9].

Compared to paper, screens are also more demanding than paper, from the cognitive and physical point of view, and make it a little harder for us to remember what we read. Displacement requires a constant conscious effort, and LCD screens on tablets, laptops and smartphones can dilate the eyes and cause headaches by directly illuminating people's faces.

People often approach computers and tablets with a mental state less conducive to learning. In addition, electronic readers can not recreate certain tactile reading experiences on paper, the absence of which is disturbing to some people. Even so-called digital natives are more likely to remember the essence of a story when they read it on paper and recognise that their attention is more focused when they read the print rather than online [10].

The 2011 study conducted by the Israel Institute of Technology [7] mentions that students who used paper approached the exam with a more studious attitude than their fellow screen readers and directed their attention and working memory more effectively.

3 Methodology

The design of the research is of a descriptive type and is based on the qualitative analysis of the messages of an electronic Forum (created in the Virtual Campus of the subject "Qualitative Methodology" of the Master "Research in Teaching and Learning of the Experimental, Social and Mathematical Sciences"), on students' opinions about studying from printed format or from digital format.

Arango [11] maintains that an electronic Forum "is a context of communication on the Internet, where debate, conciliation and consensus of ideas are promoted. By operating asynchronously, the Forum allows us to overcome the temporal limitations of synchronous communication, delaying in time the cycles of interaction that, in turn, favours the reflection and maturity of the messages" [11]. According to Martínez & Briones [12] the Forum is a space with multiple possibilities which supports the social construction of meanings through forms of interaction and collaboration.

For this study, seven research questions were delineated:

1. Are there more opinions in favour of the use of the book on paper or in favour of the use of the digital book?
2. Are there differences by gender in opinions in favour of the use of the book on paper or the digital book?
3. What reasons do students have to study from a printed book?
4. What reasons do students have for not studying with recourse to a paper book?
5. What reasons do students have to study from a digital book?
6. What reasons do students have for not studying with recourse to a digital book?
7. What are the main functionalities of the digital book mentioned by the students?

The number of participants in the study is constituted by a total of 31 students with an average age of 28 years. In relation to gender, 61% of the subjects (19 students) are female and 39% (12 students) are male. Depending on their enrollment schedule in the

Master, 13 students come from the Social Sciences, 10 from Mathematics and 8 from the Experimental Sciences.

The analysis of the messages of the electronic Forum has been carried out following a process of reduction, organization, coding, obtaining results and verification of conclusions. The content analysis methodology used is an adaptation of the sequence of phases and instructions proposed by Bardin [13], and by Fraenkel and Wallen [14], on the procedures to guarantee and confer reliability to the data structure.

To do this, we began by making an exploratory reading of each of the opinions provided by the participants in the electronic Forum, aimed at identifying the most relevant key concepts. The subsequent readings, in depth, provided the creation of an emerging system of categories (Table 1) and the definition and conceptual description of the dimensions and categories of analysis. The validation of the category system was ensured through consensus among four researchers.

Table 1. Dimensions and categories created for the analysis of the data.

Dimensions	Categories
In favor of the printed book	Health
	Accessibility
	No dependence on technogy
	Writing and underlining
	Energy saving
	Memory retention/concentration
	Senses and feelings
Against the printed book	c-transportation and storage
	c-ecology
In favour of the digital book	Diversity of functionalities
	Society's demand
	Ecology
	Price
	Finding more information
	Transportation and storage
Against the digital book	Lose or damage
	Distraction
	Dependence on technology
	Previous editions
	c-health

Subsequently, the coding of the analysis units was carried out through the establishment of relations between the text units and the categories. The procedure for verification of process reliability consisted in the application of the test-retest method, that is, in the agreement of categorisations made by the same researcher at different times, and by categorisations made by different researchers.

After the coding and interpretation of the results, we proceeded to the creation of matrices (combining the descriptive and interpretative codifications) and the descriptive analysis was carried out through frequencies, percentages and elaboration of graphs that show the characteristics of each of them, as well as as through the discussion, interpretation and narrative description of the results.

The content analysis was carried out through the webQDA software, especially for the organisation of data, the establishment of the category system, the coding of Forum messages and the obtaining of descriptive and interpretative matrices. webQDA is software designed to support the analysis of qualitative data in a collaborative and distributed environment. It is aimed at researchers, in different contexts, who need to analyse qualitative data, individually or in collaboration, synchronously or asynchronously. This type of software follows the structural and theoretical design of other programmes available in the market, differing from these by providing online and real-time collaborative work, and a support service for research.

According to Neri de Souza et al. [15], the webQDA software has many potentialities. The ones that stand out are the fact of being in the cloud (cloud computing), allowing the researcher access to his project anywhere, working in a network collaboration environment with other researchers, analysing a variety of data sources (text, audio, video and image), organise and systematise the analysis through a tree of categories, help the researcher to record the entire context of the research through "notebooks, footnotes, grouping of items, assignment of descriptive categories and, finally, help the researcher to question the data, classify relationships and build models." [15] (Fig. 1).

Fig. 1. Screen of webQDA.

4 Results

Below, an analysis of the results obtained in the study is made, in order to try to answer the research questions initially raised, and the most relevant results of the research are described and highlighted.

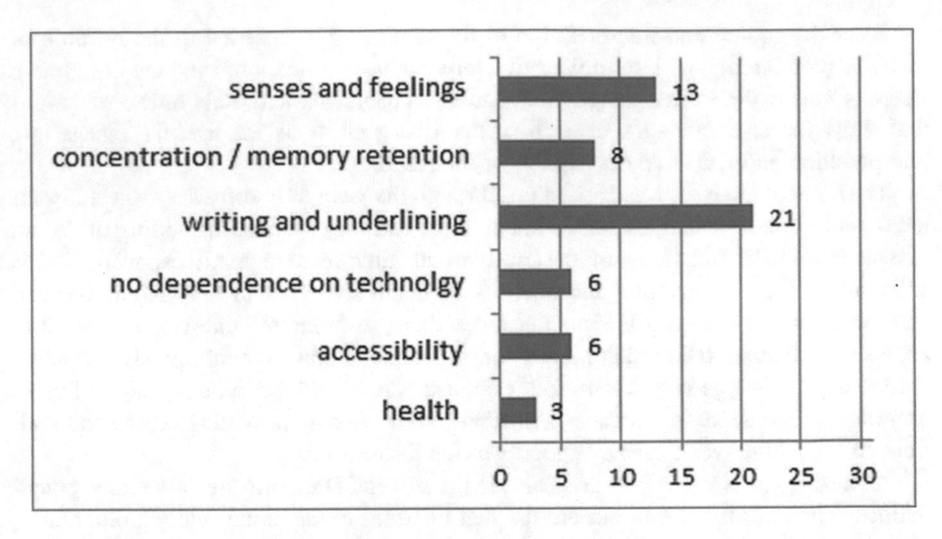

Fig. 2. Reasons to study from a printed book.

4.1 Opinions in Favour of the Printed and Digital Book

Most of the participants (74.2% - 23 students out of a total of 31 students) consider that, to study, it is preferable to use printed books, with only 19.4% (6 students) who prefer to study from books in digital format (Table 2). However, 35% of these participants (11 students) do not rule out the occasional use of the digital book, considering that both supports have their own advantages and disadvantages for studying. Each student must choose the one that best suits their needs, with which they feel more comfortable and provides more advantages and better results when studying. All the participants are aware of the emergence of the digital format, since in many classrooms tablets, smartphones and the new technologies are being used simultaneously with the paper book, the traditional notebooks and folios.

Table 2. Preference of format of the participants in the study.

Formats	Participants	%
In favour of the printed book	23	74.2
In favour of the digital book	6	19.4
In favour of combining both	2	6.5

4.2 Differences by Gender

Table 3 shows the number and the relative percentage of the participants whose opinions are in favour of reading in the two formats under analysis, depending on gender. As it can be observed, it can be said that there are more opinions in favour of the use of paper books by women (78%). In relation to the use of digital books, it can be said that men (25%) prefer the use of books in digital format.

Table 3. Preference of format of the participants, by gender.

Formats	Female	%r	Male	%r
In favour of the printed book	15	78.9	8	66.7
In favour of the digital book	3	15.8	3	25.0
In favour of both	1	5.3	1	8.3

4.3 Reasons to Study from a Printed Book

Among the main reasons to study from a book in printed format, students indicated the possibility of highlighting, annotating, crossing out, making drawings and schemes in books and even entering marks in books. They consider that these personalised actions allow them to make a more controlled reading (iR), highlight the information (mB), more easily remember the information of the text (aM), and make the search of some contents (tA) more efficient. With regard to the category "senses and feelings" students refer that reading from printed books is more pleasant, natural, intuitive, close and personal because they can feel them, handle them, turn the sheets and they even have their own smell. The same students enhance and emphasise that reading from paper, together with the actions indicated above, allows them to study in a more concentrated way and make them have a greater memory of what they read, as well as helps to memorise or better understand the topics. "The tactile sensations produced by the paper contribute to this difference." (cS).

Reasons not to Study from a Printed Book. As reasons not to study from a printed book, 4 of the participants argue that paper books require more space for storage, deteriorate over time and get lost easily, and they also consider (cS) that the manufacture of a paper book is harmful to the environment, since it produces about 2.7 kilogrammes of CO2 emissions [16]. They suggest the use of recycled paper and the option of saving ink or toner so that contamination may be lower (Fig. 3).

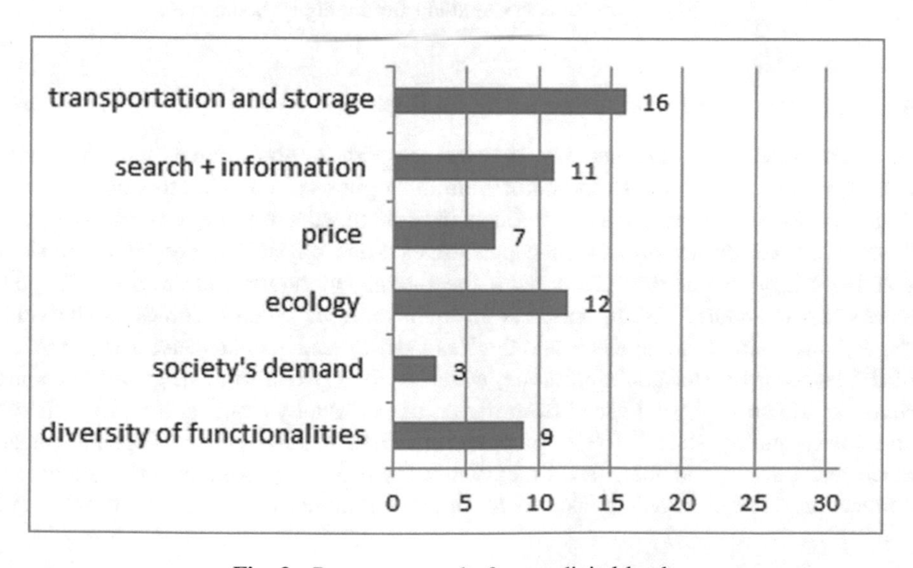

Fig. 3. Reasons to study from a digital book.

4.4 Reasons to Study from a Digital Book

The most important reasons indicated for the use of digital books to study gravitate around aspects related to their transport and storage, with the effects on living beings and the environment in which they live, and with the search for information. Students consider that digital books take up little space, are easier and more convenient to transport and store than printed books. From the ecological point of view and the sustainability of the environment, the digital book results in a considerable saving of paper. "The reading or studying from digital books results in a sustainable measure for the environment insofar as we prevent the excessive logging of trees" (jP). Students attach great importance to the links provided by digital books, which allow access to other aspects of the same topic and, in this way, better understand or expand the initial information. The digital reading devices, if they are connected to the Internet, also allow, through search engines, dictionaries databases or translators, search and expand information which complements learning (Fig 4).

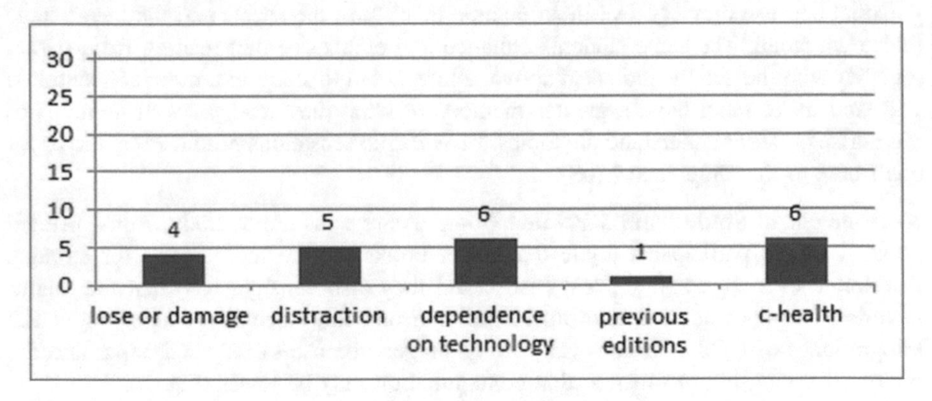

Fig. 4. Reasons not to study from a digital book.

4.5 Reasons not to Study from a Digital Book

As there were more students who showed support for the printed book to study, consequently there were also more unfavourable opinions in relation to studying from digital books. As the main reason, students indicate that the use of electronic books can have an impact on health, academic performance and physical safety. They consider that the brightness of the screen, with the passing of hours, causes visual fatigue, headaches and blurry vision, and it is an inconvenience when it comes to studying. "The lights emitted by the electronic devices used to read ebooks cause a suppression of the levels of the melatonin hormone, which is directly related to sleep and circadian rhythms. Those who use to read from digital books usually manifest a more deficient and lower quality sleep." (cS). For these students, reading always depends on an electronic device, and there are times when there is no power, battery or Internet connection. "The electronic book may at some point fail us, erase the data or run out of battery." (mG). They believe that the digital book can be a source of distraction. "It is more susceptible to fall into the temptation to be distracted by any functionality of our device (chats, browser, social networks…)." (LG) They also mention that electronic devices can fail or break down, causing the loss or damage of documents. To conclude this section, they state that "some books have not been digitised, for example, old editions, so they can not be accessed." (IR).

Functionalities of the Digital Book to Study. Considering that 9 students (Fig. 2) described some technological functionalities of digital books as specific reasons for using this type of support, we wanted to know what they are based on. The ten main functionalities described by the students are ordered and schematised in the following table (Table 4):

Table 4. Preference of format of the participants, by gender.

Funcionalities
Edit text (and associated objects) without damaging the book
Create schemes, drawings, graphs, ...
Bookmark pages
Expand and decrease the display of the pages (zoom)
Update capacity
Check links and associated objects
Include multimedia objects
Read in the dark
Copy and share books
Synchronization with other documents

5 Conclusions

Finally, the main findings of this study are summarised, pointing out contributions, limitations and perspectives of future work development.

In our society, the digital format is gaining ground on the printed format [2], but it is clear in our study that the paper format continues to be preferred by students, mainly female students. An identical result was obtained by Naomi Baron and collaborators [8] when they concluded that the paper book remains the absolute reference for university students from the countries involved in their research. However, many students emphasise that in their free time and leisure time they use the digital format for reading books, that both formats have their pros and cons and that each one must choose the format that best suits their needs, which provides more comfort and advantages when studying and through which one gets better results.

To study from a printed book, students consider that the most important reasons are the ease of annotation on its pages as a way of getting control of its content and the tactile properties of the paper that help them to concentrate and remember what they read, while among the disadvantages they mention the volume and weight of books and the expenditure of environmental or monetary resources.

As for reasons to study from a digital book, the students reason that in the same electronic support they can have several digital books, which facilitates their transport and storage, that the impact on the environment is lower and its use could help to improve the sustainability of the planet, and even that the searching functionalities allow to easily and quickly obtain contextualised information, and in different formats, on the subject under study. As for the main disadvantages, they point out that digital books are not independent physical objects and need an electronic device for reading, in addition to causing visual fatigue, mainly as a result of the brightness of the screens, and distraction resulting from the multiplicity of functionalities of the devices, especially that of Internet access.

The research team has highly valued the contribution of the qualitative methodology and its methodological approach, also considering that the use of qualitative analysis programs with webQDA greatly facilitates the achievement of conclusions.

One of the limitations of this work, which we consider pertinent, is not having started by going deeply into the concept of "studying" and the different ways of doing it. "Some people associate the term *studying* with that of memorising, but there are also those who consider that studying goes hand in hand with the term *learning*"(eB). This work opens new lines and research problems, such as knowing and understanding:

- Students' beliefs about «learning», «studying», «remembering», «knowing», «didactic resource», «text», which represent the basis of their practices of using books and which are closely linked to their conception of learning.
- The influence of age, sociocultural context and emotions in the preference to study from printed or digital books.
- Students' attitudes, practices and own intentions regarding reading from a hard copy versus reading from a screen.
- Differences in reading comprehension and ability to analyse complex ideas in educational situations, where students have to read from screens and from paper.
- If students understand and remember more in depth and in the long term what they learn through a printed book or a digital book.
- If students' expectations, opinions and learning are altered if digital books start to include more multimedia objects, diverse interactive activities, exploration guides and do not limit themselves to digitally reproduce the books on paper.
- Whether the preferences of the students for the printed format are essentially due or not to the persistence of a pedagogical paradigm based on the reception of messages or information and on the presentation and reproduction of knowledge.

Given that electronic screens have become widespread, we consider that it is still pertinent to study the relationship that students establish with new formats and reading practices, how they process, understand and use digital text in comparison with paper text, as well as its effectiveness and its implications for formal, informal and non-formal learning.

Acknowlegement. This work was supported by the European Union and Gobierno de Extremadura under Grant FEDER: Programa Operativo FEDER de Extremadura.

References

1. Area, M.: La metamorfosis digital del material didáctico tras el paréntesis Gutenberg. Realtec: Revista Latinoamericana de Tecnología Educativa **16**(2) (2017). http://relatec.unex.es/article/view/3083
2. Agencia del ISBN: Ligero aumento del número de publicaciones en 2015 (2017). https://agenciaisbn.es/web/noticias.php
3. Area, M., Pessoa, T.: De lo sólido a lo líquido, las nuevas alfabetizaciones ante los cambios culturales de la Web 2.0. Comunicar: Revista científica iberoamericana de comunicación y educación, **38**, 13–20 (2012)
4. Pérez, J.M., Pi, M.: La integración de las TIC y los libros digitales en la educación. Actitudes y valoraciones del profesorado en España. Editorial Planeta (2014). http://www.aulaplaneta.com/descargas/aulaPlaneta_Dossier-estudio-TIC.pdf
5. Del Moral, M.E., Villalustre, L.: Libros digitales: valoraciones del profesorado sobre el modelo de formación bimodal. Realtec: Revista Latinoamericana de Tecnología Educativa **13**(2), 97 (2014). http://relatec.unex.es/article/download/1291/867/0
6. Anichini, A., Parigi, L., Chipa, S.: Between tradition and innovation: the use of textbooks and didactic digital contents in classrooms. Realtec: Revista Latinoamericana de Tecnología Educativa **16**(2) (2017). http://relatec.unex.es/article/view/3061
7. Jabr, F.: Why the brain prefers paper. Sci. Am. **309**(5), 48–53 (2013). http://www.nature.com/scientificamerican/journal/v309/n5/full/scientificamerican1113-48.html
8. Baron, N., Calixte, M., Havewala, M.: The persistence of print among university students: An exploratory study. Telemat. Inform. **34**, 590–604 (2016)
9. Majul, E.: ¿Por qué el cerebro prefiere el papel?. La Voz (2014). http://www.lavoz.com.ar/temas/por-que-el-cerebro-prefiere-el-papel
10. Baron, N.: Reading in a digital age. Phi Delta Kappan **99**(2), 15–20 (2017)
11. Arango, M.: Foros virtuales como estrategias de aprendizaje. Debates Latinoamericanos **2** (2014). http://tic.sepdf.gob.mx/micrositio/micrositio2/archivos/ForosVirtuales.pdf
12. Martínez, M., Briones, S.: Contigo en la distancia: La práctica tutorial en entornos formativos virtuales. Pixel-Bit: Revista de Medios y Educación **29**, 81–86 (2007)
13. Bardin, L.: Análise de Conteúdo. Lisboa: Edições 70 (2014)
14. Fraenkel, J., Wallen, N.: How to Design Evaluate Research in Education. Mc. GrawHill, New York (2009)
15. Neri de Souza, F., Costa, A. P., & Moreira, A.: WebQDA: Software de apoio à análise qualitativa. In Rocha, A. (Ed.), 5ª Conferência Ibérica de Sistemas e Tecnologias de Informação, CISTI 2010, pp. 293–298, Santiago de Compostela, Espanha: Universidade de Santiago de Compostela (2010)
16. Wells, J., Boucher, J., Laurent, A., Villeneuve, C.: Carbon footprint assessment of a paperback book: Can planned integration of deinked market pulp be detrimental to climate?. J. Ind. Ecol. **16**(2) (2012). http://onlinelibrary.wiley.com/doi/10.1111/j.1530-9290.2011.00414.x/full

FromText: New Functionalities for a QDA Software

Gabriel da Silva Bruno[1(✉)] and Paula Carolei[2]

[1] Anhembi Morumbi University, São Paulo, Brazil
gasilvabruno@gmail.com
[2] Federal University of São Paulo, São Paulo, Brazil
pcarolei@gmail.com

Abstract. The work that resulted on this paper began as an effort to determine the functionalities required in a qualitative data analysis software created by the authors' research group as part of a study. The software is web-based, free and open-source, with a beta version available at www.fromtext.net. In accordance with Actor-Network theory, the software assists and optimizes text reading with graphs that display textual occurrences of words chosen by users and the connections established by those words in each occurrence, all without automating the processing or using algorithms to simplify text content. The functionalities to be developed for the software's second version and their main advantages were determined via experiments (in which the first usage scenarios were sketched), internal evaluations conducted by our research group and a trial of the software with students during a workshop. This paper presents a new resource for qualitative research and discusses the functionalities necessary for the type of research and software proposed by the authors.

Keywords: CAQDAS · FromText · Functionalities

1 Introduction

FromText is a tool to assist Written Text Analysis. Its first version was called WBW (Walk Between Words) and was used in all the experiments described in this paper. This paper summarizes our first impressions after developing the functionalities that turned WBW into FromText, currently available at www.fromtext.net.

1.1 About This Paper

WBW was created to assist a study conducted by the authors' research group that sought to obtain possible semantic evidence from a series of patterns we had highlighted manually in a set of texts. The development of WBW followed a design process and its functionalities were conceived in accordance with Actor-Network Theory [1].

This paper has two sections: section one presents FromText, Design Science Research and Actor-Network Theory, and section two recounts the need for new software functionalities, which are then described briefly.

A. P. Costa et al. (Eds.): WCQR 2018, AISC 861, pp. 46–57, 2019.
https://doi.org/10.1007/978-3-030-01406-3_5

Therefore, this paper is of interest to those who wish to know FromText and its many uses, as well as to those who wish to investigate matters pertaining to the sort of functionalities comprised in such tools.

1.2 FromText

WBW was designed to assist on research [2] conducted by our research group to ascertain semantic evidence of patterns and instances of methodological inadequacy in a set of texts from the Proceedings of the Brazilian Sciences and Engineering Fair. That research project sought to employ Actor-Network Theory [1] to map controversies in our corpus.

After experimenting with some of the available text analysis tools, we found that most of them perform automatic or semi-automatic calculation, categorization and groupings of data to provide better visualization of large amounts of data, which can be desirable and useful in many contexts but did not meet our specific research requirements.

Such software can be difficult to use; they may run exclusively on Linux, require complex data treatment before data can be inserted into the software, or require specific training on the user's part; some of those software generate and export complex (and often visually crowded) graphs, or are rather expensive. As a result, it seems to us that text analysis ends up being a resource/tool confined to academia and specialized users.

By contrast, we intend to make text analysis accessible even to people without any specialized training – which doesn't mean that FromText can be used without observing a few guidelines.

We did our best to ensure that the competences required to use FromText were qualitative research ones, instead of competences pertaining specifically to software usage (or to being tech-savvy in general). According to Neri de Souza et al. [3], to Linda, Kristi and Gregorio [4], analysis quality is more influenced by researchers' competences and methodological choices than by the selected QDA software [5].

Text analysis (as well as WBW/FromText) is an applicable and useful alternative to any field that relies on text to study behaviors, phenomena, reactions, patterns, dynamics etc. It falls to the researcher to determine among the software's many options which one is more suited to his/her needs and how s/he will conduct his/her research.

In short, to begin using FromText the researcher must input the text s/he wishes to analyze. The following steps depend on the researcher's goals and theoretical framework. Our research group usually works with words that are *Previously Selected*, of *Emerging Interest* and of *Correlated Emerging Interest*.

Previously Selected words are words related to user hypothesis about the analyzed text. Words of *Emerging Interest* are the ones that can come to mind while using the software, even if they are not on the analyzed text. Finally, the option *Correlated Emerging Interest* registers and seeks words in the generated graphs to check for other occurrences.

Upon input, FromText looks for those words in the analyzed text, providing an organized display of *all* their *occurrences* and *neighboring elements* as well as clickable spheres for each related word.

Graphs generated by FromText must be read from left to right, in descending lines, as shown on Fig. 1, which depicts a search for the word "conservation" on the Sciences section of the Philippine curriculum guide for the K to 12 Program, by the country's Department of Education [6].

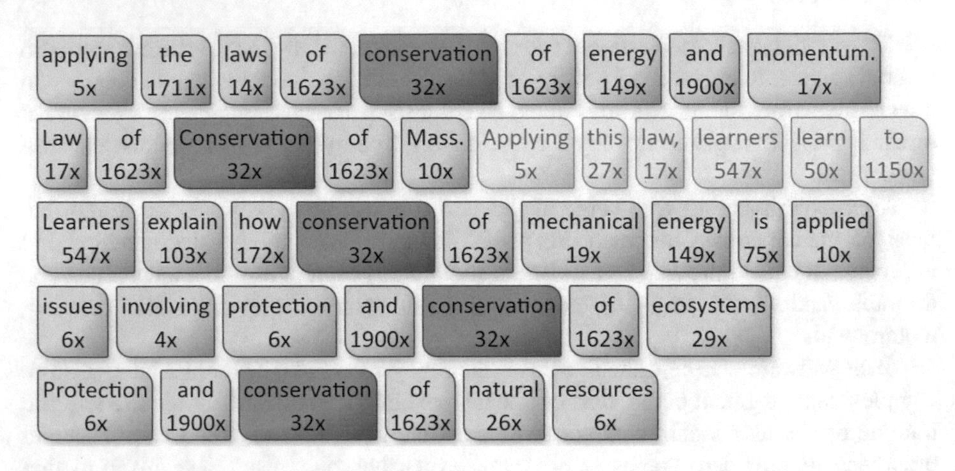

Fig. 1. Five out of thirty-two occurrences of "conservation" in the analyzed text, associating conservation to energy, momentum, ecosystems and natural resources.

A next possible step would be to click on "ecosystems" to find all occurrences of that word in the curriculum and other existing co-occurrences besides "conservation-ecosystems".

FromText was developed using JavaScript (one of the most used programming languages on the internet [7]), so anyone is able to manipulate and edit the attached files, creating their own version/edit of the software. Being web-based, FromText also works on any device connected to the internet and doesn't require much in terms of device performance.

1.3 Design Science Research

WBW (and its successor FromText) was conceived and designed using the Design Science Research methodology; in that framework, the development process required immersion in three axis: technological (programming), artifact (design) and scientific (validation and application), as discussed in further detail in another paper (in press) [5].

- The technological axis refers to computing matters, such as which programming language to use, locating and correcting errors, improving performance etc.
- The artifact axis refers to the designing aspect of the development process, especially its pragmatic validity; this axis comprehends data collection instruments, data interpretation, positing solutions and the many stages of prototyping and iteration (enhancement).

- The scientific axis refers to how our software relates to existing techniques and methods of qualitative analysis of textual data.

The Design Science framework was first conceptualized by Economy Nobel Herbert Alexander Simon in his book *The sciences of the artificial* [8]. This methodological apparatus is applicable in sciences that differ from natural and social sciences by virtue of being project-oriented; Design Sciences aim to create, develop and study man-made artifacts.

For a clearer understanding of how this paradigm relates to (and interacts with) natural and social sciences, see the following table by Dresch, Lacerda and Júnior [9] (Table 1).

Table 1. Natural sciences, social sciences and Design Science.

	Natural sciences	Social sciences	Design Science
Purpose	To understand complex phenomena, discover how things are and explain why they are like that	To describe, understand and reflect upon the human being and his actions	To project and build new systems; change existing situations to achieve better results, focused on problem-solving
Research aims	To explore, describe, explain and predict	To explore, describe, explain and predict	To prescribe. Research must be directed to solve existing problems

Regarding how Design Science relates to natural and social sciences: scientific knowledge is traditionally organized, methodic, systematic, rational and cumulative, and must be replicable, dependable, generalizable and verifiable [9]. Design Science adds a new requirement: applicability [10]. According to Dresch et al. [9], articulating scattered knowledges to develop artifacts that perform a certain function and meet a certain need is possibly what matters most in Design Science.

This (theoretical) need to be met arose from the activities of our research group as both artifact developers and users and can be approached using Actor-Network theory.

1.4 Actor-Network Theory

Pinto and Domenico [11] provide one of the clearest definitions of ANT (Actor-Network Theory): a theory that, as opposed to traditional sociology, regards the agency of human and non-human actors equally. ANT assumes that people are not the only agents to participate in, model and influence reality [12].

The association of the human and non-human elements form networks or, more specifically, actor-networks, for one must consider not only the elements but their performed agency.

It is of the utmost importance to employ the greatest possible number of approaches and observation instruments, so as to concede visibility to different points of view [11]. ANT's symmetry requirement posits that no element should be accorded less importance in the analytical process. The greatest challenge for an ANT-abiding analyst is *to build a narrative that forms a network, displaying objects, mobilizations, movements and facts*, listed or unlisted [13].

WBW/FromText's core operation (to find and display the occurrences of a given word in a text) is related to the following proposition by Latour: whenever a grouping is mentioned, its formation mechanism becomes visible and may then be tracked. According to Latour, "there is no group without a recruiting officer", a fact that becomes evident when we use FromText to check on word co-occurrences and ensuing connections.

Despite its simple purpose, our software operates considering the existence and complexity of networks and does not generate automatic representations of those systems, since it aims at enabling the interpretation of networks as real systems. Our goal is to have WBW/FromText's graphs work as a snapshot of text elements, which speak by themselves and must be heard by the researcher, who should be the one to draw relationships between them.

The last few paragraphs sum up our refusal to automate the analytical process. However, as has already been mentioned in this paper, the choice of software is less important than the choice of method; the researcher must weight available resources, the data to be analyzed and the theoretical framework structuring his/her analysis.

Considerations on the use of FromText and its main connections to ANT are presented in another paper (in press) [14], the reading of which might enable a better use of the software or even assist users in choosing the instruments best suited to their analytical needs.

1.5 Intersections

To conclude this presentation of WBW/FromText's context and theoretical framework, let us look at the following graph (Fig. 2). It presents FromText as an artifact that contributes and interacts in two directions (knowledge application and knowledge base) and is influenced by rigor and environmental needs; therefore, it can provide direct theoretical contributions either by theorization or by means of new technologies that may contribute to solve a given problem [9].

The application of the Design Science Research (DSR) methodology provides two more connections: application (DSR-environment) and contributions (DSR-knowledge base). There are also two internal movements in the DSR process: to develop/build in order to justify/evaluate and in doing so subsidize the refinement of artifacts and theories, so that the designed artifact adds value to both environment and knowledge base.

The following section discusses how we determined which software functionalities were to be developed.

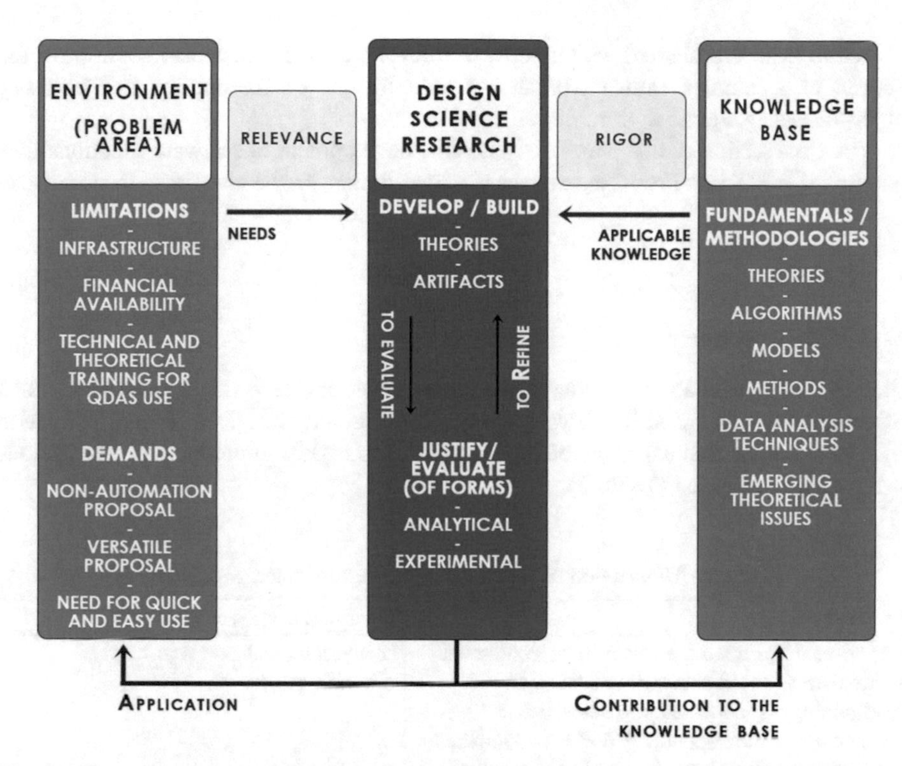

Fig. 2. The scheme displays Design Science role in the software development process and its workflow, with environmental factors (adding relevance) and knowledge base factors (adding rigor). This scheme was adapted and translated from "Design, Base and Enhancing: building a QDA software" [5].

2 Methodology

Building upon an empirical research with exploratory and descriptive objectives that used seven experiments with WBW to detect the need for new software functionalities, this paper discusses the main advantages of those functionalities.

Although our research data were first structured in tables for better visualization, the quantitative aspect of this research is negligible when compared to group discussions, technical viability assessments and other experiments with the new software functionalities that pointed at possible future FromText versions. The table used during the evaluation process is available as part of the Proceedings of the 7th Ibero-American Qualitative Research Congress [5] (in press), from which this paper derives.

Because we wished to understand different WBW usage possibilities, the seven experiments followed no standard: each one had its own goals and worked with different types of text, such as resumes, books, online forum answers etc. Since our goal was to refine the software, in each experiment we made sure to identify what could be improved (e.g. processing speed, making graphs exportable).

Aside from the charted evaluations, the development process also drew from the records of a summer course workshop taught by one of the authors to Chemistry undergraduate students at a public university.

The final section of this paper discusses the development of software functionalities that turned WBW into FromText and the steps to be taken on the next development cycle.

3 Results

3.1 The Experiments

The seven experiments were conducted by both authors, recorded and arrayed. Each experiment was assigned a keyword by which we could refer to it during evaluation [5]. However, for this paper we believe keywords to be less effective than descriptions of the usage scenarios (Table 2).

Table 2. Description of seven experiments conducted with WBW.

Experiment	Identified usage scenario
Experiment 1: First experiment to be conducted. Its goal was to find semantic evidence of methodological inadequacy in science fair projects about videogames. WBW was compared to another software and showed better results: it enabled the researchers to discern several elements that the other tool failed to make visible	Methodological issues in a set of scientific papers
Experiment 2: WBW was used to analyze the Proceedings from a Biology Science Fair, looking for indicators of projects that addressed real and local issues from their authors' respective contexts; we queried for words such as "city", "neighborhood", "school" etc.	Checking for context and environmental markers in a set of scientific papers
Experiment 3: This experiment focused on cyberspace studies and was presented at a Cyberculture Symposium. We analyzed in tandem (1) a set of user comments from a news portal story on an artistic performance in which a child interacted with the (male) artist's naked body. Several patterns were detected, such as the association between "family" and "[to] destroy"; and (2) posts from an open online forum where men talked about whether or not they would have sex with a 12-year-old girl who had "a woman's body". Our goal was to check for differences in the graphs generated by the analysis of each set of comments, focusing on childhood, nakedness and eroticization	Tensions and groupings in cyberspace

(*continued*)

Table 2. (*continued*)

Experiment	Identified usage scenario
Experiment 4: Clarice Lispector's novel *The Hour of the Star* was analyzed to check for occurrences of certain words, such as "hug", "word" and "wharf"; we found instances of polysemy and the return of some symbolic elements at the character's death scene	Analysis of a literary text
Experiment 5: Its goal was to highlight and compare thematic competences and incoherencies on governmental curriculum guidelines. WBW was used to compare different government documents and legislation on curricula, to highlight elements and problems in that set of texts. Tables proved to be a challenge, solved by establishing standards for text conversion prior to its input on the software	Analysis of government documents
Experiment 6: WBW was used to analyze the arguments presented by students discussing at an online course forum. Highlighting tensions and identifying concepts and references proved interesting. Later, our analysis subsidized the teacher's discussion summary as well as personalized student feedback	Asynchronous online education forums
Experiment 7: A classroom activity to discern relationships between texts elements, thus facilitating the creation of concept maps. This experiment was part of an educational activity with the goal of having K-12 students create concept maps from a text read for Biology class. This experiment aimed at verifying whether WBW could be used outside academia, e.g. as a pedagogic resource	Visualization for network understanding

These experiments with WBW enabled us to detect the need for new software functionalities [5], particularly the following: visualization of a greater number of spheres, simple graph export functions, all-time visible navigation, creation of content categories, simultaneous tracking of more than one word and not distinguishing between upper and lower case.

3.2 New Functionalities

Some of the functionalities that turned WBW into FromText do not result from the experiments described above, such as the option to track word parts, which was ideated during discussions between the authors.

The most important new functionalities have already been developed and their advantages are summarized in the following table (Table 3).

Table 3. New functionalities and their main advantages.

Functionality	Main advantage
Visualize original text excerpt	Enables user to better understand and check meanings
Expand reach	Better element visualization
Word gallery	Simultaneous tracking of a set of words
Simultaneously track more than one word	Optimizes word-tracking
Track for words and spaces	Enables tracking of multi-word terms
Track word parts	Includes plural forms and word variants in the tracking results
Display word frequency on graphs	Enables micro-context comprehension
Highlight a word in a previously generated graph	Displays co-occurrences
Export graphs	Records generated graphs
Navigation by icons	Optimizes usability
Breadcrumbs of previously tracked words	Enables visualization of research steps
More color choices for graph visualization and exporting	Enhances user experience
Option to remove stopwords	Affords less visually crowded results on tracking functions related to word frequency
Text reflow and left-aligned text	Better visualization; optimizes text for printing in A4-sized paper sheets
Insert metainformation and categories on text excerpts	Allows user to create, differentiate and keep track of metainformation and categories of the analyzed text

Other functionalities under consideration for future development are the possibility to upload text in different formats and better processing of tables, both of which felt necessary as we performed and discussed the experiments with WBW. Some software upgrades unrelated to the experiments and yet to be developed include interface translation to English, Spanish and French, as well as accessibility adaptations – e.g. compliance with screen reader software standards.

FromText's new interface has horizontal and vertical navigation menus, a work area and a textbox, as shown in Fig. 3 below:

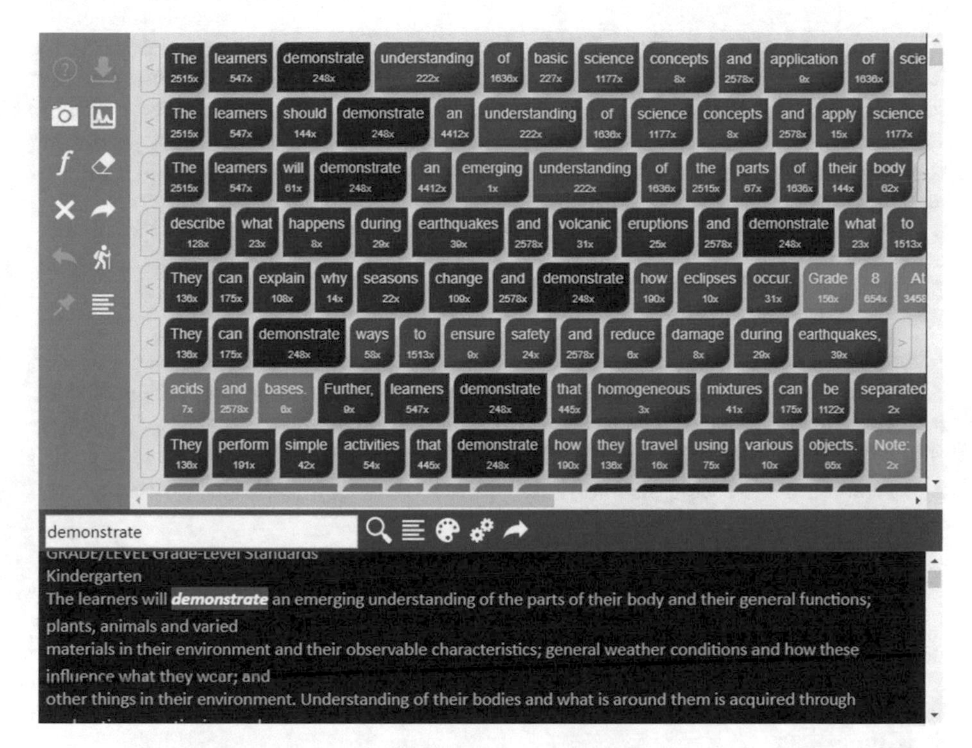

Fig. 3. New FromText interface, with horizontal and vertical navigation menus, a work area and a textbox at the bottom.

After developing these functionalities, we conducted performance tests to compare software versions. Speed improvements were negligible, but the new algorithm runs faster even though it performs several new functions. The new navigation menu provides faster access to its elements, thus optimizing software use.

3.3 Experiment Conducted During a Workshop

At a summer course workshop for Chemistry undergraduate students at a public university, the participants seemed unable to fully grasp the possibilities afforded by FromText. Out of 20 students, only 2 had any experience with text analysis. There was a grad student who pointed out the need for some software improvements, the most critical being an increase on the number of spheres (which would enable users to grasp the analyzed text as a whole). On the second day, we adjusted the code to allow for 16 spheres, but the student still felt that it should be possible to work with even more spheres if necessary [5] – the issue was solved with the "expand reach" function.

During the workshop there was no internet connection, so we had to use the software's offline version. In addition, there were not enough notebooks for all students, so some of them used WBW on their smartphones; software performance was not too different from using it on a computer. Both circumstances suggest the need to consider offline use and mobile devices for future FromText development.

Some workshop participants wanted functionalities available in other software (such as creating graphs with edges and nodes), while others saw no use for Text Analysis in their studies. Therefore, some users seem to have no need for a specific software to assist their research, while others are looking for mass data processing software (which is not the purpose of WBW/FromText and could be a source of disappointment).

4 Conclusion

Although the current aim of our software experiments and evaluations is to clarify and advance theoretical matters related to FromText, it has also been interesting to think that its technical quality may allow the software to reach a broader user-base; a possible step in that direction would be to work on compliance with ISO standards [15], which focus on aspects such as software trustworthiness, usability, efficiency and maintenance.

Faced with some technical challenges, we have considered the possibility to develop a PHP version. However, for a third software version we intend to continue with JavaScript and maybe have PHP running non-essential functions, such as the option to interact with data banks and thus enable online collaboration.

WBW and FromText certainly aren't meant to replace existing (and consolidated) text analysis techniques and software. We offer an alternative that seemed to be urgently needed.

New software functionalities are indeed necessary, although most of the essential ones were already developed and implemented for a beta version of FromText (hosted at www.fromtext.net). We believe that the lack of those functionalities has influenced user perception: certain users thought WBW was "a tool to produce text synopsis", when the software's purpose is the exact opposite: to complexify interactions with analyzed texts [5].

Following this study, we intend to finish developing additional functions and to begin a new evaluation cycle to address technical and usability matters and, more importantly, to work on software usage scenarios.

References

1. Latour, B.: Reagregando o social: uma introdução à teoria do ator-rede. EDUSC, São Paulo (2012)
2. Bruno, G.S., Carolei, P.: Analise semântica de resumos científicos como recurso para evidenciar pontos críticos. In: Atas do XVII Encontro Nacional de Educação em Ciências (2017)
3. Neri de Souza, D., Neri de Souza, F., Costa, A.P.: Percepção dos Utilizadores Sobre o Software de Análise Qualitativa webQDA. Comun. Informação **17**(2), 104–118 (2014)
4. Linda, S.G., Kristi, J., Gregorio, S.: Tools for analyzing qualitative data: the history and relevance of qualitative data analysis software. In: Spector, J.M., Merrill, M.D., Elen, J., Bishop, M.J. (eds.) Handbook of Research on Educational Communications and Technology, 4th edn., pp. 221–236. London (2014)

5. Bruno, G.S., Carolei, P.: Design, Base e Aprimoramento - O percurso de um software para QDA. In: Atas do 7° Congresso Ibero-Americano em Investigação Qualitativa (2018)
6. D. of E. of P. DepED: K to 12: Primary and Junior High School Curriculum Guides (CG) [Online] (2016). https://www.spideylab.com/primary-junior-high-school-curriculum-guides-2017/. Accessed 2 July 2018
7. Flanagan, D.: Javascript: o guia definitivo, 6th edn. Bookman, Porto Alegre (2013)
8. Simon, H.A.: The Sciences of the Artificial, 3rd edn. MIT Press, Cambridge (1996)
9. Dresch, A., Lacerda, D.P., Júnior, J.A.V.A.: Design Science research: método de pesquisa para avanço da ciência e tecnologia. Bookman, Porto Alegre (2015)
10. Appolinário, F.: Dicionário de metodologia científica: um guia para produção do conhecimento científico. Atlas, São Paulo (2011)
11. Pinto, C.C., De Domenico, S.M.R.: Teoria Ator-Rede em Estudos Organizacionais: Encontrando Caminhos via Cartografia de Controvérsias. In: VIII Encontro Estud. Organ. da ANPAD, p. 16 (2014)
12. Montenegro, L.M., Bulgacov, S.: Reflections on actor-network theory, governance networks, and strategic outcomes. BAR - Brazilian Adm. Rev. **11**(1), 107–124 (2014)
13. Silva, F.A.R.e., Lisboa, D.do P., Oliveira, D.do P.L., Coutinho, F.Â.: Teoria Ator-Rede, Literatura e Educação em Ciências: Uma proposta de materialização da rede Sociotécnica em Sala De Aula, pp. 47–64 (2016)
14. Bruno, G.S., Carolei, P.: A influência da Teoria Ator-Rede no desenvolvimento e uso de um software web para pesquisa qualitativa. In: Semana de Ciências Sociais PUC-Rio (2017)
15. ISO: ISO/IEC 25010:2011 - Systems and software Quality Requirements and Evaluation (2011). https://www.iso.org/standard/35733.html. Accessed 2 July 2018

IRAMUTEQ Software and Discursive Textual Analysis: Interpretive Possibilities

Maurivan Güntzel Ramos(✉), Valderez Marina do Rosário Lima, and Marcelo Prado Amaral-Rosa

School of Sciences, Postgraduate Program in Science and Mathematics Education, PUC – Pontifical Catholic University, Porto Alegre/RS 90619-900, Brazil
mgramos@pucrs.br

Abstract. The presence of specialized software is becoming more and more intertwined with the procedures of data analysis in qualitative surveys. Therefore, the guiding question of this text was: *In what way can the software analysis of qualitative data, IRAMUTEQ, contribute to the process of Discursive Textual Analysis?* The topic was focused on technology in the context of the classroom. The objective was to identify analytical possibilities existing in the relationship between the IRAMUTEQ software and the method of Discursive Textual Analysis, with a view to the emergence of new interpretative pathways using technology in the classroom. The participants, subjects, were teachers joining the post-graduate degree (n = 40). The data collection instrument was an open questionnaire. The method for analyzing the data was Discursive Textual Analysis, in conjunction with an exploratory analyses using the IRAMUTEQ software. The software identified five textual classes based on the *corpus* of analysis, which in turn resulted in three definitive categories: (i) technologies for increased understanding between teachers and students; (ii) technologies to modify teaching practice; and (iii) technology for changes in school. Lastly, it is confirmed that the IRAMUTEQ software contributes to the methodological process of Discursive Textual Analysis, offering agility, new perspectives and rigour to qualitative textual data analysis.

Keywords: Discursive Textual Analysis · IRAMUTEQ software
Qualitative analysis

1 Introduction

The use of specialized software is increasingly present in the analysis of data in qualitative surveys [1, 13–15, 19, 31]. Today, the virtual characteristics of technology are inseparable from the day to day functioning of society [8, 24], and thus, they are found in the processes of designing, processing and understanding research data in the field of Education in Science and Mathematics.

The crux of this text is the perceptions [23, 26] concerning technologies in the context of the classroom. Discussions of this nature and their implications are timely and, it seems, will be opportune for many years [13]. Therefore, those interested in

A. P. Costa et al. (Eds.): WCQR 2018, AISC 861, pp. 58–72, 2019.
https://doi.org/10.1007/978-3-030-01406-3_6

research in education and/or teaching and learning should do likewise, giving attention to the methodological possibilities offered by such applications [3, 11, 12, 22].

Before proceeding, it should be noted that this text is a development of the work presented in the 7th CIAIQ – Congresso Ibero-Americano em Investigação Qualitativa (Iberian-American Congress in Qualitative Investigation) and published in the procedings, *Contribuições do IRAMUTEQ para a Análise Textual Discursiva* (IRAMUTEQ contributions to Discursive Textual Analysis). It is emphasized that, in this version, the treatment of the analyses gains greater prominence and depth.

For the desired understanding of the theme under study, the processing of data was achieved through the use of Discursive Textual Analysis. This is a method of data analysis with the purpose of producing new understandings, presenting a hermeneutic interpretative process [20], since it considers the context of those who speak. It is organized using four essential foci, being: (i) dismantling of the texts (unitarization); (ii) establishment of relations (categorizations); (iii) the capture of new findings (production of a metatext); and (iv) communication [18, 20, 21, 25]. Thus, it permits a deep understanding of research topics and with immersion in the data, new reflections are formed [18].

The relevance of this work is centered on the use of the IRAMUTEQ software [6, 29] in the partial treatment of the data using the steps of Discursive Textual Analysis [20]. We emphasize, it is neither intended to mechanize the analytical steps of Discursive Textual Analysis nor to exclude the interpretive responsibilities of the researcher during the process [31]. If it were so, then we would not be doing qualitative research and we would not be following the steps of Discursive Textual Analysis. Thus, as previously stated, we draw to the attention of the reader the new aspect of this work [12] for the teaching of Science and Mathematics [6, 16, 17, 28].

In the face of the above, the guiding question was formulated: *In what way can the IRAMUTEQ software analysis of qualitative data, contribute to the process of Discursive Textual Analysis?* In this sense, the objective was to identify analytical possibilities existing in the relationship between the IRAMUTEQ software and the method of Discursive Textual Analysis, with a view to the emergence of new interpretative pathways using technology in the classroom. To this end, the individual statements [23, 26] of the teachers (n = 40) regarding the use of digital technology in teaching practice, were submitted to the IRAMUTEQ software [6, 29].

Regarding the organization, this article is presented in four sections: (i) *presentation of the IRAMUTEQ software in the Analysis of the Data*, in which the central point is the basic characteristics of the text analysis software in question; (ii) *methodological procedures*, which aims to clarify the established techniques and procedures, presenting three subsections: (ii.i) *object of study*; (ii.ii) *subjects and procedures*; and (ii.iii) *data collection and analysis*; (iii) *results and discussions*, the crux is the results obtained in the IRAMUTEQ in discourse with the literature; and finally, (iv) *conclusions*, a return to the guiding question and a consideration of the main contributions of the IRAMUTEQ software when used in order to meet the principles of Discursive Textual Analysis.

2 The IRAMUTEQ Software and Analysis of Qualitative Data

The presence of technology in present day society is considerable [8, 24], and the trend will increase in all sectors in the coming years [8]. Thus, the area of research in teaching and learning and/or education, whilst it allows understanding about social dynamics, is also directly affected by several technological advances. In this section, one of the multiple possibilities available on the market for qualitative data analysis will be presented in a succinct manner: the IRAMUTEQ software [6, 29].

There are various software options available related to the processing of qualitative data, for example, WebQDA [22], SPHINX [11], Nvivo [3], ALCESTE, among others, each with its resources and purposes [1, 12, 14]. In Brazil, the use of this type of software for the analysis of data in qualitative research began in the 1990s [6].

The IRAMUTEQ software was developed using the French language in 2009 [6, 29]. It is an acronym for *Interface de R pour les Analyses Multidimensionnalles de Textes et de Questionnaires* (Interface of R for Multi-dimensional Text and Questionnaire Analysis) [6, 29]. The textual statistical analyses are rooted in the same algorithm of ALCESTE, which allows for the recovery of the context in which the words belong, a possibility that made it very popular among researchers in the humanities, and more specifically of the area of Social Representations [6, 28], and Health [16, 17].

It is free and was developed under the free software license and open source, being linked to statistical software *R*, and the programming language is *Python*[1] [6, 7, 29]. The importance of IRAMUTEQ is also linked with its license type and programming codes, since it permits: (i) any user to run the program; (ii) the study and adaptation of the programme as required by individuals; (iii) the distribution of copies and to help others; and finally, (iv) the perfecting and sharing of ones findings with the community [30].

In Brazil, the IRAMUTEQ software began to be used in 2013 [6]. It is a recent software to the National Academy, so academic work arising from its use is scarce and confined to specific areas such as Social Psychology [6]. Directly responsible for the testing, development of dictionaries and dissemination, is LACCOS - Laboratório de Psicologia Social da Comunicação e Cognição (Laboratory of Social Psychology of Communication and Cognition), da UFSC – Universidade Federal de Santa Catarina (Federal University of Santa Catarina, state of Santa Catarina, southern region of Brazil); CIERS-Ed – Centro Internacional de Estudos em Representações Sociais e Subjetividade – Educação (International Center for Studies in Social Representations and Subjectivity – Education), da FFC – Fundação Carlos Chagas (Carlos Chagas Foundation, state of São Paulo, Southeast region of Brazil); and the Grupo de Pesquisa Valores, Educação e Formação de Professores (Research Group Values, Education and Teacher Training), da UNESP – Universidade Estadual Paulista Júlio de Mesquita Filho (State University of São Paulo Júlio de Mesquita Filho, state of São Paulo, Southeast region of Brazil) [6].

[1] Created by Guido van Rossum in 1991, with the aim of being a language programming that produces agile maintenance codes. For more information, access: www.python.org.

Among the features of the software is the possibility of executing textual data analyses at different levels [6, 7, 16, 29]. There are five possibilities for textual data analysis, namely: (i) classical textual statistics - identifies quantity and frequency of words, single words (Hapax coefficient), identifies and searches words according to the grammatical classes and search root-based words (stemming); (ii) search for group specifications; (iii) descending hierarchical classification (by consequence of Factorial Correspondence Analysis); (iv) analyses of similarity; and (v) word cloud [6, 7, 16, 29].

While parsing simple textual data, for example, it is possible to perform *word frequency calculations* that can be presented in a cloud of words. And, for multivariate analyses, more sophisticated than the first, it is possible to carry out the *Descending Hierarchical Classification* (DHC), *Factorial Correspondence Analysis* (FCA) and Analyses of Similarity [6, 7, 29].

In this study, *Descending Hierarchical Classification* (DHC) and *Factorial Correspondence Analysis* (FCA) will be dealt with. IRAMUTEQ performs a subdivision, by means of statistical calculations, of the corpus of analysis to reach the *text segments* TSs [6, 7, 16, 29]. After, the *Text Segments* (TSs) are sorted based on their vocabulary, the sets are divided based on the frequency of the words already lemmatized, in order to create classes similar to each other and different from others [6].

The *Factorial Correspondence Analysis* (FCA) is achieved as a result of the analysis of *Descending Hierarchical Classification* (DHC), approaching a kind of "internal function" of DHC. In the FCA, in a Cartesian plan, the different grouping of words or subjects that constitute each of the classes proposed in the DHC are presented [6, 7, 16]. It is also possible to know the intensity of each word next to the class set and to access the text segments of each subject entered in the classes for a more qualitative interpretation of the data by the researcher.

With respect to the installation, downloading of *Software R* (www.r-project.org) and installation, without the need to carry out specific procedures. After that, downloading and installing the IRAMUTEQ software (www.iramuteq.org). This order of installation is of paramount importance, since IRAMUTEQ uses the *Software R* in the process of textual analysis and will need to have the codes that act with its installation [6, 7]. It is advisable, before attempting to insert textual data into IRAMUTEQ, that the user makes a detailed reading of the manual [7]. The preparation of the necessary encodings in the texts must be accurate, thorough and planned. Thus, one may avoid mistakes in the generation of data in the analyses, distorted interpretations and annoyances when inserting the files from the *corpus*, since it is not possible to insert saved files with the same name more than once.

Finally, it is underlined that the approximation with Discursive Textual Analysis [20] is possible because the software is only a tool to support the method and does not form a method p*er se*. In this way, it is worth noting that IRAMUTEQ [6, 7, 29] does not conclude the analysis, with the role of the respondent being the interpretive agent of the data [18, 20, 21].

3 Methodological Procedures

3.1 Object and Context of the Study

The object of the study is the contributions of the features of the IRAMUTEQ software [6, 29] applied to Discursive Textual Analysis [20]. The basic texts used for the analyses are the views of teachers on digital technologies in the context of teaching practice. The intention was to capture the positions pertaining to a topic imposed by contemporary society in the classrooms that may (or may not) be of personal interest to these teachers: digital technologies in a didactic context [2]. It is important to understand such a point, since the individual's view of something [23] is often not evident [2].

Therefore, they can determine the adoption or exclusion of technology in their daily pedagogical activities [3]. The views of these subjects are formed based on their experience of the world, and thus, they are powerful enough to assume an important explanatory role of that which is, in fact, real, thus giving meaning to the world that surrounds the individual [26]. Thereby, in this article, the crucial point was the possibilities offered by technology, both in terms of the academic context that encompasses instruments for research and in the generation of hypotheses regarding the school context by the participants (subjects) themselves.

3.2 Subjects and *Corpus* of Analysis

The subjects of the research were 40 teachers joining the post- graduate degree in Education in Science and Mathematics, being 30% (12) male and 70% (28) female, with an average 32 years of age. Of these, 37.5% (15) are professors of Science (Chemistry, Physics and Biology), 55% (22) are Mathematics and 7.5% (03) are from other areas. All participated voluntarily, filling out the information and consenting to the use of the data. Anonymity was guaranteed by means of the designation of *Subject* with sequential numbering (*Subject1, Subject2, ..., Subject40*). In IRAMUTEQ [6, 29] the global analysis of the *corpus* was made up of 446 Text Segments (TS), with use by the software of 77.8% (347 TSs). The textual segments present, on average, 3.25 lines, with ~36 words per TS. The total lexicon set has 16,119 occurrences (average ~400 words by text), being 1,691 distinct and 761 of single occurrence.

3.3 Data Collection and Analysis Tools

The data collection used was carried out through an open questionnaire [1, 14, 31]. The instrument was ample, containing eight dimensions, each with internal questions, which dealt with the experiences, personal theories and classroom reality in the school context, being: (i) teaching and learning; (ii) experimenting in the teaching of Science and Mathematics; (iii) Internet and social networks in the teaching and learning in Science and Mathematics; (iv) environmental education; (v) evaluation in the teaching of Science and Mathematics; (vi) the function of language in the teaching of Science and Mathematics; (vii) inter and transdisciplinarity in the teaching of Science and Mathematics; and finally (viii) research in the classroom. The intention was to generate a robust database with a holistic perspective on the global representation of the subjects with the view of echoing the topics that would be addressed during the academic semester within the post-graduate course.

Attention falls on the third topic of the questionnaire: *internet and social networks in the teaching of science and mathematics.* The questions of the target topic were: (i) how do you perceive the Internet and social networks in the context of teaching and learning in Science and Mathematics?; (ii) what is the role of the Internet and social networks in the classroom?; (iii) how do you use these resources?; (iv) what are the main difficulties that you perceive in your practice in relation to the Internet and social networks in teaching and learning in Science and Mathematics?; (v) what solutions do you propose for these difficulties?; and (vi) as an example, explain some situation in your experience as a teacher or as a student which shows the relationship with the use of the computer in teaching and learning.

The questionnaire was applied in two separate groups. The first group responded in the first quarter of the Year of 2013 and the second in the same period of the following year. The intention is not a comparison between the groups, but a general view of the understandings as teachers joining the post-graduate course in Education in Science and Mathematics. The number of subjects (n = 40) provided data saturation for analyzes against the teacher understandings [10, 14], a difficult issue to be achieved in similar studies.

IRAMUTEQ software presents in its programming structure a quantitative analytical basis [6, 29]. However, the macro analysis of the data is, in essence, qualitative [1, 14, 20, 31] since it is founded on the theoretical and practical assumptions of Discursive Textual Analyses [18, 20, 21]. In this way, it is underlined that the quantitative analysis was maintained as a feature of a background analysis in an automated way through the resources of the software itself, without the need for direct application of its procedures [6, 7, 29]. Quantitative analyses were therefore used as a first resource to base and understand the desired qualitative analysis in greater depth [1, 10, 14].

The organization of the material and the analyses were carried out according to the requirements and possibilities of IRAMUTEQ [6, 7, 29]. The character of the analyses is exploratory, with a view to enhancing the process, manually, through the use of Discursive Textual Analysis [18, 20, 21, 25]. It is underlined that the intention is not to question the merits of the procedures developed, nor provoke controversy between the processes of manual analysis *Versus* digital [3, 12, 14, 22]. Efforts are focused solely on identifying potential contributions in the processing of data through exploration of the resources [3, 11, 12, 22] offered by IRAMUTEQ [6, 29].

4 Results and Discussions

In order to obtain results in the IRAMUTEQ software, a group of texts (n = 40) was used in a single file, which is called the *corpus* [6, 7, 16]. The data resulted from the Descending Hierarchical Classification analysis (DHC) and, consequently, the Factorial Correspondence Analysis (FCA). In such a way, it is pointed out that they were considered only one of the five textual data analyses available in the software (see Sect. 2).

The *corpus* of analysis is made up of 446 text segments (TSs), with use by the software of 77.8% (347 TSs). It should be noted that in IRAMUTEQ, due to the programming employed, the context of the words is considered, being the total of TSs called *Initial Context Units* (ICU) and the TSs utilised are the *Elementary Text Units* (ETU) [6, 7].

In DHC analysis, a dendrogram was generated with five classes (see Fig. 1[2]). From this data, it is possible to draw up interpretations of the formations of each class, as well as to understand the similarities and differences between them.

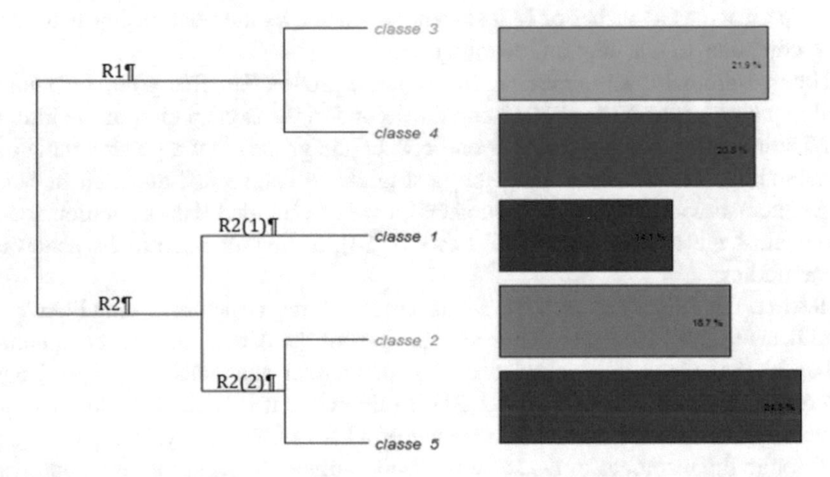

Fig. 1. Descending hierarchical classification of the representations of the subjects.

With respect to the five classes generated for the *corpus* in analysis, reading from left-to-right, two ramifications are evident: (i) ramification 1 (R1), with 3 classes 3 (21.9%) and 4 (20.8%); and (ii) ramification 2 (R2), with the class subdivision (R2 (1) and R2 (2)) 1 (14.1%), 2 (18.7%), and 5 (24.5%). This is the final configuration of the DHC, given that IRAMUTEQ provides the subgrouping until the classes present themselves in order to have the *Elementary Context Units* (ECU) in their most similar vocabulary densities, becoming distinct, one from the other [6, 7, 29], thus becoming established (see Fig. 2).

With the dendrogram of the global DHC referring to the construction of the classes, it is possible to perceive that the stabilization of the classes occurs as follows: three macro ramifications (R0, R1 and R2) and seven internal ramifications, all belonging to R2. In R1, classes 3 and 4 demonstrate superior affinity with each other, demonstrating a *strong interaction*, in addition to a *strong distance* from the other classes, with 42.7% of the total on the *corpus*. Already in R2, there have been subdivisions on three internal

[2] All analyses in the IRAMUTEQ software was performed in the Portuguese language. It is fundamental, since the subjects of the research are Brazilian teachers and the method of Discursive Textual Analysis is based on the interpretation, including that which is not is not said.

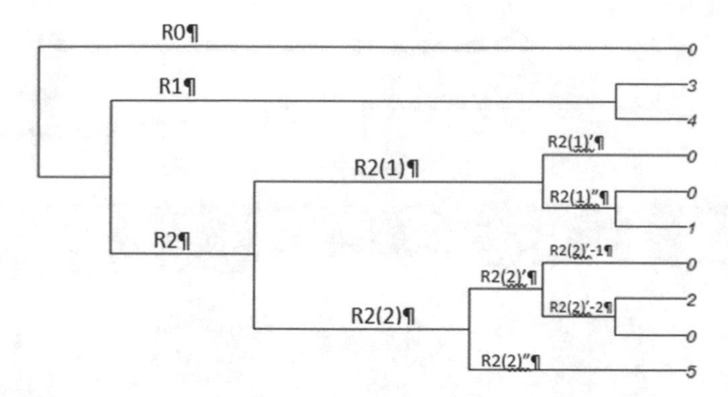

Fig. 2. Global Descending Hierarchical Classification of class constructs.

levels. There is the formation of a solitary class (Class 1 with 14.1% of the *corpus*) and the approximation of classes 2 and 5 (43.2%), with *average interaction*, since it is arising after irrelevant class 2 relations with discourses of *low relevance* to IRAMUTEQ.

Furthermore, concerning Fig. 2, the number zero at the end of the horizontal lines means that there is a discourse that distorts the classes, however, without sufficient strength to exist as a class. Therefore, one may observe that there are ten *areas of discourse*, however, only five with a force of approximation or distance for IRAMUTEQ to consider relevant to the point of turning them into classes. However, it can be interpreted that the constructions of the classes will be *more consistent* with *fewer ramifications*, because they are linked to more clear and declared representations on the theme.

From the disposition of classes presented, they are considered as *intermediate categories* under the view of Discursive Textual Analysis [18, 20, 21] and so, the interpretations begin. There are two ways to begin: (i) when there is *solitary class*, because of the single context and little likelihood of connection with other classes in the interpretative act; and (ii) classes grouped with higher percentage, since they form the majority of the *Corpus*. In the dendrogram of DHC (see Fig. 1), at first glance, identified in R2, class 1 with an isolation profile, but it is cautioned that to confirm the characteristic, definitively, we need careful interpretative analyses based on all the possibilities of analysis offered in IRAMUTEQ (see Sect. 2).

The representations of the subjects of classes 3 and 4, because they are in R1, present approximations between each other and distances against classes 1, 2 and 5, respectively, since the farther away from the DHC switching, the fewer the relationships between the words in the context of the classes. The closer the classes are, the greater the contextual affinity and the likelihood of future collations in the construction of the final categories (or definitive) according to Discursive Textual Analysis [18, 20, 21, 25].

Figure 3 demonstrates the main words with similar vocabulary between each other and different vocabulary to the other classes [6]. The DHC with Phylogram of the words favors viewing on the part of the researcher of the main words that make up each class constructed by the IRAMUTEQ software [6, 7, 22].

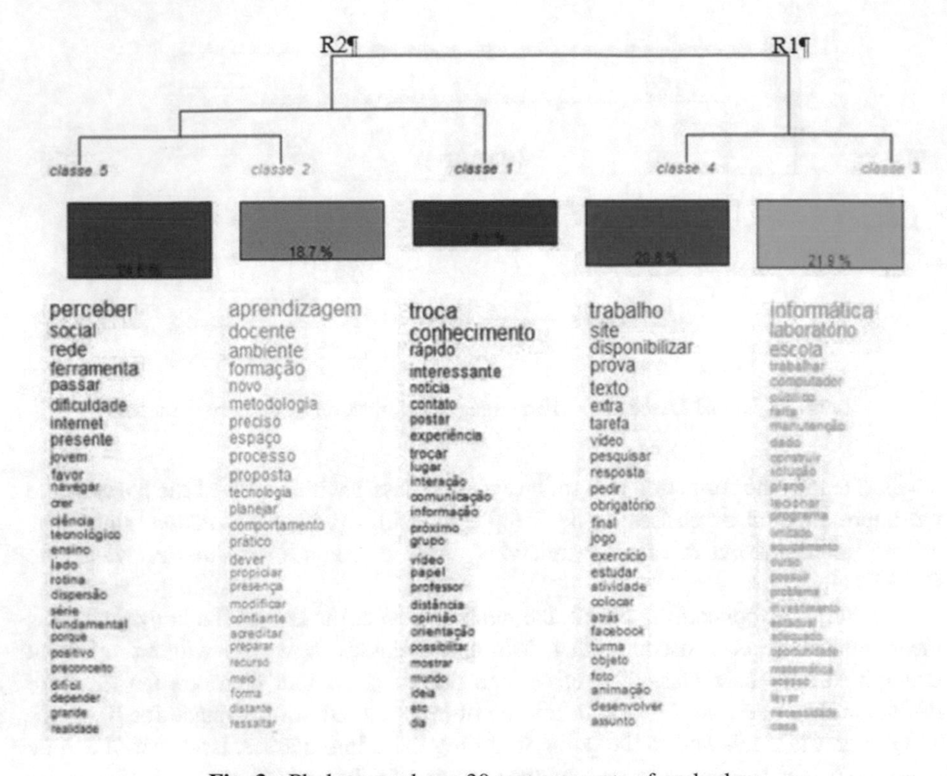

Fig. 3. Phylogram above 30 text segments of each class.

Teachers tend to form their beliefs about technologies in the school context, based on experiences of particular success through trial and error [4]. There are conflicts between personal professional skills *versus* pedagogical requirements, having implications on how technologies will be (or not) used in teaching and learning in the classroom [27]. Therefore, it is essential for teacher training, and also in more robust public policies [4], to understand the representations of teachers concerning good and bad uses in the face of digital technologies.

By deepening the understanding of the class formation – or intermediate categories [18, 20, 21] – which emerged in IRAMUTEQ, we progressed to Factorial Correspondence Analysis (FCA), which is possible after Hierarchical Classification Analysis of data (HCA). This is internal data because it is not possible to apply the FCA only to the textual data without considering the data of DHC [6, 7]. In Fig. 4, FCAs of the data pertaining to classes and subjects of each class are presented.

In the Cartesian plane the approximations/distances between the classes can be identified accurately according to the positioning in the quadrants. The interpretation of the image on the left is as follows: (i) the isolation of class 1 (red) is attested by the position on the vertical axis, since the representations of subjects do not relate to any other class, however, there is a *smaller* distance to Class 4 (blue); (ii) in the lower left quadrant (Q3) the relationship is represented between classes 2 (grey) and 5 (purple), very close to the horizontal axis demonstrating homogeneity in the representations between these classes and distance to others; and (iii) class 4 (blue) is isolated in Q1

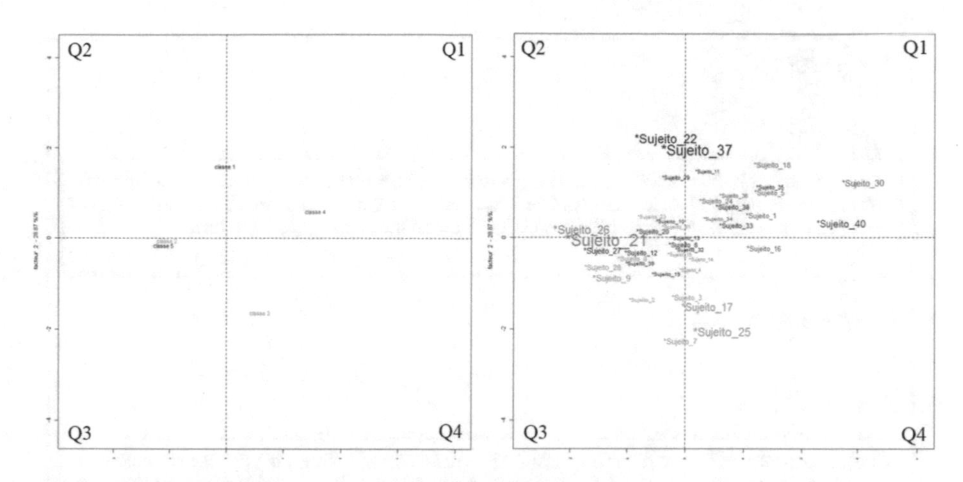

Fig. 4. FCA class (left) and FCA of subjects (right).

and next to the horizontal axis whereas Class 3 (green) is isolated in Q4 and next to the vertical axis, meaning that the representations of the subjects are related by the context and not by the words, thus justifying their approximation in R1 (see Figs. 1 and 2).

Already in the cartesian plane in the right image, the composition of each class of the subjects is demonstrated. The highlights in each class are, namely: (i) Class 1 (red), subjects 22 and 37; (ii) class 2 (grey), subject 21 and 26; (iii) class 3 (green), subjects 17 and 25; (iv) class 4 (blue), subjects 30 and 40; and (v) in class 5 (purple), subject 20 and 27. We highlight the majority position of the representations of the subjects on both axes, which demonstrates a homogeneous positioning of the group's representations as a whole. Thus, IRAMUTEQ attests that there are no completely dissonant discourses on the topic and that there is probably a standardized understanding of the large group on the topic, an aspect that does not occur with the FCAs performed in the area of health [16].

Entering the detailed interpretation of the classes or intermediate categories [18, 20] it is also possible to select the segments of the text of each subject. Figure 5 demonstrates some prominent excerpts from the subjects for the classes, extracted from IRAMUTEQ itself [6, 7, 29].

In the statement of excerpts, an example was captured from each of the five classes: (i) class 1, subject 22 (score 145.23); (ii) class 2, subject 21 (score 197.36); (iii) Class 3, subject 17 (score 363.41); (iv) class 4, subject 40 (score 121.39); and (v) class 5, subject 20 (score 130.60). By dividing the texts into Text Segments (TS), IRAMUTEQ assigns a value (*score*) for each unit of direction [18, 20, 21].

The higher the value of this *score*, the greater the density of the text segment, having higher importance within the class. In the five classes the values for the units of meaning vary in the following way: (i) class 1: 28.39 to 240.67; (ii) class 2: 23.65 to 211.13; (iii) class 3: 2.66 to 363.41; (iv) class 4: 19.13 to 230.92; and (v) class 5: 11.88 to 175.39. Between all the text segments the peak score value was 363.41 and the lowest value was 2.66, which attests that IRAMUTEQ considers all segments contextually for the construction of classes (intermediate categories).

****** *Sujeito_22**

score : 145.23

to access the information for the construction and reconstruction of knowledge to associate the internet to learning can make classes more interesting with the use of videos games images and other resources allow students a wider perspective beyond the traditional class format.

****** *Sujeito_21**

score : 197.36

it is not possible to think of internet or social networks without first thinking that the presence and use of the computer in the school environment directly changes the teacher's practice in the class and also the relation between teacher and pupil and the concept of school space.

****** *Sujeito_17**

score : 363.41

the schools in which I had the opportunity to work do not have adequate equipment or when they have they are without the maintenance they need when some school has an adequate computer lab the number of computers per student is not enough to supply the need of each class.

****** *Sujeito_40**

score : 121.39

without doubt the internet is a powerful tool for learning when used well through the social networks it is possible to offer activities and create interest for you pupils to follow also it is possible to create forums for discussion and research about specific topics.

****** *Sujeito_20**

score : 130.60

to offer many discussions with respect to a topic, the internet and social networks after we become familiar with these tools and of their proper use must have a fundamental role in our classrooms.

Fig. 5. Excerpts from the representations of the subjects.

Teachers need to acquire technological skills that enable them to generate opportunities to learn the technologies [32]. The integration of technologies into the teachers' school routines is a challenge [9], whereas use among students is a growing reality [2], unceasingly so [8], and that opens up unprecedented possibilities for teaching and learning [5]. Thus, researching the function that technologies have played in the present, implies the need to understand the social, cultural and educative conditions of their contexts [24].

Based on the analyses and interpretations of the five classes generated from the IRAMUTEQ software, only three possible categories emerged, being: (i) *technologies to bring teachers and students closer together*, deriving from class 1 (solitary class); (ii) *technologies to modify teaching practice*, consequence of the union of classes 3 and 4; and finally (iii) *technologies for changes in school*, resulting from the harmonization of classes 2 and 5.

Thus, it is confirmed that the development of the emerging categories, in their completeness with the metatexts, is an aspect of Discursive Textual Analysis that requires breath and space [18, 20, 21]. This will be a focus in future texts, because the central point here was to understand the point of view provided by IRAMUTEQ [6, 7, 22] on the analyses. Finally, it stressed that these technologies enable forms of teaching and learning based on interaction, and in autonomous and collective constructions, which can favour both teachers and pupils [13], and understanding how teachers think is one of the steps in achieving efficient, reflective and enduring activities within the walls of schools.

5 Conclusions

The purpose of this article was to identify the contributions of the IRAMUTEQ software to the Discursive Textual Analysis method. To this end, the answer to the following guiding question was sought: *How can IRAMUTEQ software contribute to Discursive Textual Analysis procedures?* Considering the textual data analysis performed by the IRAMUTEQ software, the main contributions to the Discursive Textual Analysis are:

(i) all analysis of possible textual data is performed with extreme agility. In particular, this assists the qualitative researcher, who as a rule, spends countless hours/days with large volumes of information to extract data capable of interpretation;

(ii) the classes that emerge in IRAMUTEQ, can be considered *intermediate categories* in the Discursive Textual Analysis, streamlining the analysis process and offering new positions, interpretations and relationships that, in general, would go unnoticed in manual application of the construction of the final or emerging categories;

(iii) the resources concerning the generation of intermediate categories, are a viable, safe and free alternative for qualitative researchers, since they shed light on data construction, as is the case with the categories in Discursive Textual Analysis, using instruments such as questionnaires and interviews.

It is emphasized that the software in question is limited in generating information about intermediate-categories, since the final categories require texts described in a gradual dialogue between the tacit knowledge of the researcher, with the theory and empirical data. In this way, this article reports three possible contributions of the IRAMUTEQ software applied to Discursive Textual Analysis method, with regards to the use of technologies in the context of the classroom.

With a view to fomenting discussions in future work, the application of the IRAMUTEQ software is suggested for the analyses of data in research in the area of Education in Science and Mathematics, with the intention of clarifying the interpretations presented. In particular, for the improvement of this research, it is necessary to deepen understanding of the internal aspects of each of the five intermediate categories generated in IRAMUTEQ, and in addition to the need to complete the Discursive Textual Analysis with the metatexts of each emerging category.

Finally, it should be reaffirmed that the point here is not to rival manual against digital procedures for analyzing qualitative data. The crux is the effort to consider the limits of contribution of the software for the analysis processes, previously performed successfully, as is the case with the use of Discursive Textual Analysis, and point out new perspectives for the area of teaching and learning and/or education.

Acknowledgement. CAPES – Coordenação de Aperfeiçoamento de Pessoas de Nível Superior, Brazil: for awarding a post-doctoral grant to the third author.

References

1. Amado, J.: Manual de Investigação Qualitativa em Educação (Handbook of Qualitative Research in Education). Imprensa da Universidade de Coimbra, Coimbra (2013)
2. Amaral Rosa, M.P., Eichler, M.L., Catelli, F.: "Quem me salva de ti?": representações docentes sobre a tecnologia digital ("Who Saves Me from You?": Teaching Representations on Digital Technology). Revista Ensaio **17**(1), 84–104 (2015). https://doi.org/10.1590/1983-211720175170104
3. Amaral-Rosa, M.P., Eichler, M.L.: O Software QSR Nvivo: utilização em pesquisas no ensino de Química (QSR Nvivo Software: Us in Research in the Teaching of Chemistry). Chem. Educ. Point View **1**(1), 120–143 (2017). https://doi.org/10.30705/eqpv.v1i1.895
4. Amaral-Rosa, M.P., Eichler, M.L.: Brazilian teacher's beliefs about technologies in a training program in Portugal. Acta Scientiae **19**(4), 679–692 (2017)
5. Arias-Ortiz, E., Cristia, J.: El BID y la tecnologia para mejorar el aprendizaje: ¿Cómo promover programas efectivos? (IDB and Technology to Improve Learning: How to Promote Effective Programs?). IDB Techinal Note (Social Sector. Education Division), IDB-TN-670 (2014). http://goo.gl/PGuPFr. último acesso 18 Mar 2018
6. Camargo, B.V., Justo, A.M.: IRAMUTEQ: um software gratuito para análise de dados textuais (IRAMUTEQ: A Free Software for Textual Data Analysis). Temas em Psicologia **21** (2), 513–518 (2013)
7. Camargo, B.V., Justo, A.M. (S.d.). Tutorial para uso do software IRAMUTEQ (Tutorial for Using the IRAMUTEQ Software). https://goo.gl/22jP4X. Accessed 20 Mar 2018
8. Castells, M.: A sociedade em rede. (A era da informação: economia, sociedade e cultura; v.1) (The Network Society). (The Age of Information: Economics, Society and Culture; v. 1). Paz e Terra, São Paulo (1999)

9. Chai, C.S., Koh, J.H.L., Tsai, C.C.: A review of technological pedagogical content knowledge. Educ. Technol. Soc. **16**(2), 31–51 (2013)
10. Creswell, J., Miller, D.: Determining validity in qualitative inquiry. Theory Pract. **39**(3), 124–130 (2000)
11. De Paula, M.C., Lori, V., Guimarães, G.T.D.: Análise Textual Discursiva com apoio do software SPHINX (Discursive textual analysis with support of the SPHINX software). Investigação Qualitativa em Educação - Atas CIAIQ **2**, 352–357 (2015)
12. De Paula, M.C., Lori, V., Guimarães, G.T.D.L.: A pesquisa qualitativa e o uso de CAQDAS na análise textual: levantamento de uma década (Qualitative research and the use of CAQDAS in textual analysis: a decade survey). Internet Latent Corpus J. **6**(2), 65–78 (2016)
13. García-Valcárcel, A., Basilotta, V., López, C.: Las TIC em el aprendizaje colaborativo em el aula de Primaria y Secundaria (ICT in collaborative learning in the primary and secondary classroom). Comunicar **42**, 65–74 (2014). https://doi.org/10.3916/C42-2014-06
14. Gray, D.: Pesquisa no mundo real (Real-World Research). Penso, Porto Alegre (2012)
15. Johnston, L.: Software and method: reflections on teaching and using QSR NVivo in doctoral research. Int. J. Soc. Res. Methodol. **9**(5), 379–391 (2006)
16. Kami, M.T.M., Larocca, L.M., Chaves, M.M.N., Lowen, I.M.V., Souza, V.M.P., Goto, D.Y. N. (2016). Trabalho no consultório na rua: uso do software IRAMUTEQ no apoio à pesquisa qualitativa (I work in the office on the street: use of the IRAMUTEQ software in support of qualitative research). Escola Anna Nery **20**(3). https://doi.org/10.5935/1414-8145.20160069
17. Leonidas, S.R.: As crenças dos gestores, profissionais e usuários sobre o Centro de Referência em Saúde do Trabalho Cearense (The beliefs of the managers, professionals and users of the Centro de Referência em Saúde do Trabalho Cearense). (Dissertação de mestrado). UNIFOR – Universidade de Fortaleza (2016). https://goo.gl/kn311G. último acesso 10 Feb 2018
18. Lima, V.M.R., Ramos, M.G.: Percepções de interdisciplinaridade de professores de Ciências e Matemática: um exercício de Análise Textual Discursiva (Interdisciplinarity Perceptions of Professors of Science and Mathematics: An Exercise of Discursive Textual Analysis). Revista Lusófona de Educação **36**, 163–177 (2017)
19. Mayring, P.: Qualitative Content Analysis: Theoretical Foundation, Basic Procedures and Software Solution. Klagenfurt, Austria (2014). https://goo.gl/pNjubm. Last accessed 20 Feb 2018
20. Moraes, R., Galiazzi, M.C.: Análise Textual Discursiva (Discursive Textual Analysis). Ed. Unijuí, Ijuí (2011)
21. Moraes, R., Galiazzi, M.C., Ramos, M.G.: Aprendentes do aprender: um exercício de Análise Textual Discursiva (Learners of Learning: An Exercise in Discursive Textual Analysis) Indagatio Didactica **5**(2), 868–883 (2013)
22. Neri de Souza, D., Neri de Souza, F., Costa, A.P. Percepção dos utilizadores sobre o software de análise qualitativa WebQDA (Users perception of WebQDA qualitative analysis software). Comunicação & Informação **17**(2), 104–118 (2014)
23. Olsen, W.: Coleta de dados: debates e métodos fundamentais em pesquisa social. (Data Collection: Debates and Fundamental Methods in Social Research). Penso, Porto Alegre (2015)
24. Porto, T.M.E.: As tecnologias de comunicação e informação na escola; relações possíveis… relações construídas (Communication and information technologies at school; possible relationships… relationships built). Revista Brasileira de Educação **11**(31), 43–57 (2006)
25. Ramos, M.G., Ribeiro, M.E.M., Galiazzi, M.C.: Análise Textual Discursiva em processo: investigando a percepção de professores e licenciados de Química sobre aprendizagem (Discursive textual analysis in process: investigating the perception of teachers and graduates of chemistry on learning). Campo Abierto **34**(2), 125–140 (2015)

26. Pesavento, S.J.: História & História Cultural (History and Cultural History). Autêntica, Belo Horizonte (2008)
27. Prestridge, S.: The beliefs behind the teacher that influences their ict practies. Comput. Educ. **58**, 449–458 (2012)
28. Pryjma, L.C.: Ser professor: representações sociais de professores (Being a Teacher: Social Representations of Teachers). (Tese de doutoramento). UNESP – Universidade Estadual Paulista, Faculdade de Ciências e Tecnologia (2016). https://goo.gl/aRJLZf. ultimo acesso 15 Feb 2018
29. Ratinaud, P.: IRAMUTEQ: Interface de R pour les Analyses Multidimensionnelles de Textes et de Questionnaires - 0.7 alpha 2 (IRAMUTEQ: R Interface for Multi-dimensional Analysis of Texts and Questionnaires - 0.7 Alpha 2) (2014). http://www.iramuteq.org. último acesso 16 Feb 2018
30. Silveira, S.A.: Software livre: a luta pela liberdade do conhecimento (Free Software: The Struggle for Freedom of Knowledge). Ed. Fundação Perseu Abramo, São Paulo (2004). https://goo.gl/PWHkv9. último acesso 26 Jan 2018
31. Stake, R.E.: Pesquisa qualitative: estudando como as coisas funcionam (Qualitative Research: Studying How Things Work). Penso, Porto Alegre (2011)
32. Unesco: Padrões de competências em TIC para professores. Organização das Nações Unidas para a Educação, a Ciência e a Cultura (ICT skills standards for teachers. Organization of the United Nations for Education, Science and Culture) (UNESCO) (2009). https://goo.gl/R8zoAq. Accessed 10 Jan 2018

Risk Management of STEM Education - The Strategic Risk: Teachers - Opportunities, Training and Social Status in Israel

Anat Even-Zahav(✉)

Talpiot Academic College of Education, Holon, Israel
anatez@gmail.com

Abstract. A doctorate study conducted in the Faculty of Education Science and Technology at the Technion – Israel Institute of Technology, offers an innovative implementation of the risk management process for STEM (Science, Technology, Engineering and Mathematics) Education in Israel. A three-phase risk-management process employed: Risk Identification, Risk Assessment & Prioritizing, and Risk Response. The **research goal** was to outline **a risk-management plan** to STEM education in Israel based on the conceptions of five stakeholder groups: Educators, academics, industrials, military and philanthropy actors. **Research findings** presented according to the three phases of the risk management analysis performed in the research: SWOT analysis aimed at identifying risks, a Delphi method for the purpose of risks rating, and a response plan aimed at mitigating risks faced by STEM education in Israel. The findings presented in the article focus on one risk category: Teacher - opportunities, training and social status, as this risk category was ranked as strategic risks in terms of the effect on the objectives of STEM education.

The response plan of my research highlights the cooperation of the stakeholders in the risk management process of STEM education as a way to mitigate strategic risks. It should be noted that, in reality, cooperation already takes place between the education system and other stakeholders – academia, military, industry, and philanthropy institutions. Such cooperation is much desired, but the research suggests it should be carefully examined, so that the best mode of cooperation is adopted, one which accounts for the weaknesses and strengths of all stakeholders.

Keywords: STEM teachers · Risk management · STEM education
SWOT analysis · Delphi method

1 Introduction

This article presents a doctorate study conducted in the Faculty of Education Science and Technology at the Technion – Israel Institute of Technology. The study implied a Risk Management process to STEM (Science, Technology, Engineering and Mathematics) education in Israel's high school education system. Much like in a business organization, a three-phase, risk-management process was employed: **Risk Identification, Risk Rating,** and **Risk Response.**

A. P. Costa et al. (Eds.): WCQR 2018, AISC 861, pp. 73–89, 2019.
https://doi.org/10.1007/978-3-030-01406-3_7

Strategic objectives for STEM education in Israel emphasize the importance of building scientific and technological human resources, in order to maintain the high-tech industry's role as a central driver of the country's economy. STEM education can promote equal opportunity and enable potential-realization across population groups and strata. Science and technology knowledge is part of the general education necessary for the contemporary citizen, and even more so for the future one. STEM education also contributes to the development of learner functioning, adopted to the 21st century.

The main goals of STEM education in Israel are, therefore, to increase the numbers of students who choose studying STEM subjects in high school on the highest level, to institutionalize excellence programs, and to attract as many students as possible to those programs. Those goals become ever more challenging, as one witnesses the expected shortage of STEM teachers in Israeli high schools in the coming years.

Facing this reality, and the challenges it presents, professionals across the board – both in Israel's education system, the academia, the military, and the industrial sector – are encouraged to realize the need for action. It is necessary - if Israel wants to preserve its relative edge in research, if it wants to remain a high-tech superpower, if it seeks to maintain its human capital, which is a direct result of wise investment in education.

This research performs a strategic analysis of high-school STEM education. The **research goal** was to outline **a risk-management plan** to STEM education in Israel based on the conceptions of five stakeholder groups (below, research participants): Educators, academics, industrials, military and philanthropy actors, who all have vested interest in STEM education in general and in promoting STEM education in high school education system in Israel in particular. **Research question** was derived from the research goal. We asked: Can we draw a risk-management plan to STEM education in Israel?

The risk management process performed in the research included SWOT analysis aimed at identifying risks, a Delphi method for the purpose of risks prioritizing, and a response plan aimed at mitigating risks faced by STEM education in Israel.

In what follows the **Theoretical background** reviews concepts of strategic analysis. The **Research framework** will describe research tools, research participants and research process applied in the study. The **Research findings** will present the three phases of the risk management analysis performed in the research: (A) SWOT analysis enabled risks identification; (B) Delphi Method produced risks rating; (C) Risk response planning. This article highlights only one risk category: Teacher - opportunities, training and social status, as this risk category was among the four **strategic risks** ranked as high-level risks in terms of the effect on the objectives of STEM education.

The Summary and discussion chapter highlights of the response plan expressed: cooperation of the stakeholders in the risk management process of STEM education in Israel to mitigate strategic risks.

2 Theoretical Background

This chapter will present three concepts of strategic analysis that were used in the research for strategic analysis of STEM education in Israel: SWOT analysis, Delphi Method and Risk Management.

2.1 SWOT Analysis

SWOT – Strength, Weaknesses, Opportunities and Threats – analysis is a methodological examination of the environment in which an organization operates. It is based on the examination of (a) internal characteristics of the organization (strengths and weaknesses) and (b) characteristics of the external environment of the organization (opportunities and threats). SWOT analysis allows the organization choosing operational strategies that foster its strengths and opportunities and protect it from its weaknesses and threats (Barney 1995).

Though the origin of SWOT analysis is at the business sector, it has been used also for the analysis of public sector organizations, e.g., schools and hospitals (Rego and Nunes 2010). For example, in the field of education, institutions of higher education carried out SWOT analysis for the evaluation of educational initiatives, such as the integration of information technologies (Sabbaghi and Vaidyanathan 2004).

SWOT analysis was applied in the case presented in this research for the identification of risks that the system of STEM education in Israel should prepare itself to face with.

2.2 Delphi Method

Delphi Method is based on expert evaluation of the topic under discussion. The process takes place in several rounds, in each of them a set of questions is answered by a group of experts. The Delphi procedure was first introduced by Olaf Helmer (1966) and included the following steps (usually called rounds):

- First round
  - Gathering a group of experts from the said field
  - Presentation of a set of questions about future trends to each expert separately
  - Each expert answers the questions individually and confidentially without any direct contact with the other experts.

  This stage is usually implemented by interviews or questionnaires, which, in most cases, include open questions relevant to the study. The participants are asked to identify topics that will be discussed in the next rounds.
- Second round
  - The experts' answers gathered in the first round are presented to each expert, who is now asked to express his or her opinion about each of them.
- Additional rounds take place in a similar manner until an agreement with respect to the desirable directions is reached.

The working assumption is that each round decreases the level of disagreement between the experts and eventually it is possible to formulate a strategy which is agreed upon all experts. The Delphi method attempts to avoid group thinking in which one expert opinion affects the other experts' perspective as it sometimes happens during brainstorming sessions (Linstone and Turoff 1975).

2.3 Risk Management

Risk is an internal or external event that has the potential to affect the implementation of the organizational strategy and the achievement of the objectives it sets for the future. The risk severity level is determined according to its (a) likelihood – the probability of its realization and (b) impact – the damage that the risk realization can cause (ISO Guide73 2009). The event, that is, the risk realization, may deviate the organization from achieving its desired orientation, either positively (upside) by enabling the organization to exhaust an opportunity, or negatively (downside), by threating the achievement of the desired results.

The following events are commonly conceived as risks for different kinds of organizations: nature disasters, security holes (e.g. cyber-attacks), shortage or failures of human resources, financial crisis, unstable business environments, and project failures. On the one hand, in the field of accidents and safety at work, for example, only events that have negative consequences are considered as risks, and therefore, risk management as the field of safety focuses solely on the prevention of damage and the reduction of the intensity of the risk impacts. On the other hand, events which are recognized as an opportunity for the organization reflect a positive future. For example, an unexpected business opportunity may evolve as a result of a change introduced into the tax policy that may enable the organization to expand its markets.

Risks are classified in different ways, e.g., different organizational concerns: strategic risks, financial risks, operational risks, political risks, and hazard risks (related to facilities or human lives) (IRM 2002). Based on Mikes and Kaplan's (2014) terminology, the research examination categorized risks according to three resources: operational risks, strategic risks and external (political and financial) risks.

Bruckner *et al.* (2001, in Hosseinzadehdastak and Underdown 2012) defines risk management as follows:

> Risk management refers to strategies, methods and supporting tools to identify, and control risk to an acceptable level. Additionally, all events that may prevent an organization from realizing its ambitions, plans and goals are known as **risks**. In other words, **risks are potential problems that might happen**. As a result, identifying risks, assessing them and estimating their impacts can help to mitigate negative their effects (p. 2).

Bruckner *et al.*'s definition (2001, in Hosseinzadehdastak and Underdown 2012) has been adopted for the research. Specifically, in the process of risk management, we used methods and tools to identify and control risks; e.g., SWOT analysis was applied for the risk identification. Thus, the identified risks represent weaknesses and threats for STEM education in Israel, that their existence in the future endanger the desired achievements of STEM education (for example, the need to increase the number of high school graduates in the STEM subjects on the highest level and respectively, the number of qualified STEM teachers). The risk rating according to their severity level led to the formulation of a response plan which lays out thirteen courses of action to alleviate (mitigate) the negative impact of the highly ranked risks, in order to reduce the severity of their impact on STEM education in Israel.

3 Research Framework

This chapter describes the research framework of the risk management process presented in this article. It reviews the research methodology, the research participants, the research tools, and the research process.

3.1 Research Methodology

The research method was mixed and used both qualitative (the main one) and quantitative research tools. The qualitative method allowed describing the perspective of the study participants and was used to analyze the data gathered by the Delphi survey.

3.2 Research Participants

The research participants participated in the three phases of the study (several of them participated in the three phases): Phase A: Risk identification; Phase B: Risk rating; and Phase C: Risk response. They belonged to the following two groups.

Group I: Employees of the education system

- STEM Teachers who teach in the secondary school the following subjects: Mathematics, Physics, Chemistry, Biology, Computer science and the Technology education subjects. They all have a teaching certificate, at least a Master's degree, and a teaching experience of at least five years. They all prepare their students towards the matriculation exams in all levels.
- High school principals.
- Administrative executive role holders in the education system, local councils, and school chains.

These participants had a direct contact with the education system. They all work in the field of STEM education in Israel and face challenges while performing their role in the system.

These research participants were interviewed in an in-depth interview designed as a SWOT analysis. At the following phases, the risk rating (Phase B) and risk response (Phase C), additional data were collected from this group by the SWOT interview, questionnaires and focus groups.

Group II: Managerial role holders in organizations involved in STEM education

These research participants belong to four groups of stakeholders: academia, industry, military and non-profit NGOs. Data were collected from this group at an advanced stage of Phase A of the study, in Phase B by questionnaires, and in Phase C by focus groups. This group represented a wide range of expertise from various fields and, thus, provided a broad and diverse perspective on STEM education in Israel. In this research, they contributed to the building of a strategic plan that copes with current and future challenges of STEM education in Israel. We selected stakeholders who are familiar with STEM education, serve in key roles, and cooperate with the education system in order to promote STEM education.

3.3 Research Tools

Data was collected using the following research tools: SWOT interviews, questionnaires, focus groups, and documents.

SWOT Interview. The interviews designed as a semi-structured, in-depth interview in order to reveal the perception of the research participants with respect to STEM education in Israel.

The SWOT interview as a research tool was especially designed by the researcher uniquely to this research to relate to the research question. In our attempt to characterize the of five stakeholders' perception of STEM education in Israel, we constructed an interview along the lines of the SWOT analysis, which enables an analysis of the Education system according to their **s**trengths, their **w**eaknesses, the **o**pportunities that are open, and the **t**hreats it faces. The interviews were conducted according to the four SWOT dimensions as they apply to four different levels: the individual, the school, the curriculum, and the education system (see Table 1). Thus, each interview in essence constituted both a SWOT analysis of the participant, as well as a SWOT analysis of the system, on three levels: the school, the curriculum, and the education system.

The analysis of the interviews showed that this framework allowed the interviewees to elaborate beyond the focus of the question and to express their perspective on additional related issues. In addition, the specific designed interview using the four components enabled to reveal the perceptions regarding the Opportunities and Threats which otherwise were missing.

Table 1. SWOT interview structure

Strengths	Weaknesses
Individual level School level Curriculum level Education system level	Individual level School level Curriculum level Education system level
Opportunities	**Threats**
Individual level School level Curriculum level Education system level	Individual level School level Curriculum level Education system level

Questionnaires. Questionnaires were designed mainly as a Delphi survey to collect data in the process of risk rating (Phase B) and in the formulation of the response plan (Phase C).

Based on the data analysis carried out in Phase A, a 'Risk management – Risk rating' questionnaire was designed in Phase B. The purpose of the questionnaire was to validate the data analysis of Phase A, as well as to assess and prioritize the risks and their implications for STEM education in Israel.

The questionnaire asked the research participants to rate the risks on a Likert scale. The data was analyzed by descriptive statistics (mean, standard deviation, etc.) and statistical tests.

Focus Groups. Focus group were used in Phase C - risk response plan. Focus groups emphasize the interaction between group members depending on the issues brought to them, when the researcher plays as a mediator.

We facilitated nine focus groups in which fifty practitioners, from the two groups of the research population described above, participated. Each focus group included representatives from the five groups of stakeholders – education, academia, industry, military and non-profit NGOs. In addition to reaching a consensus related to the risk response, the participants examined strategies to mitigate the identified risks.

Documents. Documents were analyzed to further validate our findings by formal publications, such as, regulations published periodically by the Ministry of Education, committee reports, school websites, principal messages to teachers, etc.

3.4 Research Process

The study was carried out for four years (2012–2015) in three phases: Risk identification (Phase A), risk rating (Phase B) and response plan (Phase C). The three phases will be elaborated in the next chapter.

4 Findings

4.1 Phase A: Risk Identification

The identified risks present the different perspectives of the research participants, and outline weaknesses and threats faced by the STEM education system, the existence of which endanger reaching central objectives. The risks identified by STEM teachers are ones directly regarding teachers, bearing the marks of possible ramifications of those risks on them **as teachers** (e.g. risks associated with professional opportunities, training and professional status). Risks identified by **other stakeholders** express the risks they perceive as having potential effects of their organizations. Thus, for instance, failings in forming a clear learning and pedagogical path from high school to academia to the work market, can trigger manpower shortages in the military, the industry and the academia; high school pupils' self-perceptions with respect to scientific subjects, and social perceptions regarding diminished status of technology professions may effect the work market.

Forty-three **risk factors** (see Table 2) were identified and grouped into seven **risk categories:** (a) Sectors in STEM education, (b) Teacher - opportunities, training and social status, (c) Curriculum of STEM subjects; (d) Study sequences - school, higher education, military, labor market, (e) Management of STEM education, (f) Social perceptions, and (g) National programs in STEM education.

It this article I will elaborate only on one risk category: (b) Teacher - opportunities, training and social status. This risk category was among the four **strategic risks** ranked

Table 2. Risk categories, conflicts and barriers: the 43 risk factors of STEM education in Israel[a]

Risk categories	Conflicts[b]	Barriers[c]	Risk factors
1. Sectors in STEM education		Psychological-personal barrier Socio-cultural barrier Procedural barrier	1. Most of the ultra-orthodox female students do not study science courses on the advanced level in high school
			2. Male students in ultra-orthodox schools do not study STEM subjects at all
			3. Many students attending military[d] classes are associated with social-economic periphery
			4. Many students attending vocational education are associated with social-economic periphery
			5. Percentage of students studying advanced level science subjects among male students is higher than among female students
			6. If a female student faces difficulty in learning advanced science subject, there is a tendency to suggest her to study the subject on a lower level
2. Teachers - opportunities, training and social status	Professional opportunities conflict Teacher status conflict	Procedural barrier	7. Lack of promotion tracks for STEM teachers*
			8. The teaching profession is not appreciated by the public*
			9. STEM teachers' salary is low relative to alternatives jobs in the industry*
			10. The working conditions of STEM teachers are not attractive*
			11. Placement and guidance of novice STEM teachers is not provided
			12. Many teachers who teach science in elementary school lack education in the field
			13. Only small number of junior high school teachers who teach "Science and Technology" are Physics or Chemistry teachers
			14. Number of students who study STEM education in universities is low
			15. Lack of a collaborative professional community for STEM teachers

(*continued*)

Table 2. (*continued*)

Risk categories	Conflicts[b]	Barriers[c]	Risk factors
3. Curriculum of STEM subjects	Profession perception conflict		16. Technological aspects are not sufficiently integrated in science studies in the elementary school
			17. Structure and content of STEM education in junior high schools: the subject "Science and Technology" is taught as one block of four disciplines (Biology, Chemistry, Physics and technology)
			18. Frequent changes in the curricula of STEM subjects
			19. The curriculum of the vocational education subjects does not match the needs of the labor market*
			20. The teaching of STEM subjects focuses mainly on the preparation towards the matriculation exams
4. Study sequences - school, higher education, military, labor market		Procedural barrier	21. Long training is needed to prepare high school graduates for technological jobs in the military service
			22. Shortage of engineers in the industry*
			23. Shortage of practical engineers and technicians in the industry*
			24. Shortage of skilled low-tech professionals in the industry
			25. Lack of career plan programs for high school students
			26. Universities are required to offer preparatory courses for high school graduates before accepting them to undergraduate science and engineering studies to close gaps in the STEM subjects
5. Management of STEM education	Academic freedom conflict Discourse on STEM education conflict	Procedural barrier Economic barrier	27. Tight supervision over teachers and principals
			28. Bureaucratic work is required from principals and teachers
			29. School success is measured by the eligibility of a matriculation certificate and not by the subjects included in the certificate
			30. Lax discipline in schools
			31. Parents' involvement in schools

(*continued*)

Table 2. (*continued*)

Risk categories	Conflicts[b]	Barriers[c]	Risk factors
			32. Involvement of military and industry in STEM education in the high school
			33. Limited budget to open technology classes
6. Social perceptions		Psychological-personal barrier Socio-cultural barrier	34. Negative labeling of vocational education
			35. Vocational jobs are not valued
			36. The perception "I am not qualified for studying science subjects" is common among students
			37. The perception "It is better to study easier subjects" is common among students
			38. The perception "Vocational education is associated with social gaps in the Israeli society" is common in the public
7. National programs in STEM education		Procedural barrier Political barrier	39. The introduction of new national programs reduces the importance attributed to previous national programs
			40. Lack of teacher training process prior the implementation of a new national program
			41. Lack of persistence in the implementation of long-term national programs
			42. Budgets for infrastructure, teaching and training hours are increased especially when a crisis is identified
			43. Implementation of national programs without appropriate infrastructure

[a]The reviewed risk factors are denoted by*

[b]In the Research we identified five conflicts STEM teachers face whose essence is the teachers as professionals vs. the system in which they work: (a) the professional opportunities conflict, (b) the teacher status conflict, (c) the academic freedom conflict, (d) the profession perception conflict, and (e) the discourse on STEM education conflict.

[c]In the Research we identified five barrier which introduce difficulties to implement changes in the system. In particular, the stakeholders' perceptions reflected the following five barriers: (a) economic barrier, (b) procedural barrier, (c) psychological-personal barrier, (d) socio-cultural barrier, and (e) political barrier.

[d]The students in these classes learn professions required by the army.

as high-level risks in terms of the effect on the objectives of STEM education (see next section, Phase B).

Risk category (b) Teachers - opportunities, training and social status. This category is presented from two perspectives: the perspective of the STEM teachers and the viewpoint of the other stakeholders of STEM education. **Risk factors** 7, 8, 9, and 10 (see Table 2) will be reviewed below.

Risk Factor 7: Lack of Promotion Tracks for STEM Teachers. Promotion and advancement tracks for STEM teachers, as for all other teachers, lie primarily in the administrative track; therefore, the STEM teachers face a professional-opportunity conflict with the education system, which, they believe, fails to provide them with adequate opportunities for advancing their *teaching* careers. Specifically, the STEM teachers are mostly interested in the development of the professional aspects of their careers, rather than in the promotion of the administrative aspects of their position. In particular, the STEM teachers expressed the desire to fill research and development (R&D) positions, in which they would engage in the development of learning and teaching curricula, serve as the teachers of teachers, study towards their PhDs, and fulfill leadership roles within the education system.

The STEM teachers referred to this particular weakness of the system, emphasizing the importance they attribute to academic studies in the STEM subjects and to the need to belong to a broader *professional* community. For example, A., a Physics teacher said:

> The development horizon for teachers is very limited… [it is] a critical point: The professional horizon. There's a difference between a Math, Physics or Chemistry teacher and the other teachers, no offence intended… [but] those Math and Physics teachers are a kind of people who don't go into management roles… **a different way must be found. Things like Hemda[1], doing some research, experiments.**

The teachers' words, as expressed in the interviews, raised the question of whether or not STEM teachers only rarely hold administrative positions. We found a partial answer to this question by an online questionnaire distributed to the schools whose teachers participated in the study. It was found that only a small percentage of STEM teachers fill educational and administrative positions. Specifically, on average, 28% of all teachers in the school are STEM teachers, but only 19% of all administrative role-holders in the school are STEM teachers. It is important to mention that even if teachers wish to promote their career in the administrative path, the number of such roles is limited (Barak 2011).

Risk factors 8, 9 & 10: The teaching profession is not appreciated by the public, STEM teachers' salary is low relative to alternatives jobs in the industry, and the

---

[1] Hemda is a Science Education Center located in Tel Aviv. Students from schools in the Tel Aviv-Jaffa area come to this center to study Physics, Chemistry and Computer science. Most of the teachers in Hemda hold a Ph.D. in Physics or Chemistry. The work environment is modern and includes laboratories, computer equipment and demonstration equipment that enable hands-on experience for each and every student. For more details, see Hemda website: http://www.hemda.org.il/english/.

working conditions of STEM teachers are not attractive. The teacher status in the public conflicts their own self-perception as professional practitioners who have been trained for their jobs in academic institutions, hold academic degrees in the subjects they teach, and wish to continue their professional development. According to the STEM teachers who participated in the study, teachers' employment conditions (their salary and work environment) reflect the way they are perceived by the system. The STEM teachers mentioned the salary and pension as personal threats that jeopardize the chances they will continue working as teachers.

I., a math teacher, who studied in a second-career program for academic high-tech employees, describes the situation:

> The main threat is the economic issue… If my economic situation is such that I cannot afford it, then I will simply not be able to afford it. I really don't know whether the "Courage for Change" [in Hebrew, Oz Latmura - a reform whose target is to increase teacher salary and status] will make the difference. Thanks to the career-change program in which I participated, I receive a very considerable subsidy on a quarterly basis, … and I still subsidize the education system on a monthly basis.

4.2 Phase B: Risk Rating

Delphi survey reached out to 186 research participants. The survey's findings indicate **strategic risks,** represent **social perceptions in relation to STEM education.** Strategic risks ranked as high-level risks in terms of the effect on the objectives of STEM education. Among those risks: high school pupils' self-perceptions with respect to scientific subjects; social perceptions regarding diminished image of technology studies; diminished public recognition granted to teachers in Israel; and sectorial gaps resulting from deeply rooted perceptions as well as historic processes which generated them.

4.3 Phase C: Risk Response

Response plan reveals opportunities to mitigate strategic risks. ***A reaction plan for strategic risk mitigation*** that proposes action items to *mitigate* strategic risks, by *avoiding* operational risks and *accepting* external risks was formulated. **Thirteen courses of action** were proposed: five regarding internal action within the education system, eight involving cooperation with stakeholders of STEM education in Israel.

Five courses of action regarding internal action within the education system are: (1) Teacher training, career development and guidance: proposals to strengthen the training component among STEM teachers; opening professional tracks to promote STEM teachers career and constructing accompanying programs for novice teachers (will be elaborated below); (2) Autonomy, trust and supervision discharge: The granting of academic freedom for teachers while releasing supervision of the teacher's work (will be elaborated below); (3) Creating STEM education policy managed by the education system to encourage students to choose STEM studies in high school; (4) focusing on developing capability and competence among students and avoiding referring them to low level courses in STEM subjects; (5) Equalizing the wage to novice and senior STEM teachers.

Eight courses of action involving cooperation with stakeholders of STEM education in Israel are: (1) Seeking ways to enable integration of STEM teachers in the industrial sector; (2) Designing curriculum of STEM studies and career planning programs, which are future-work-market-oriented; (3) Exposing students to the high-tech industry; (4) Creating cross-sector cooperation to advance populations under-represented in STEM education; (5) Improving teachers' income through part-time employment in the industry; (6) Generating budget growth for STEM education through business and philanthropy involvement; (7) The establishment of a National Council for STEM education, thus joining stakeholders in a common national effort; (8) Creating Institutionalized Collaboration with business sector, NGO (Non-Government Organizations) and military as a way of coping with challenges and conflicts in multi-sectoral relations.

I will elaborate on two courses of action: (1) Teacher training, career development and guidance; (2) Autonomy, trust and supervision discharge. These two courses of action relate to STEM teachers and were found among the **strategic risks** ranked as high-level risks in terms of the effect on the objectives of STEM education (see section above: Phase B, Phase C).

Course of action 1: Teacher training, career development and guidance. Study participants pointed out a clear association between the social status of the teaching profession and the need to improve teachers' training and advancement. Three themes brought up in this context.

a. *Increment in the requirements for STEM teachers training, specifically for professional training of elementary school science teachers and STEM teachers in junior-high schools*. The need for enforcement the training component among STEM teachers was mentioned mainly in context of elementary school teachers who teach science professions without expertise in this field and in the context of STEM teachers in junior high school.

   Study participants noted also a difference in teacher training between teachers who have completed their training in colleges and teachers who were trained in universities. It was argued that the admission conditions for teacher colleges in Israel are low, which may harm the professional skills of the graduates and accordingly the status of the teacher profession.

   The participants argued that an increase in the level of admissions, not only will contribute to the quality of teaching, but will also improve teachers' status both among students and among their parents.

   In Finland, the consideration of teacher status as a risk that may actualize one day was taken into account at the beginning of the 70 s. As a response, the admission requirements to the teaching profession were raised and teacher training was reshaped, turning into a research-oriented study program. Eventually, the risk has been removed. Sahlberg (2011) describes the current situation in Finland as follows:

   > Teacher training in Finland is also recognized because of its systematic and research-based structure. All graduating teachers, by the nature of their degree, have completed research-based masters' theses accompanied by rigorous academic requirements of theory, methodology and critical reflection equal to any other field of study in Finnish universities at that level (p. 94).

b. *Professional development options for post-elementary teachers*. The need to provide teachers options for professional development in the teaching career is supported by the theory of personal well-being at work, which attributes importance to the fulfilment of the individual's aspirations in the organization (Deci and Ryan 2008). In the context of STEM teachers, our study identified their desire for advancement in the *professional* aspect of their teaching career, rather than in the administrative/managerial aspect (as is possible by dual ladder promotion processes). L., a Chemistry teacher describes her personal ambitions for professional advancement:

> To do something beyond teaching… do things that promote… if I could, I would have done more things that effect… trying to do something that will bring additional solutions on the national level but also to improve teaching methods and ways of thinking that are important to me… to go out to colleges, become a head of a department, these options are very interesting.

c. *Implementation of mentoring programs for teachers in their first teaching years in order to cope with the high rate of teacher drop out*. In order to reduce teacher drop out in the first years of teaching, it was suggested to provide pedagogical support and guidance system for new teachers who join the education system.
Recently, efforts are being made in Israel to include a guidance program for new teachers as part of the teacher training programs. In the 2015–2016 academic year, a national program, called "Academia-Class" was launched. In this program, prospective teachers in their third year of study, join an experienced teacher in the school for 12–16 weekly hours, serve as additional classroom teacher and work side-by-side with the Mentor Teacher. The program is based on an Irish program and is one of the recommendations of the EU for improving quality of teachers and their training (European Commission 2013).
In conclusion, according to the research participants, teachers' status maybe improved by raising the requirements of STEM teachers training program, opening professional development options for post-elementary STEM teachers, and guiding new teachers in the first years of teaching.

Course of action 2: Autonomy, trust and supervision discharge. Providing academic freedom and autonomy to teachers, while decreasing the supervision of their work, may increase teachers' trust and public recognition both by the education system and by the society in Israel.

S., a principal of a post-elementary school, describes her expectation for trust which will be reflected in reduced level of supervision when new programs are implemented:

> There are too many supervisors and implementers of different programs… Say what your expectations are and leave us. If we need, we will ask for help… believe in us. Help a school that does not meet the requirements, that doesn't have what you consider as essential.

Indeed, studies identified that social rewards, such as the society's recognition of the professional status of teachers, and granting autonomy related to their professional freedom, are top priority motives for choosing the teaching profession (Oplatka 2007).

5 Summery and Discussion

The response plan highlights the cooperation of the stakeholders in the risk management process of STEM education in Israel as a way to mitigate strategic risks. These positions of the study participants' reflects the fact that more and more organizations, both for-profit and nonprofit, cooperate with the education system in different forms: development of educational content in STEM education, implementation of projects in science and technology in secondary schools, budget assistance, construction of educational technology centers, etc.

Such forms of cooperation are desired also from the perspective of the education system. Dr. Ofer Rimon, Head of the Science and Technology Administration at the Ministry of Education, presented this perspective at the Knesset (the Israel representative house) Committee of Science and Technology in 2015 in a discussion on STEM education:

> Many of our partners are sitting here around the table. The technology chain of school, the Manufacturers Association, the professional unions, the IDF, which … is the first professional organization that our students, many of them, meet. So, we are closely coordinated with the IDF.

So far, we can learn that opportunities for cross-sectoral organization exist, when the risks are managed by the system of STEM education. However, alongside the desire for cooperation, weaknesses were also emphasized in the cooperative work with the education system. The research participants, representatives of for-profit and non-profits organizations, whose activities are limited by contract with the education system, pointed to bureaucratic processes and expressed the desire to cooperate in professional decisions making and not just being their operators.

Thus, the study's findings indicate the need to examine optimal ways to involve the stakeholders of STEM education and cooperate with them in STEM education processes.

The research findings specified the *institutionalized cooperation* (Schiffer *et al.* 2010) as the preferable collaboration pattern. This pattern of cooperation enables to create an infrastructure for joint activities, when the education system is the responsible agent for what goes inside the education system, while the second and third sector organizations operate through philanthropic assistance in fields which are determined by the education system. In the implementation process of this pattern of collaboration, a regulation system is established that preserves the power balance in favor of the education system.

However, this type of partnership does not provide freedom to the second and third sectors in determining the educational content. In practice, the education system is the body which initiates the creation of new programs and is just assisted by the second and third sectors with their implementation. In addition, the education system operates control mechanisms over the partners' activities to ensure maximum efficiency of the allocated budget.

Johnston *et al.* (2011) explain that successful collaborative governance is reflected in a climate that fosters trust, mutual responsibility, and a willingness to share risks. The stakeholders are required to trust each other for a long-term, which is a profound conceptual change that may take place only in an open and trustworthy environment. This idea is further developed by the economist Robert Reich (the Minister of Labor of the United States under Bill Clinton administration) in general, and in the Israeli context by Tamir (2015). These ideas present a new public dialogue. A dialogue that reflects the society's desire to reduce social gaps for the benefit of everyone. If the public conception will undergo a conceptual change process and the second and third sectors will operate as ideal philanthropy[2], different or other collaboration patterns may fit.

For example, in the case discussed in this Brief, stakeholders' cooperation for the mitigation of the strategic risks may be more efficient if the cooperation will also transfer the responsibility. Examples for such a process are the proposed courses of action for dealing with the salaries of STEM teachers: 'STEM teachers' mobility' (Course of action 3) and 'Integrated earning' (Course of action 5). These actions suggest the recruiting of the industry and the education system to enable STEM teachers to be engaged in both occupations at the same time and to allow teachers' mobility between sectors. If the responsibility of managing this process had been transferred to the stakeholders, and the cooperation was not perceived as help provided to the education system to "solve" its problems, the process might have been more effective.

It should be noted, in conclusion, that in reality cooperation already takes place, between the education system and other stakeholders – academia, military, industry, and philanthropy institutions. Such cooperation is much desired, but the research suggests it should be carefully examined, so that the best mode of cooperation is adopted, one which accounts for the weaknesses and strengths of all stakeholders.

One possible mode of action is the creation of **institutionalized cooperation, cross-sectors**, one which preserves the role of the education system as the carrier of primary responsibility for public education, and yet enables other sectors and stakeholders to act in the service of one important goal: advancing STEM education, for the benefit of society, and future generations.

References

Barak, M. Improvement of teacher quality by organizational-oriented policy, the whole (school) is bigger than the sum of its parts (teachers). Hed Hachinuch **86**(1), 29−30 (2011). (Hebrew) http://portal.macam.ac.il/ArticlePage.aspx?id=4471

Barney, J.: Looking inside for competitive advantage. Acad. Manag. Exec. **4**, 49–61 (1995)

Deci, E.L., Ryan, R.M.: Hedonia, eudaimonia, and well-being: an introduction. J. Happiness Stud. **9**(1), 1–11 (2008)

European Commission: Key data on teachers and school leaders in Europe (2013). http://eacea.ec.europa.eu/education/eurydice/documents/key_data_series/151EN.pdf

Helmer, O.: Social Technology. Basic Books, New York (1966)

Hosseinzadehdastak, F., Underdown, P.E.: An analysis of risk management methods. In: Proceedings of the IIE Annual Conference, p. 1. Institute of Industrial Engineers-Publisher (2012)

[2] Ideal philanthropy is an altruistic approach that focuses on the contribution to social needs that are not provided by the government.

IRM: Risk Management Standard. The Institute of Risk Management (IRM), The Association of Insurance and Risk Managers (AIRMIC) and the National Forum for Risk Management in the Public Sector (ALARM) (2002). https://www.theirm.org/media/886059/ARMS_2002_IRM.pdf

Johnston, E.W., Hicks, D., Nan, N., Auer, J.C.: Managing the inclusion process in collaborative governance. J. Public Adm. Res. Theor. **21**(4), 699–721 (2011)

Linstone, H., Turoff, M.: The Delphi Method. Addison-Wesley Publication, Boston (1975)

ISO Guide 73: Risk management – vocabulary: a collection of terms and definitions relating to the management of risk. International Organization for Standardization (2009). http://www.iso.org/iso/home/standards/iso31000.htm

Mikes, A., Kaplan, R.S.: Towards a Contingency Theory of Enterprise Risk Management. Harvard Business School, Boston (2014)

Oplatka, Y.: The Foundations of Education Administration: Leadership and Management in the Educational Organization. Pardes Publications (Hebrew) (2007)

Rego, G., Nunes, R.: Hospital foundation: a SWOT analysis. IBusiness **2**, 210–217 (2010)

Sabbaghi, A., Vaidyanathan, G.: SWOT analysis and theory of constraint in information technology projects. Inf. Syst. Educ. J. **2**(23), 1–19 (2004)

Sahlberg, P.: Finnish lessons. The series of school reform (2011)

Schiffer, et al.: The involvement of third sector organizations in the education system. White Paper, Van-Leer Institute, Jerusalem (Hebrew) (2010). http://www.vanleer.org.il/sites/files/product-pdf/73-sFileRedir.pdf

Tamir, Y.: Who's afraid of equality? On education and society in Israel. Yediot Achronot, Hemed Books (2015)

The Main Features of Nvivo Software and the Procedures of the Grounded Theory Methodology: How to Implement Studies Based on GT Using CAQDAS

Jakub Niedbalski(✉) and Izabela Ślęzak

Faculty of Economics and Sociology, University of Lodz, Lodz, Poland
jakub.niedbalski@gmail.com, iza.slezak@gmail.com

Abstract. The aim of the article is to introduce the reader to the possibilities offered by software supporting the analysis of qualitative data in research projects based on qualitative methods. The analysis is based on NVivo software and its functions. In this article we present the way in which the options of this program should be used to constitute an effective means to conduct research in accordance with the procedures of grounded theory methodology. The aim of the article is to introduce the reader to the possibilities offered by the software supporting the analysis of qualitative data in research projects based on qualitative methods.

Keywords: Computer-Assisted qualitative data analysis software (CAQDAS) NVivo software · Grounded theory methodology

1 Introduction

Technology plays an increasingly important role in today's world. One of its faces is the use of common present computers and applications, both universal and specialized, aimed at a narrow audience of scientists and researchers. Computer software has been used in social sciences and humanities for numerous years now [1–3]. We selected NVivo from plenty of qualitative data analysis applications, based on our knowledge and long-standing experience in CAQDAS.

It is emphasized in the literature on the subject that NVivo is a tool that aids research projects based on discourse analysis, grounded theory methodology, conversational analysis, ethnography, studies based on phenomenology as well as other methods, including mixed research methods [4]. Nevertheless, as we ourselves employ GT in our research, we look at the described application from the perspective of this research method. Therefore, we would like to present the most important functions that NVivo has to offer and indicate their applicability within the qualitative analysis process compliant with grounded theory methodology.

A. P. Costa et al. (Eds.): WCQR 2018, AISC 861, pp. 90–101, 2019.
https://doi.org/10.1007/978-3-030-01406-3_8

2 Grounded Theory Methodology - Basic Assumptions and Procedures

The choice of methodology always involves specific ontological and epistemological assumptions, which constitute the research framework; if a researcher aims to learn the perspective of the researched social actors and capture the process-oriented dimension of the researched phenomena, the grounded theory methodology seems particularly useful [5]. According to its assumptions, the theory emerges during systematic field research as a result of the empirical data analysis, directly referring to the observed fragment of social reality [6, 7]. It is a middle range theory that includes terms derived from observations and the empirical area specified for the research [8]. Generation of theories is a procedure where the stage of data collection is not explicitly separated from the phase of establishment of hypotheses. In case of GT, empirical data are collected not in stages or phases, but alternately with analysis and interpretation carried out in parallel [9]. Both stages entangle, and researchers' continuous getting back to the collected material guarantees that the developed theory is really derived from the empirical data [7].

Thus, a theory is created in an inductive way, based on the analysis of collected data; only according to the categories "induced" in this manner is a deduction analysis carried out, indicating subsequent groups of cases for comparative analyses, as a result of which a more abstract level of theoretical connection of the categories can be achieved. We deal here with an abduction procedure, namely going from induction to deduction [7]. The theory is therefore not free literary creation, but laborious, controlled, and "objective" research methodology [5]. One of the assumptions underlying its application is to minimize research pre-conceptualization, in order to maintain openness, the chances for discovering phenomena that we did not seek and were not aware of at the beginning of our research. These actions are also accompanied by other procedures of methodological correctness, which in the case of MTU include, among others, theoretical sampling, constant comparative method, coding, writing memos. Sampling is of processual character and it takes place until theoretical saturation is achieved, i.e. to the moment, when no new data appears, and subsequent cases are similar to the previous ones, and they may be analyzed based on the already existing categories [5].

It is worth stressing that the grounded theory methodology is not a uniform proposal, not differentiated internally. Due to the fact that it has been critically reflected upon by its creators for decades, and it has been employed up to now by successive generations of researchers, who often introduce certain modifications to its assumptions and applied analytical procedures, it can be referred to rather as methodologies of grounded theory [6, 7, 11]. Without considering the differences between these approaches in more detail we would like to point out, however, that the paper adopts the approach represented by A. Strauss, drawing on the tradition of symbolic interactionism, which is thus a coherent whole with the ontological and epistemological representations we assume.

3 Starting to Work with the Application

NVivo's history dates back to the early 1980s. The application was developed by Tom Richards to support social research by Lyn Richards and became one of the first qualitative research applications. The application versions before NVivo had been released under the designation of NUD*IST (*Non-Numerical Unstructured Data Indexing Searching and Theorizing*), followed by software called QSR - from N4 to N6 versions. Since then, the program has undergone numerous modifications, and its successive versions have been enriched with newer functions that were better tailored to the needs of qualitative researchers. The dynamics of change causes that a new version of the application is launched every 2–3 years.

Despite its numerous options, NVivo is quite intuitive. Its interface and the arrangement of individual functions allow even an inexperienced user to get familiar with the application quite quickly. There are drop-down menus and toolbars at the top of the window. There is a search engine bar below - it allows a user to search through the data from all available sources and materials stored in the database. Another area is the navigation menu located on the left side of the interface window. Definitely the largest part of the interface is occupied by the work window that displays the project elements selected by the user.

Work in the program is started with the creation of a project, which *de facto* constitutes a database that is gradually filled with materials. NVivo offers a lot of possibilities in this scope, as it allows to import text files and photos, as well as audio and video materials. It is important here that the application support not only different types of data, but also plenty of data formats what makes their applicability very broad.

It is also worth stressing that the application has a specific naming system. The change in the term "codes" in NVivo to "nodes" is particularly noteworthy. For beginners, this idiosyncratic difference may be slightly problematic, but it can be accustomed to relatively quickly.

4 Functions of NVivo and Implementation of a Project Based on GT

Knowing how to start working with NVivo we can begin with strictly analytical activities related to the process of generating theory in compliance with GT. Therefore, what will be of particular interest to us is the possibility of using selected functions available in the application in such a way that the researcher can apply specific research procedures of the grounded theory methodology.

4.1 Coding

The basic elements of the theory are *categories* and their *properties* (the most specific, features of a given category that can be conceptualized, specific enough to be useful in further analysis) and *hypotheses* (links between categories, which the researcher discovers in the course of analyses). Categories and their properties are generated during the coding procedure.

A key role is played in this context by the process of coding, i.e. ascribing batches of material with particular labels that reflect their sense and meaning allocated by social actors, and reflected by the researcher [9]. Two basic types of coding are distinguished in the grounded theory methodology: substantive and theoretical coding. Substantive coding means conceptualizing a given area of research by using (often alternately) three types of coding: open, axial and selective [7].

The open coding forms the basis for the other types of coding; it involves assigning labels (in all possible ways) to the collected empirical material, identifying the categories, their properties and consequently moving from data to the conceptual level. In other words, it is about assigning a word or phrase to one or more paragraphs. Therefore, the code is a concept that ascribes meaning to descriptive information. Strauss and Corbin [7] define them as "an analytical process where concepts are defined and developed in terms of their properties and dimensions". This goal is achieved by searching for meanings hidden in data, comparing them, as well as developing and creating more general categories for similar phenomena [7].

Such a coding performed in NVivo takes place through marking a fragment after a fragment of a text, which is afterwards ascribed with certain labels [12]. Importantly, the generated codes may be subject to modification, depending on the researcher's subsequent findings and evolving ideas. It means that a once ascribed code could change its name, become merged or replaced with another code, along with advancement of the theorization process. In NVivo we can copy, cut, delete, rename or merge one code with another. This allows us to change, add or delete selected codes on an ongoing basis, thus redefining and reconstructing the categories gradually as the researcher's ideas change.

4.2 Coding Paradigm

Encoding cannot be restricted only to the detection of categories and their names. The elements that make up the theory are, among others, categories and their properties. Categories are presented as "conceptual elements of theory", i.e. some basic components that they are built from. Categories are conceptual equivalents of phenomena [13] and, as Glaser and Strauss put it, "are based on themselves" [6]. According to the mentioned authors, this is the basis for distinguishing categories from properties. The latter only make sense if they are conceptual aspects or elements of a category. There is therefore a difference in generality between categories and properties [13]. As Marek Gorzko argues [13], the authors of grounded theory assumed that general categories form the core of the theory and are constructed of elements of a lower order. Furthermore, categories need to be developed within the definitions of their properties. The more of these conceptual elements (components, properties) are included in the analysis, the better the categories are described and the "denser" the theory based on those. Therefore, another step in the analysis based on GT is the axial coding, when connections between categories and subcategories are determined. A coding paradigm may be helpful in finding them. It consists of causal conditions of occurrence of the analyzed phenomenon, the context in which it occurs, the conditions intervening in its course, as well as actions, interaction strategies that accompany it and the consequences of the occurrence of these factors [7]. An additional analytical tool is the conditional matrix, which facilitates encoding and description of casual and intervening conditions as well as the context of a given phenomenon [7].

Within the course of work, NVivo allows to create an ordered code structure, i.e. a category tree. Moreover, the category tree, as the analytical process progresses, allows to make certain changes and modifications. These may include, for instance, changing the positions of categories that remain at the same level in the hierarchy.

The tree structure is also highly helpful in creating a core category that emerges during axial coding. The core category is usually the basic process, and its essence is the fact that other categories are associated with it and that the theory builds upon it [5].

For example [14], during the analysis process, we developed and used almost 500 codes, which were gradually combined, grouped or simply excluded, ultimately providing 40 categories. The interviews and notes from the observations were coded as a "stage" in relation to the course of the story, which is composed of the following phases: "psychosocial disintegration", "discovering sport", "conversion of position", "skills testing", "progress verification", "final rehearsal", "fight for a win" and "specifically." Fragments related to turning points identified on the basis of the data analysis within particular phases of the process of becoming an athlete were marked within particular stages. During the analysis, there was a key for material codes developed for each phase in the scope of the main process, that is, the development of a sports career. Grouping codes into a hierarchical code took place from the bottom. Each of the above-mentioned phases was allocated several dozen codes which were later arranged into more general categories until seven main groups were achieved.

4.3 Generating Hypotheses

Selective coding can be carried out on the basis of the axial coding, where the collected material is analyzed only in terms of those variables and codes that are related to the core category. Within the course of research, the respondents' statements revealed numerous *in vivo* codes that express the conceptual work of the subjects, their thinking about their key issues; they are included into the theory under development directly from the colloquial language and the substantive area of research [5].

Since the researcher's aim is not only to codify, but also to systematize the categories and to organize them in relation to each other, they will try to place the codes in a specific structure resembling the category tree that imitates their hierarchy. The category tree itself allows to make certain changes and modifications as the analytical process progresses. These may include, for instance, changing the positions of categories that remain at the same level in the hierarchy. This function may be used to determine the "significance" of a category within the analytical determinations by moving it above or below other categories.

Despite reflecting a code structure in a form of the category tree, NVivo allows to create relationships between particular categories. While using this function, you can indicate the type of a relationship and whether it is uni- or bidirectional, or simply establish the very existence of a relationship without indicating its direction. Hence, we obtained a tool supporting the process of hypotheses development, thus enabling to reach a higher level of conceptual analysis, according to the principles of the grounded theory methodology. At the same time, a hypothesis in grounded theory methodology is not a statement that we want to verify within the course of research; it is rather a thesis that shows the grounded empirical relationships between concepts, for which the conditions of occurrence have been established [7]. Very often during field studies such relationships between variables are discovered *in vivo* [6]. It is also worth noting that the categories can also be borrowed from already existing theories. However, this requires careful consideration by the researcher and taking into account the different meanings that they may have [15]. The final criterion determining their application is the requirement that they emerge directly from the collected data [6, 7].

4.4 Writing Theoretical Memos

The process mentioned in the previous section is theoretical coding, which allows to discover the mutual relations and interdependencies between categories [5]. Theoretical coding takes place through preparation of theoretical memos, i.e. thoughts of a researcher considering the encoded categories and hypotheses, written in a theoretical language. The memos are a reflection of analytical thoughts related to codes, and they are adopted to make the applied categories more specific and provide the coding process with a certain direction. What is more, they also pose a certain connector between two stages of an analysis - coding and writing a report. According to Marek Gorzko [13] the written memos are a kind of an analytical tool, creating certain thought and theoretical space, where the researcher can conceptualize data.

Writing of memos accompanies an analyst applying the grounded theory methodology since the beginning of the research process. These memos may be related to the whole project, collected data in general or any source of data individually, also subsequent stages of analysis and particular codes [16]. Therefore, preparation of notes is of crucial meaning on each level of the coding and data analysis process [7].

In the NVivo application, a role of theoretical notes is played by memos, i.e. records of theoretical thoughts and concepts by a researcher. Memos are concise notes drawn up by a researcher, including information on concepts regarding the whole project, particular material or issues to be discussed or interpreted in the future. The concept of memos is analogical to the procedure of generating notes in the grounded theory methodology. They are intended to help the researcher to move to a higher conceptual level, and they serve the generation of theories as tools of theoretical coding. Hence, apart from carrying out actions in the field, we also went back to our "old" notes with a new research "objective" to confirm or reject the theses from the presented story. Each thesis must be supported with an accurate description that complies with the facts. In other words, it must be enriched with details and episodes that show the reader that the research is solid and rich. Thus, we extended the story, composed a more elaborate model, and made the theory a little more "sophisticated" [14].

4.5 Theoretical Sampling

When collecting, encoding and analyzing materials, the researcher decides on an ongoing basis where and what kind of data to collect further, where the theoretical sampling procedure plays a key role. Hence, it is possible to select the right cases for comparison. They are chosen based on their conceptual similarity, which allows to saturate the categories with their properties (so as to develop as complete a list as possible) and to build new hypotheses with a higher level of generalization. Therefore, the researcher aims to collect both very similar cases (minimum comparison strategy) and definitely different ones (maximum comparison strategy), in order to observe some possibly different conditions of categories occurrence and their mutual relationships [6, 11]. It is worth noting that although selection of samples to be compared is open and processual in character, it is not chaotic, but controlled through an emerging theory; the procedures of the grounded theory methodology, such as the requirement to write theoretical and methodological memos that justify the researchers' choices and help identify the moment of theoretical saturation, provide systematic and methodological guidance [5, 6]. It means that no new data are produced during the collection of the material, that emerging examples are similar to those already collected and that they can be analyzed using existing categories. Saturation is therefore achieved by supplementing the properties of the category with various data [6, 7], which can be derived from observations, interviews, and from existing documents, and it can also be statistical data [5, 6]. In such cases, we can talk about application of the triangulation procedure, especially the triangulation of data.

In NVivo, cases are created through a corresponding function. It allows to add and specify attributes and properties. The cases may serve as a representation of particular persons or organizations that pose a subject of research interest. Additionally, using the *Attributes* option allows to assign the properties of metric data to specific cases. This is how the researcher working with the application can use one of the key procedures of grounded theory - theoretical sampling - where these cases and their attributes are of great help.

For example [14], we wanted to specify the scope of similarities and differences regarding the situation of disabled women and men who practice sport, respectively. Hence, we excerpted fragments from particular texts where women spoke about the manner in which they are perceived by those around them, and how – according to them – others see their dysfunctions, and we compared them to the segments of texts raising the same topic, but coming from interviews with men. Such parameters related to the comparison of particular data allowed to put forth several initial hypotheses related to the gender of disabled athletes. An example may be provided by the following hypothesis: "Men may count on greater approval from those around them as regards their bodies, which, despite certain dysfunctions, usually do not pose such a great barrier in relationships with the environment as they do in the case of women."

4.6 Constant Comparative Method

The logic of the research process in the grounded theory methodology is based on seeking an increasingly higher conceptual level, and, as a result, dropping data and turning to theorization. In this context, the constant comparative method is of great significance, consisting in the search for differences and similarities between fragments of data, codes or cases [5, 6, 17]. In other words, the constant comparison method is about confronting various components of the project in order to check the similarities between them or to emphasize certain features that differentiate them. Increasingly more general categories revealing underlying uniformities are generated on the basis of similarities and differences analysis (Strauss, Corbin 1990, 2008: 86). The constant comparative method is based on three mutually dependent and common types of comparison:

Firstly, to compare individual cases in order to identify both their common features and their strong differentiation, and to specify the conditions under which they occur. This is the basis for generation of concepts;

Secondly, the concepts are then compared with the subsequent cases observed, thus generating new theoretical properties of the concepts and new hypotheses, and saturating the properties of a given concept-category;

Thirdly, within the course of the first- and second-type comparisons, the researcher begins to compare concepts, which makes it easier to integrate them into hypotheses [5, 6, 17]. This procedure also avoids bias [5].

NVivo facilitates activities related to the constant comparative procedure which, by providing the researcher with a basis for discovering general mechanisms or the extent to which certain social phenomena and phenomena occur [18], is one of the main procedures of the grounded theory methodology. In practice, while using the NVivo application, the comparative procedure is performed with the data search option. This process consists in reviewing fragments of a text and other data, which were coded with a particular code (or codes). Hence, we obtain knowledge on opinions on a given topic - which we grasped in an analytical category - expressed by particular speakers.

For example [14], the hypothesis of the relationship between sport being practiced by a disabled individual and the impact other people have on the process may be studied through searching for all elements of the text coded under "sport interest" and fragments coded with "persons from their surroundings," located a certain distance from the first time it is mentioned (expressed by the number of paragraphs). On the basis of a search presented in such a manner, it might be agreed that interviews carried out with various individuals were coded in such a manner that there are text fragment codes that represent the distance between fragments. If such an interdependence is of a repeatable character, and it can be observed in various interviews, the researcher may claim that the hypothesis is corroborated in the current pool of data.

4.7 Matrices and Diagrams

Tables and matrices are good means for data comparisons. They allow text extracts from the entire data set to be presented in a form that facilitates systematic comparison. They may result in generated typologies - dividing the set of all analyzed cases into subsets in such a way that each subset is assigned only to one of the selected types. Use of this tool allows to condense large materials and segregate them thematically into more accessible and clearer fragments that can be compared with each other in a much simpler manner.

While generating theories, it is also helpful to create visual representation of the established theories, picturing relationships between them, and allowing to distance oneself towards the collected data [5, 7]. From the GT methodology perspective, the most significant are the models which form the basis for diagrams that integrate data.

NVivo enables to create models, among others, to determine and review the initial concepts and ideas on the questions that are interesting for the researcher, visual representation of relationships between the project elements, identification of emerging patterns, theories and explanations, as well as documentation and recording of subsequent stages of work over the project. From the practical and technical perspective, the models can be created from various geometric shapes available in the application (e.g. square, rectangle, rhombus), which may (but do not need to) be additionally combined with a given project component. The above-mentioned geometric shapes can represent some specific project components. At the same time, different types of lines and arrows present in the model visualize the kind of connections existing between the project components. If a given geometric figure represents an existing project element (e.g. a code, memo, data set), it is possible to easily display its content by right-clicking on the shape [19, 20]. Thanks to this function, a researcher is allowed to develop a project draft, and a vision of own ideas related to development of material [12, 18]. The model may reflect some Kaefer current modifications that result from the analytical process, and they may present the current analyst's ideas.

4.8 Other Functions Supporting the Work Process in the Application and Generation of Theories

In addition to features that support the process of generating theories, NVivo also offers other numerous interesting and useful options. We could mention here the complex data search tool, employed both for querying the information contained in texts (words, phrases) and the categories generated by the researcher. NVivo has various search options available, which are included in the entire battery of query tools. Text Search, i.e. a dictionary search tool, is one of the search tools based on the codes generated by the researcher that can be used. Thanks to them, the researcher (apart from the data comparison described above) receives assistance in the process of formulating ideas, verifying their intuition, checking the previous result of analysis (e.g. coding of material) or discovering completely new relationships, dependencies, relations, etc. between the existing project components.

A useful feature offered by NVivo is also the ability to publish the results of analyses carried out in the application. The researcher can use one of numerous available options to export data in a format that can be open in common office applications [9]. Data can also be published by creating some research reports, which can illustrate a certain stage of analysis or final processing of research results, as well as by exporting charts and models created in the application in the form of images.

These and plenty of other functions in NVivo represent some significant potential that, when used appropriately by the researcher, can be of great help in the process of data analysis, regardless of the research method employed. Nevertheless, taking into account the implementation of research based on grounded theory procedures, the functions indicated in the article and the ways they are used constitute the basis of tools, especially for a beginner analyst. Over time, it can be expanded with additional options that will certainly aid the researcher's work to an even greater extent.

5 Summary

NVivo is a universal tool, equipped with a range of helpful features that should meet the expectations of researchers representing a variety of theoretical research schools and methods. Some of them are especially significant and useful for researchers who work with GT, while others will be helpful regardless of the analytical approach adopted in the project [12, 21].

From the perspective of the grounded theory methodology, the NVivo's advantage is certainly the possibility to organize a single database from different types of source materials [19]. An important feature of NVivo is also the fact that it allows to order various project components, both the source materials and all information developed by the researchers involved in data analysis, which actually is the analysis results. Researchers who apply GT will surely appreciate the possibility to write various memos and create the code tree, supporting a systematic analysis. NVivo enables to move freely between the open and axial coding, writing memos and modeling.

It is also worth emphasizing that the application architecture itself somehow forces the researcher to continuously think about the relations between codes and categories, to compare and modify a system created by the researcher. It also encourages concentration on the core category, which should be the focus of the researcher's activities.

On the other hand, despite numerous advantages in favor of compliance of the internal software architecture and GT procedures, the application structure causes some researchers to accuse it of excessive stiffness and the necessity to subordinate the analysis to solutions implemented by software engineers. The use of computer software also requires the researcher not only to have methodological and theoretical skills, but also to become familiar with the specifics of a particular application and its functions [22, 23]. It is also sometimes stressed that the application developers, trying to make it universal, have created a tool that is so complex that it no longer meets the requirements of any research method at all. However, it seems that in the NVivo's case it is more a matter of how the user exploits the capabilities of the software than of the architecture of the software itself. First of all, it should be kept in mind that computer software is only a tool, but it is the researcher who decides how to use it. Eventually, the researchers and their analytical abilities and skills related to operation of a given application will be crucial in the context of quality of conducted research and its results.

References

1. Gilbert, L.S., Jackson, K., di Gregorio, S.: Tools for analyzing qualitative data: the history and relevance of qualitative data analysis software. In: Spector, J.M., Merrill, M.D., Elen, J., Bishop, M.J. (eds.) Handbook of Research on Educational Communications and Technology, pp. 221–236. Routledge, London (2014)
2. Woods, M., Paulus, T., Atkins, D.P., Macklin, R.: Advancing qualitative research using qualitative data analysis software (QDAS)? Reviewing potential versus practice in published studies using ATLAS.ti and NVivo, 1994–2013. Soc. Sci. Comput. Rev. **34**(5), 597–617 (2016)

3. Zamawe, F.C.: The implication of using NVivo software in qualitative data analysis: evidence-based reflections. Malawi Med. J. **27**(1), 13–15 (2015)
4. Schönfelder, W.: CAQDAS and qualitative syllogism logic—NVivo 8 and MAXQDA 10 compared. Forum Qualitative Sozialforschung (Forum: Qual. Soc. Res.) **12**(1), (2011). http://www.qualitative-research.net/index.php/fqs/article/view/1514
5. Glaser, B.: Doing Grounded Theory: Issues and Discussions. Sociology Press, Mill Valley (1998)
6. Glaser, B., Strauss, A.: The Discovery of Grounded Theory. Strategies for Qualitative Research. Aldine Publishing Company, New York (1967)
7. Strauss, A., Corbin, J.: Basics of Qualitative Research. Grounded Theory Procedures and Techniques. Sage, Newbury Park, London, and New Delhi (1990)
8. Merton, R.: Teoria socjologiczna i struktura społeczna. Wydawnictwo Naukowe PWN, Warszawa (2002)
9. Niedbalski, J., Ślęzak, I.: Analiza danych jakościowych przy użyciu programu NVivo a zastosowanie procedur metodologii teorii ugruntowanej. Przegląd Socjologii Jakościowej **8** (1), 126–165 (2012)
10. Hammersley, M., Atkinson, P.: Metody badań jakościowych. Wydawnictwo Zysk o S-ka, Poznań (2000)
11. Charmaz, K.: Grounded theory. Objectivist and constructivist methods. In: Norman, D., Lincoln, Y. (eds.) Handbook of Qualitative Research, Advances, pp. 213–222. Sage, Thousand Oaks (1994)
12. Kaefer, F., Roper, J., Sinha, P.A.: Software-assisted qualitative content analysis of news articles: example and reflections [55 paragraphs]. Forum Qualitative Sozialforschung (Forum: Qual. Soc. Res.) **16**(2), Art. 8 (2015). http://nbn-resolving.de/urn:nbn:de:0114-fqs150283
13. Gorzko, M.: Procedury i emergencja. O metodologii klasycznych odmian teorii ugruntowanej (Procedures and Emergence. About Methodology of Classical Variants of the Grounded Theory). Wydawnictwo Naukowe Uniwersytetu Szczecińskiego, Szczecin (2008)
14. Niedbalski, J.: From a qualitative researcher's workshop—the characteristics of applying computer software in studies based on the grounded theory methodology. Przegląd Socjologii Jakościowej **13**(2), 46–61 (2017)
15. Clarke, A.: Social worlds/arenas theory as organizational theory. In: Maines, D.R. (ed.) Social Organization and Social Processes: Essays in Honor of Anselm Strauss, Advances i., pp. 119–158. Aldine De Gruyter, New York (1991)
16. Saillard, E.K.: Systematic versus interpretive analysis with two CAQDAS packages: NVivo and MAXQDA. *Forum* Qualitative Sozialforschung (Forum: Qual. Soc. Res.). **12**(1), (2011). http://www.qualitative-research.net/index.php/fqs/article/view/1518
17. Konecki, K.: Studia z metodologii badań jakościowych. Teoria ugruntowana (Studies on Methodology of the Qualitative Research. The Grounded Theory). Wydawnictwo Naukowe PWN, Warszawa (2000)
18. Seale, C.: Wykorzystanie komputera w analizie danych jakościowych. In: Silverman, D. (ed.), Prowadzenie badań jakościowych, Advances i., pp. 233–256. Wydawnictwo Naukowe PWN, Warszawa (2008)
19. Wiltshier, F.: Researching with NVivo. Forum: Qual. Soc. Res. **12**(1). http://www.qualitative-research.net/index.php/fqs/article/view/1514 (2011)
20. Bringer, J.D., Johnston, L.H., Brackenridge, C.H.: Maximizing transparency in a doctoral thesis1: the complexities of writing about the use of QSR*NVIVO within a grounded theory study. Qual. Res. **4**(2), 247–265 (2004)

21. Paulus, T., Woods, M., Atkins, D.P., Macklin, R.: The discourse of QDAS: reporting practices of ATLAS.ti and NVivo users with implications for best practices. Int. J. Soc. Res. Methodol. **20**(1), 35–47 (2017)
22. Houghton, C., Murphy, K., Meehan, B., Thomas, J., Brooker, D., Casey, D.: From screening to synthesis: using NVIVO to enhance transparency in qualitative evidence synthesis. J. Clin. Nurs. **26**(5–6), 873–881 (2017)
23. Salmona, M., Kaczynski, D.: Don't Blame the Software: Using Qualitative Data Analysis Software Successfully in Doctoral Research. Forum Qualitative Sozialforschung (Forum: Qual. Soc. Res.). **17**(3), (2016). http://www.qualitative-research.net/index.php/fqs/article/view/2505/4009

Curriculum Co-design for Cultural Safety Training of Medical Students in Colombia: Protocol for a Qualitative Study

Juan Pimentel[1,2,3(✉)], Germán Zuluaga[3], Andrés Isaza[3], Adriana Molina[2], Anne Cockcroft[1], and Neil Andersson[1,4]

[1] McGill University, Montreal, QC, Canada
juan.pimentel@mail.mcgill.ca
[2] Universidad de La Sabana, Chía, Cundinamarca, Colombia
[3] Universidad del Rosario, Bogotá, Cundinamarca, Colombia
[4] Centro de Investigación de Enfermedades Tropicales (CIET), Universidad Autónoma de Guerrero, Acapulco, Guerrero, Mexico

Abstract. Cultural safety in medical training encourages practitioners, in a culturally congruent way, to acknowledge the validity of their patients' world-views. Lack of cultural safety is linked to ethnic health disparities and ineffective health services. Colombian medical schools currently provide no training in cultural safety. The aim of this qualitative study is to: (i) document the opinions of stakeholders on what a curriculum in cultural safety should teach to medical students; and (ii) use this understanding to co-design a curriculum for cultural safety training of Colombian medical students. Focus groups will explore opinions of traditional medicine users, medical students, and cultural safety experts regarding the content of the curriculum; deliberative dialogue between key cultural safety experts will settle the academic content of the curriculum. The research develops participatory methods in medical education that might be of relevance in other subjects.

Keywords: Cultural safety · Participatory research · Medical education Colombia · Thematic analysis

1 Introduction

In 1977, the WHO called for collaboration of Western and traditional medicine [1], in its view of Primary Health Care [2] recognising an inextricable relationship between culture and health outcomes. Yet, this international recognition does not guarantee acknowledgment in everyday medical practice [3]. Western physicians continue to receive medical education and to be presented with role models that do not emphasise culture as a positive resource in health outcomes.

Medical education curricula in most Western countries still focus on biomedical content and perspectives, reducing the chances that the next generation of physicians will acquire the skills and mindset to provide culturally congruent health services. This is compounded by differences in cultural background between physicians and their patients that accentuates the shortfalls of Western medical education, hindering

A. P. Costa et al. (Eds.): WCQR 2018, AISC 861, pp. 102–109, 2019.
https://doi.org/10.1007/978-3-030-01406-3_9

accessibility, acceptability, and effectiveness of health services in the intercultural context [4]. At worst, these differences lead to confrontation with, discrimination against, and even harm to patients, with racial/ethnic health disparities as the outcome [5]. In terms of economic impact, the combined cost of these disparities was estimated at $1.24 trillion between 2003 and 2006 in the US [6].

There is growing agreement about the need to train medical students to provide culturally congruent services [7, 8]. For instance, the 2015 Standards for Accreditation of Medical Education Programs in Canada [9] call for training on "the basic principles of culturally competent health care" (p. 21) and "the manner in which people of diverse cultures perceive health and illness and respond to various symptoms, diseases, and treatments" (p. 21). Some medical curricula have implemented "cultural competence" training with positive results including reduction of health care disparities [10], increased satisfaction, increased adherence to prescribed treatments [11], a healthier doctor-patient relationship [12], and even improved physiological and biochemical indices of disease [13].

These improvements, notwithstanding, some authors criticize cultural competence as a concern that improves Western service delivery/supply without dialogue about demand, a new form of colonialism [14], leaving the *power* relations between the patient and professional unaffected [15]. The relatively newly popularised concept of "cultural safety" [16] goes beyond cultural competence, insisting that the patients should have an opportunity to "comment on practices and contribute to the achievement of positive health outcomes and experiences" [17]. Cultural safety embraces *dialogue* between patients and physicians to make joint decisions and especially to judge whether the interaction is culturally safe or not [18].

Increasing awareness of cultural safety in medical education would yield the benefits of cultural competence, but also acknowledge the power relationships that occur in practice while "accepting the legitimacy of difference and diversity in human behavior and social structure" [17]. Such a shift in practice would facilitate the transition to a more equitable and client-centred provision of health services, simultaneously reaffirming the communities' right to self-determination and providing respectful services free of colonized perspectives.

Multicultural Colombia is an ideal setting for implementing cultural safety in medical training. In the country, the government supports health services firmly based on the Western biomedical model. In contrast, up to 40% of the population seek care in traditional medicine [19], creating a care gap between the community expectations and needs, and the physician's knowledge and skills. Unfortunately, at present Colombian medical schools provide no cultural safety training.

In light of this, the purpose of our study is two-fold: (a) to examine the opinions of several stakeholders on what a curriculum in cultural safety should teach to medical students so they can provide a culturally safe practice when interacting with traditional medicine users in Colombia; and (b) to use this understanding to co-design a curriculum for cultural safety training of Colombian medical students.

2 Research Question

What are the opinions of stakeholders on what a co-designed cultural safety curriculum should teach to medical students so that they can provide culturally safe services in Colombia? In this study, the stakeholders include key-informant traditional medicine users and medical students from Colombia, as well as key cultural safety experts from Colombia and Canada.

3 Methodology

3.1 Research Design

We will use a qualitative research design that uses a sequence of qualitative research methods aimed at producing data "with adequate generalizability, (…) to influence public health programming and clinical work" (p. 417) [20]. The goal of this methodology is to produce relevant knowledge to generate social change of stakeholders, end-users, and their communities.

3.2 Participants

We will invite three different groups of stakeholders: (a) traditional medicine users from the "Seed of life" (Semilla de vida) community organization at Cota, Colombia; (b) senior medical students from *La Sabana* University (Colombia); and (c) cultural safety experts from the Center for Intercultural Medical Studies (CEMI) and the Research Group on Traditional Health Systems (GESTS) in Colombia, as well as from the Participatory Research at McGill (PRAM) and the McGill Institute for Human Development and Well-being (IHDW) in Canada.

According to Israel, a participatory research expert, "building upon prior positive working relationships is a viable strategy for conducting participatory research" (p. 187) [21]. This project is based on previous partnerships between CEMI, GESTS, McGill PRAM and the Seed of Life organization. Collaborating for more than 13 years in participatory initiatives to protect traditional cultures [22, 23], these stakeholders have developed reliable relationships that will facilitate the progress of the project.

3.3 Sampling Strategy

We will use a purposive sample of key informants [24]. The Seed of life organization is comprised of 10 key traditional medicine users and community leaders. They are key informants because they have been recognized by the community as knowledgeable about traditional medicine and also have 20 years of experience working in community-based projects to protect their culture. The 25 medical students that we will invite are former research assistants in community-based intercultural health interventions conducted by CEMI and GESTS. They are key informants because they have experience in community-based interventions aimed at strengthening traditional medicine. Finally, CEMI, GESTS, and McGill PRAM bring together 20 cultural safety experts with nearly 30 years of experience in intercultural health projects in Latin America, Canada, and Africa. We will mail/email invitations to all these stakeholders to participate in the project.

3.4 Methods for Collecting and Analyzing Data

The qualitative study will have two phases:

Phase One. In this phase, individual self-administered structured qualitative questionnaires and focus group discussions will explore the opinions of the stakeholders on what a co-designed cultural safety curriculum should teach to medical students in order for them to provide culturally safe services in Colombia. The questionnaires will gather individual opinions, enabling us to capture and compare what has been said in public and in private, as is proposed by Green [25]. The stakeholders will complete the questionnaires before participating in the focus groups. The questionnaires will inform the focus group discussion.

Phase one will use inductive semantic thematic analysis following the six steps proposed by Braun and Clarke [26]. With the consent of all participants, we will audio-record, transcribe, de-identify and safely hold the data produced by questionnaires and focus groups. We will invite two end-users (medical students from Colombia) to analyze the transcripts. In participatory research, hiring staff from the community is a way of increasing the ownership of the research process and capacity building [27].

Using AtlasTi V8.0, two research assistants will code the transcripts separately using an inductive approach. Subsequently, they will meet, compare their individual analysis, and create themes and sub-themes. Here, the research assistants will implement two levels of analysis. Firstly, they will look at the quotations and codes within each theme, looking for consistency and internal coherence. Secondly, they will look at the validity of the theme in relation to the data set. Finally, we will generate a visual representation of the themes using a thematic map to display the relationships between themes.

Phase Two. In phase two, two expert panels comprised of cultural safety experts, one in Colombia and one in Canada, will use the results of phase one to decide on the learning goals of the co-designed curriculum. The panels will follow a deliberative dialogue format [28].

Deliberative dialogue is a "a group process that emphasizes transformative discussion and may be informed by research evidence" (p. 1939) [28]. This process has recently received attention in health policy and systems research. Deliberative dialogue supports the use of evidence for decision making by: (i) using evidence as an input for discussions; (ii) providing an opportunity for stakeholders to discuss, contextualize, and determine the meaning of research evidence in light of their real-world experiences; and (iii) equipping decision-makers with decision-relevant knowledge in a format they can use [28].

The expert panels will use formal group facilitation techniques [27] as a way of creating a safe environment to maximize the effectiveness of the meeting. Firstly, we will present the results of phase one, using short and easy to read visual representations of the data. Secondly, we will provide the experts with materials (boards, paper, post-its, etc.) to work together to decide on the learning goals and academic content of the co-designed curriculum. The objective is to reach a consensus among experts.

We will use Bloom's revised taxonomy of educational objectives [29] as a framework for creating the learning goals. One research assistant will transcribe and organize the proposed learning goals and academic content. We will share the proposed curriculum via email with the experts who will suggest adjustments and give the final approval. Finally, we will share the co-designed curriculum with the traditional medicine users and medical students, who will modify and approve the final version.

3.5 Rigor

We will follow the strategies for ensuring trustworthiness in qualitative research projects proposed by Shenton [30]. *Credibility* will be assured by adopting validated research methods to gather the data (semi-structured questionnaires, focus groups, deliberative dialogue) and analyze the data (inductive thematic analysis).

The inclusion of different methods to collect data (questionnaires, focus groups, expert panels), stakeholders (traditional medicine users, students, experts) and sites (Colombia and Canada) will ensure good triangulation. Our key informants' universe is comprised of 55 individuals, and we will invite all of them in order to ensure a maximum variation sample. Similarly, we plan to undertake at least two-member checks to ensure that we correctly report what stakeholders want to say. There will be ongoing debriefing sessions with the research team every two months as well as continuous feedback provided by an experts committee in the Department of Family Medicine at McGill University.

Secondly, we will ensure *transferability* by implementing qualitative methods sequentially. In qualitative designs, the combination of different qualitative methods used at various stages of the research project strengthens the external validity of the data, thus helping to shape the opinions of decision makers [20]. Although the specific results of this project will not be generalizable to other settings, as traditional medicine is context and culture specific, the research design and methods we will employ will be transferable to other settings in Latin America and beyond.

Thirdly, we will ensure *dependability* by providing an in-depth methodological description that will allow researchers to replicate the study in the future. Moreover, we will use "overlapping methods" such as individual qualitative questionnaires and focus groups.

Finally, we will ensure *confirmability* by disclosing the researchers' background and other predispositions that may influence the analysis of the data, as well as by recognizing the limitations of the study.

4 Expected Research Contributions for Theory and Practice

This participatory research project will produce the first co-designed curriculum on cultural safety in medical education in Colombia. This curriculum will integrate the perspectives of different stakeholders, such as traditional medicine users, medical students, and cultural safety experts. The curriculum will inform future cultural safety training in medical and health sciences education. Ultimately, the results of this project will yield evidence to develop participatory methods to co-design medical training programs.

Indirect outcomes include: (1) capacity building: involved stakeholders will learn about cultural safety in medical education, participatory research, and qualitative inquiry; (2) strengthened partnership between stakeholders that will facilitate future projects; and (3) finally, given that cultural safety is a new concept to medical education in Colombia, this project will bring awareness of it to academia, thus facilitating its potential acceptance in the future.

Long-term potential benefits for stakeholders include enhanced quality of delivery of healthcare services (higher patient satisfaction, improved doctor-patient relationship, increased patient adherence) and reduced health disparities in communities of Colombia. The results of this study will be relevant to Canada and other multicultural settings.

5 Conclusion

The research supports cultural safety in medical education. It will develop participatory methods in medical education that might be of relevance in other subjects. The co-designed curriculum can be used to inform medical education interventions to foster cultural safety skills for medical students, improving quality of health services, and enhancing overall population health.

Acknowledgments. This study is funded by the CEIBA Foundation (Colombia) and the Fonds de recherche du Québec – Santé (Canada). The traditional medicine users from the "Seed of Life" community organization, the cultural safety experts at the Group on Traditional Health Systems Studies and the Center for Community Health Studies, and medical students from La Sabana University supported the project. Cass Laurie helped proofread the final version of the manuscript and supported its write-up.

References

1. World Health Organization: The Promotion and Development of Traditional Medicine. http://apps.who.int/medicinedocs/en/d/Js7147e/. Accessed 30 July 2018
2. World Health Organization: Declaration of Alma-Ata. http://www.who.int/publications/almaata_declaration_en.pdf. Accessed 30 July 2018
3. Williamson, M., Harrison, L.: Providing culturally appropriate care: a literature review. Int. J. Nurs. Stud. **47**(6), 761–769 (2010). https://doi.org/10.1016/j.ijnurstu.2009.12.012
4. Hester, R.J.: The promise and paradox of cultural competence. HEC Forum **24**(4), 279–291 (2012). https://doi.org/10.1007/s10730-012-9200-2
5. Chomat, A.M., Kring, B., Paiz, L.: Approaching maternal health from a decolonized, systemic, and culturally safe approach: case study of the Mayan-indigenous populations of Guatemala. In: Maternal Death and Pregnancy-Related Morbidity Among Indigenous Women of Mexico and Central America, 1st edn, pp. 483–511. Springer, Augusta, GE (2018)

6. LaVeist, T., Gaskin, D., Richard, P.: Estimating the economic burden of racial health inequalities in the United States. Int. J. Heal. Serv. **41**(2), 231–238 (2011). https://doi.org/10.2190/hs.41.2.c
7. Liaison Committee on Medical Education: Functions and Structure of a Medical School: Standards for Accreditation of Medical Education Programs Leading to the MD Degree. http://lcme.org/publications/. Accessed 30 July 2018
8. General Medical Council: Tomorrow's Doctors - Outcomes and Standards for Undergraduate Medical Education. http://www.gmc-uk.org/Tomorrow_s_Doctors_1214.pdf_48905759.pdf. Accessed 30 July 2018
9. Committee on Accreditation of Canadian Medical Schools: Standards for Accreditation of Medical Education Programs Leading to the MD Degree. https://www.afmc.ca/pdf/CACMS_Standards_and_Elements_June_2014_Effective_July12015.pdf. Accessed 30 July 2018
10. Brusin, J.H.: How cultural competency can help reduce health disparities. Radiol. Technol. **84**(2), 129–147 (2012)
11. Starr, S.S., Wallace, D.C.: Client perceptions of cultural competence of community-based nurses. J. Community Health Nurs. **28**(2), 57–69 (2011). https://doi.org/10.1080/07370016.2011.564057
12. Cai, D.Y.: A concept analysis of cultural competence. Int. J. Nurs. Sci. **3**(3), 268–273 (2016). https://doi.org/10.1016/j.ijnss.2016.08.002
13. Hawthorne, K., Robles, Y., Cannings-John, R., Edwards, A.G.K.: Culturally appropriate health education for type 2 diabetes in ethnic minority groups: a systematic and narrative review of randomized controlled trials. Diabet. Med. **27**(6), 613–623 (2010). https://doi.org/10.1111/j.1464-5491.2010.02954.x
14. Pon, G.: Cultural competency as new racism: an ontology of forgetting. J. Progress. Hum. Serv. **20**(1), 59–71 (2009). https://doi.org/10.1080/10428230902871173
15. Brascoupe, S., Waters, C.: Cultural safety: exploring the applicability of the concept of cultural safety to aboriginal health and community wellness. J. Aborig. Heal. **5**, 6–41 (2009)
16. Dyck, I., Kearns, R.: Transforming the relations of research: towards culturally safe geographies of health and healing. Health Place **1**(3), 137–147 (1995). https://doi.org/10.1016/1353-8292(95)00020-M
17. Nursing Council of New Zealand: Guidelines for Cultural Safety, the Treaty of Waitangi and Maori Health in Nursing Education and Practice. http://www.nursingcouncil.org.nz/Publications/Standards-and-guidelines-for-nurses. Accessed 30 July 2018
18. McEldowney, R., Connor, M.J.: Cultural safety as an ethic of care. J. Transcult. Nurs. **22**(4), 342–349 (2011). https://doi.org/10.1177/1043659611414139
19. World Health Organization: WHO Traditional Medicine Strategy 2002–2005. http://www.wpro.who.int/health_technology/book_who_traditional_medicine_strategy_2002_2005.pdf. Accessed 30 July 2018
20. Groleau, D., Zelkowitz, P., Cabral, I.E.: Enhancing generalizability: moving from an intimate to a political voice. Qual. Health Res. **19**(3), 416–426 (2009). https://doi.org/10.1177/1049732308329851
21. Israel, B.A., Schulz, A.J., Parker, E.A., Becker, A.B.: Review of community-based research: assessing partnership approaches to improve public health. Annu. Rev. Public Health **19**(1), 173–202 (1998). https://doi.org/10.1146/annurev.publhealth.19.1.173
22. Zuluaga, G., Andersson, N.: Initiation rites at menarche and self-reported dysmenorrhoea among indigenous women of the Colombian Amazon: a cross-sectional study. BMJ Open **3**, e002012 (2013). https://doi.org/10.1136/bmjopen-2012-002012

23. Sarmiento, I., Zuluaga, G., Andersson, N.: Traditional medicine used in childbirth and for childhood diarrhoea in Nigeria's Cross River State: interviews with traditional practitioners and a statewide cross-sectional study. BMJ Open **6**, e010417 (2016). https://doi.org/10.1136/bmjopen-2015-010417
24. Marshall, M.N.: Sampling for qualitative research. Fam. Pract. **13**(6), 522–526 (1996). https://doi.org/10.1093/fampra/13.6.522
25. Green, J., Thorogood, N.: Qualitative Methods for Health Research, 3rd edn. Sage, London (2014)
26. Braun, V., Clarke, V.: Using thematic analysis in psychology. Qual. Res. Psychol. **3**(2), 77–101 (2006). https://doi.org/10.1191/1478088706qp063oa
27. Salsberg, J., Parry, D., Pluye, P., Macridis, S., Herbert, C.P., Macaulay, A.C.: Successful strategies to engage research partners for translating evidence into action in community health: a critical review. J. Environ. Public Health **2015**, 1–15 (2015). https://doi.org/10.1155/2015/191856
28. Boyko, J.A., Lavis, J.N., Abelson, J., Dobbins, M., Carter, N.: Deliberative dialogues as a mechanism for knowledge translation and exchange in health systems decision-making. Soc. Sci. Med. **75**(11), 1938–1945 (2012). https://doi.org/10.1016/j.socscimed.2012.06.016
29. The Center for Teaching and Learning: Bloom's Taxonomy of Educational Objectives. https://teaching.uncc.edu/services-programs/teaching-guides/course-design/blooms-educational-objectives. Accessed 30 July 2018
30. Shenton, A.K.: Strategies for ensuring trustworthiness in qualitative research projects. Educ. Inf. **22**(2), 63–75 (2004). https://doi.org/10.3233/EFI-2004-22201

Contribution of Textual Analysis by ALCESTE Software to Determine Dimensional Publicness and Public Values: An Application on Two Banks of the French Local Authorities

Muriel Michel-Clupot[1(✉)] and Serge Rouot[2]

[1] IUP Finance Nancy, CEREFIGE (EA 3942), Université de Lorraine, 54000 Nancy, France
muriel.michel@univ-lorraine.fr

[2] IAE Nancy, CEREFIGE (EA 3942), Université de Lorraine, 54000 Nancy, France
serge.rouot@univ-lorraine.fr

Abstract. This paper presents an original qualitative empirical approach in local management. It is indeed based on a textual analysis (using a French software, named ALCESTE) of financial reports of two major banks of the French local sector: Dexia CL, as the market leader until 2011 and AFL, as the new entrant into the market in 2013. This research reveals public values contained in their both communications and proposes a measurement of their publicness (that is to say the importance of political constraint compared to economic constraint, weighing on the organization). The textual analysis shows vocabulary used and significant contents. First results are somewhat surprising: AFL, the bank for and made by local authorities, presents no more political constraints or more public values, despite the nature of its shareholding and its objective of general interest. A compared literature on companies per share and on cooperative banks could be used to explain the future funding sector of local governments in France.

Keywords: Textual analysis · Publicness measure · Public values

Local public investment represents almost 70% of the public investment in France. A financing sector was developed specially for local authorities. It proposes both bank loans and bond issues. It has long been dominated by a private corporation, the Dexia Group, which disappeared because of a subprime crisis. In 2013, a new bank, called "Agence France Locale" (AFL) was created, dedicated to the funding of local governments. AFL is a bank, for and made by local authorities, who are also shareholders. The bank will raise funds on markets and lend to its members.

Would the financing of local public sector be characterized by public values and political constraints? What will be the impact on these public values and the considerations of political constraints due to this change? Do shareholders who are also customers influence a change on the values framework?

A. P. Costa et al. (Eds.): WCQR 2018, AISC 861, pp. 110–116, 2019.
https://doi.org/10.1007/978-3-030-01406-3_10

The two banks produce a financial communication for their shareholders. Their reports reflect their own public values. To establish an inventory of these public values [3], we suggest an original study using a qualitative empirical approach, by means of a textual analysis of these financial discourses. This article will present Dexia and AFL (1), then publicness and public values (2). It will detail the qualitative empirical approach (3) and present some first results (4) and perspectives for further research (5).

1 From Dexia to AFL

The financing of the local authorities was characterized by a diversity of lenders and a diversity of products. "Dexia Crédit Local" remained the first bank of the local governments (40% of market share), until the international financial crisis. The aforementioned and a bad risk management created serious difficulties in 2008 for the group.

Without being the cause of bankruptcy in 2011, the scandal of structured debts (or toxic debts) seriously affected Dexia's reputation. During the subprime crisis and its impact on interest or change rate, the derived products contained in structured loans induce great increases of the cost of many territorial debts [4].

There is a temptation to say that the maximum profit for the shareholders would be turned over against customers. "The shareholders indeed should rather increase the risk of assets because their possible losses are limited in their contribution, whereas their potential earnings are unlimited" [6, p. 127].

With the end of Dexia, the question of the financing of the local public investment is made apparent. There is no banking leader, so bond issues become necessary. The idea of joint issues appears to make the markets available to smaller local authorities.

The concept of the "Agence France Locale" (AFL) is born [8]. The AFL is a specialized credit institution, in which local governments are both shareholders and loan recipients. Adhesion is opened to any local community, whatever its size. It depends on a contribution in start-up capital, proportional to its debt stock. Nevertheless, it is subordinated to the preliminary financial analysis of the candidate: the decision is based on healthy finances. Even if the local authorities are shareholders, the AFL objective is not profit-making. Benefits would not be distributed, but would be incorporated in the share capital, to increase its financial strength. Its credit terms would then be improved and benefit all the local communities' borrowers. Usually, the financing of the local public sector was characterized by public values and by the respect for political constraints.

Does this evolution from Dexia to AFL have an impact on these two dimensions?

2 Public Values and Publicness

According to the theory of publicness [1], any organisation is constrained by two authorities: one political, one economic, in which it will find the resources it needs to exist. The challenge then is to determine their respective proportions. This is the "dimensional publicness" [1] that is useful to describe organizations as "public" or "private" in their operation and governance. Indeed, the legal status of an entity or the identity of its owners have become partial elements of a more complex reality, combining the two authorities in proportions which allow them to be measured. This measurement of publicness offers an assessment of the political influence and the influence of the market, a balance between "political" values and "private" values. The latter are voluntarily apprehended only through the economic values which constrain policy choices.

This raises the question of the link between theory of publicness and the study of public values, defined as the "rights, advantages and prerogatives to which citizens should (and should not) be entitled; the obligations of citizens towards society, the State and others; and the principles on which governments and policies must be based" [1, p. 189]. Public values, carried by elected representatives, civil servants and citizens, are the cornerstone of the administration and public policy. The "balance" of the political and the economic directs and shapes public values; "we need to understand the mix of the political and economic authority of institutions and policies if we are able to understand their potential to achieve public values and to work towards public interest ideals" [1, p. 186].

These public values are mapped out on the basis of public administration literature and the theory of organisations [3]. "The inventory" thus established is based on an approach that brings out the different spheres affected by the value in the public administration: public sector's contribution to society; transformation of interests to decisions; relationship between public administrators and politicians; relationship between public administrators and their environment; intra organizational aspects of public administration; behaviour of public-sector employees; relationship between public administration and the citizens.

3 Qualitative Approach: Textual Analysis for a Rating of Publicness and the Detection of Public Values

Our goal here is to identify public values in the financial discourses of the two banks: Dexia and AFL. Our field of research is their financial communication. As far as they finance public local authorities, these discourses should convey public values [3]. Hence, this reading of the reporting of the two banks through the prism of public values and the measurement of publicness is a type of confrontation with a "market logic", a real challenge for these public values [3]. Indeed, financial markets constraints are numerous in local public financing, as in any debt management. We explore how local authorities who become shareholders will build the AFL corporate strategy.

There are several kinds of textual analysis software. Resulting from European research of linguists and the French school of statistical analysis, some allow a further study of a corpus [2]. In order to show differences or proximities of the financial communication of these two banks, we chose to use ALCESTE (in French, Analyse des Léxèmes Co-occurents dans un Ensemble de Segments de TExte: analysis of lexemes co-occurring in a set of text segments). Developed in partnership with CNRS (French National Center for Scientific Research), ALCESTE (www.image-zafar.com) is frequently used in management sciences, in particular in marketing, strategy and more recently, finance. What are the principles of this method? The corpus is cut out in a succession of segments of text and one observes the distribution of the words in these segments [7]. The automated textual analysis consists in a “destructuring” of the corpus [2]. What retains our attention is the repetition of the words and the proximity of certain words, ones compared to the others (we speak about co-occurrences).

A corpus treated by the ALCESTE software is segmented in Elementary Context Units (noted ECU). At this step lies the question of cutting out the corpus in text segments: with relatively arbitrary length but of the same size. Resulting from the factorial analysis of the correspondences, this original method is one of the characteristics of ALCESTE. On the contrary, to an ascending classification, it will not classify words, but ECU. Moreover, it carries out a hierarchical downward classification. In the ECU, the full words will be detected to define classes. So, ALCESTE calculates the decomposition of the corpus in various classes. Among all the possible decompositions, it maximizes the oppositions, i.e. in technical terms, khi2 of the margins. At the end of the treatment, the corpus is thus “arranged” in various classes. The work of the researcher then consists in defining, naming and qualifying these classes, according to the vocabulary which they integrate. The analysis focuses on the content of the financial discourses, which requires access to drafted supporting material, a non-standardized document, written by Dexia and AFL.

In this research, two corpora are analyzed: a first one made up with the two last financial reports of Dexia CL before its phase of liquidation (years 2010 and 2011) and a second one made up with the two first reports of AFL (years 2015 and 2016). We then have several classes, with significant words, named by the searcher. Thus, present words and headings of classes lead to attach the different classes to rather “economic” or rather “political” constraints of the publicness theory. Consequently, we propose a measure of publicness [5] for these two corpora, by giving points to each category (0 for economic; 1 for political; 0.5 for a mix), on all classes. From the words significantly present in discourses, we create an original “rating of publicness”, estimating the weight of the political authority according to Bozeman [1], conveyed by the communication.

In a second time, financial communication words are brought closer to a representation of the public values, according to Jorgensen and Bozeman [3]. The ambition of this research lies in the detection of the values contained in Dexia’s and AFL’s reports. We propose to compare and discuss them too. This is made possible by the detailed study of the words classified by the software. The results are shown in the next part.

4 First Results: Publicness Rating and Public Values Detection

The results of the textual analysis on the financial discourses of the two banks are shown in Figs. 1 and 2. The corpora are cut out in 3 classes for Dexia and 6 classes for AFL.

A first analysis of the results reveals the following points. Its synthesis appears in Fig. 3.

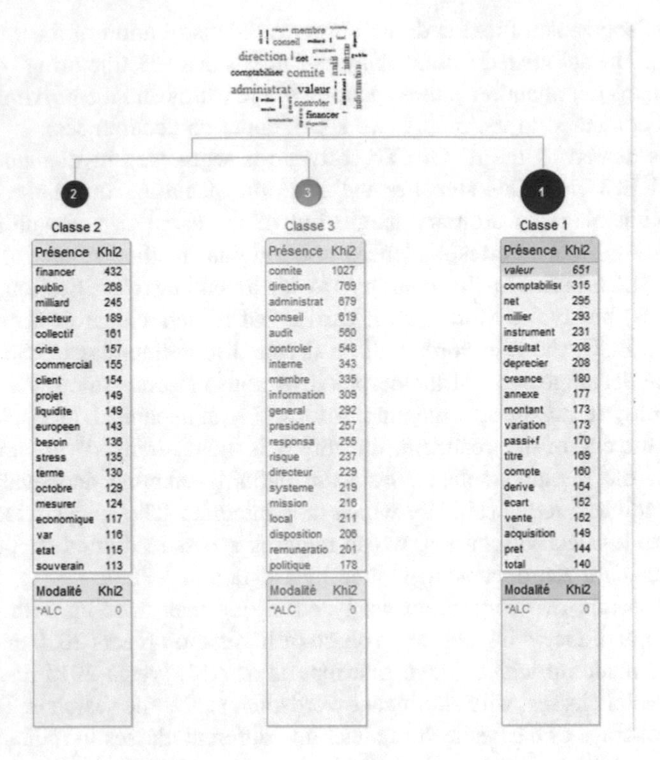

Fig. 1. Results from the ALCESTE on the Dexia corpus.

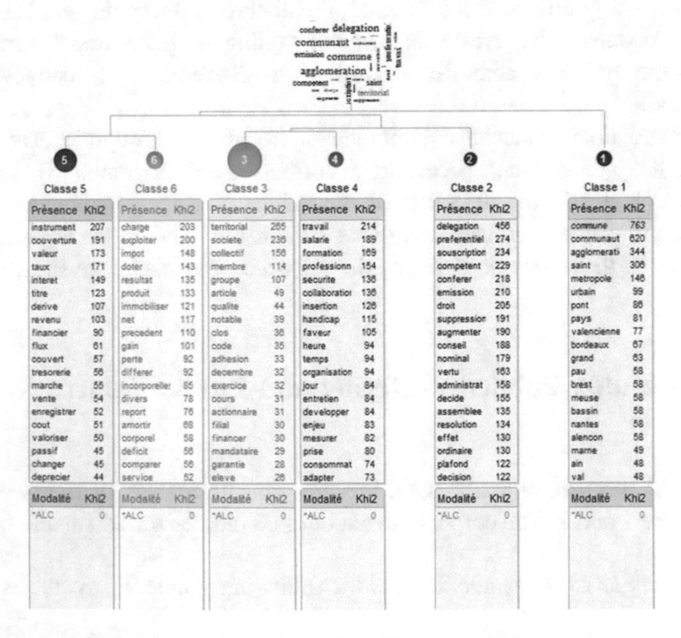

Fig. 2. Results from ALCESTE on the AFL corpus.

	Dexia			AFL					
Classes	**2**	**3**	**1**	**5**	**6**	**3**	**4**	**2**	**1**
Thematics	banking environment	Governance	accounting	debt management	accounting	Organization	mission of local authorities	corporate law	adherents
Public values spheres and *words (in French)*	**Public sector contribution to society** *public* *collectif* *projet* **Inter organizational aspect of public administration** *commercial* *client*	**Public sector contribution to society** *local* *politique* **Inter organizational aspect of public administration** *audit* *contrôler*	-	-	-	**Public sector contribution to society** *territorial* *collectif* *société* *membre* *groupe* *adhésion*	**Public sector contribution to society** *formation* *sécurité* *collaboration* *intention* *handicap* *entretien*	-	-
Publicness and rating	eco and political 0,5/1	eco and political 0,5/1	eco 0/1	eco 0/1	eco 0/1	eco and political 0,5/1	eco and political 0,5/1	eco and political 0,5/1	political 1/1

Fig. 3. Public values and publicness in the financial discourses of Dexia and AFL.

In the two banks' financial reports the software shows the significant presence of words which one can attach to the same public value: "Public sector contribution to society". It could be assumed that Dexia, a private corporate, whose shareholders expect profits, should develop fewer public values than AFL, built by and for local authorities, therefore concerned by the pursuit of shared interest. In a paradoxical way, we observe more public values in the Dexia's discourse ("Inter organizational public aspect of administration") than in the AFL's one.

Lastly, the measures of publicness show a very close rating: 33% of political constraint for Dexia and 41% of political constraint for AFL. Once again, we could have believed that a clearly stronger political constraint would be applied on AFL. But this is not the case: AFL does not present a major political constraint. Concerning public values, these results show after all a small difference between the financial discourses of the two banks of the French local public sector. This conclusion is surprising because of their operating processes and the nature of their respective shareholding.

5 Concluding Remarks and Perspectives for Further Research

This research is an original empirical approach which carries out the textual analysis of the financial communications of the two major banks of the French local sector. Using ALCESTE software, the analysis reveals the public values which financial reports convey and measures their dimensional publicness. The first results are relatively surprising. There are a few differences between the public values in the two banks and in the constraints, which are mainly economic, while the nature of their shareholdings could have an assumption: political constraint and public values are not more significant in the AFL's discourses.

These first results raise many questions. Following the bankruptcy of Dexia, the reorganization of the territorial financing system in France didn't lead to the creation of a new bank, with more public interest concerns. This first study requires exploring and these results suggest several methodological tracks of qualitative nature. Indeed, the ALCESTE software produces "networks of forms" which reveal the proximity between words. On these first corpora, a fine analysis of the associations of significant words could show new public values. It can also improve the rating of the dimensional publicness. Moreover, we intend to pursue a longitudinal study of the financial discourses of Dexia, including older periods, to analyze the impact of the crisis on the public values. In 2008, the financial crisis has surely modified the financial discourses of Dexia.

Lastly, AFL is a very recent creation. These results have perhaps not revealed the future of AFL. It appears necessary to integrate in the research the last financial report of the AFL (not yet available), in order to evaluate if the new bank of the French local sector will progress to more political and public interest concerns.

References

1. Bozeman, B.: La publicitude normative: comment concilier valeurs publiques et valeurs du marché. Politiques et Manag. Public **25**, 179–211 (2007)
2. Gavard-Perret, M., Gotteland, D., Haon, C., Jolibert, A.: Méthodologie de la recherche. Pearson Education, Paris (2008)
3. Jorgensen, T., Bozeman, B.: Public values. Inven. Adm. Soc. **39**(3), 354–381 (2007)
4. Michel-Clupot, M., Rouot, S.: Communication financière des collectivités locales françaises et crise internationale: des pratiques en mutation. Politiques et Manag. Public **29**(3), 343–367 (2012)
5. Michel-Clupot, M., Rouot, S.: Public values and function creep of the financial rating by local authorities. Int. Rev. Admin. Sci. **8**(2), 303–325 (2015)
6. Ory, J.-N., Jaeger, M., Gurtner, E.: La banque à forme coopérative peut-elle soutenir durablement la compétition avec la banque SA. Finan. Contrôle Stratégie **9**(2), 121–157 (2006)
7. Reinert, M.: Contenu des discours et approche statistique In: Analyse statistique de données textuelles en sciences de gestion, Editions EMS, Collombelles (2007)
8. Saoudi, M.: Un nouvel instrument financier de développement des territoires. L'Agence France locale. In: 3e colloque international du CIST (2016). https://hal.archives-ouvertes.fr/hal-01353682/

Generating Empirically Grounded Typology from Narrative Data

Victoria Semenova(✉)

Institute of Sociology of FCTAS RAS, State Academic University for the Humanities, 24/35, Krzhizhanovskogo Str., Moscow, Russia
victoria-sem@yandex.ru

Abstract. The paper is devoted to the methodological aspects of the analytical interpretation of qualitative data as the process of constructing an empirically grounded typology from the textual information provided by biographical interviews. The main questions for discussion are how the researcher can generalize from individual data and what the step-by-step procedure of going "up" from empirical data is. The author argues for the possibilities of combining subjectivity and objectivity in this process, as well as the role of empirical, theoretical, and contextual knowledge in this process. An additional aspect to discuss is the procedure of interpretation of narrative texts and their use as the basis for constructing a typology. As a result, the author presents the entire procedure for constructing a grounded typology and illustrates it with the example of individual career strategy. Finally, the article debates the risks and advantages of the proposed research strategy. The objective of the paper is achieved through the following scheme: 1. theoretical approaches and methodology towards subjective/objective typologies in biographical research; 2. an analytical model for constructing empirically grounded typologies; 3. an empirical step-by-step procedure for elaborating a typology from biographical data; 4. interpretation and some conclusions on typology-construction-oriented biographical research.

Keywords: Empirically grounded typology · Biographical research
Narrative analysis

1 Theoretical Approaches and Methodology Towards Subjective/Objective Typology in Biographical Research

Generating a typology seems to be the final analytical stage of all qualitative research [3, 6, 8, 9, 13, 15] but as a rule this poses questions for researchers in the process of realization. Our main objective in this paper is to propose a model for the analytical interpretation of unstructured biographical texts developed from empirical subjective information. This aim leads to a number of research questions about an "empirically

'Social mobility in four generations: from XXth to XXIst century' (project supported by the Russian Scientific Foundation, no. 14-28-00217).

A. P. Costa et al. (Eds.): WCQR 2018, AISC 861, pp. 117–135, 2019.
https://doi.org/10.1007/978-3-030-01406-3_11

grounded typology." What could be the starting point for such a construction? A set of objective facts or perhaps the respondent's subjective position? How does one combine a theoretical model with firsthand data? Moreover, how does one present it publicly in short to make it sound both theoretically "grounded" and empirically valid? These questions will be discussed here as a possible step-by-step research strategy and will be proven as a valid procedure by the example of the researcher's experience.

As a rule, the construction of a typology as a "constructed typology" [17] is a way of describing groups of respondents according to different clusters of behaviors, their attitudes, or their views of the world that are used for comparison and measurement of empirical data. According to John McKinney [17], it is the process of 'purposive, planned selection, abstraction, combination and (sometimes) accentuation of a set of criteria with empirical referents, that serves as a basis for the comparison of empirical cases" [17: 25]. It starts as a set of descriptive names or "types" of behavior and/or attitudes for each group, but can go much further, claiming to categorize far broader attitudes and lifestyles [17: 25]. McKinney also emphasizes that "on the one hand, the type is related to a conceptual scheme and hence implicated in theoretical context more broadly than any concrete problem under the empirical consideration. On the other hand, it serves as the unit for comparison and statement of empirical occurrence" [17: 41]. In this way, the typology bridges the gap between a system of categories and empirical events. Therefore, in his view, the most important points of the strategy of "constructed typology" are: (1) initial description and comparison of cases as behavior patterns, and (2) the combination of empirical data with theoretical scheme at different steps with the aim of developing wider attitudes or lifestyles according to the conceptual scheme preliminarily developed by the researcher.

Though there appeared some additional aspects to this general strategy, if the researcher wants to develop biographical generalization. This is why the analysis in biographical research usually ends with text interpretation and only seldom goes further to behavior patterns; only then is it possible to find both subjective and objective information. That is why the main strategy in biographical research is the biographical interpretative method (BIM) [4, 22, 32].

The biographical researcher's interest, however, is often exclusively on subjectivity in order to understand and compare motivation for action, compare different lifestyles and identities, or investigate other forms of subjectivized differentiation within one community. Therefore, we should search for some typological strategy that could at the same time use the general principles of "constructed" typology and generalize from subjectivized information. McKinney remarks on this issue that every typology construction process should be focused on definite problematic areas that are of most theoretical significance in concrete research situations [17: 41].

Another analytical strategy, "grounded" typology, derives from B. Glaser and A. Strauss's concept of "grounded theory" [8], and could be most helpful here, as it is mostly close to subjective data and the researcher starts from "below," using the inductive approach to collected data; categorization comes later and emerges step-by-step from firsthand types. Dealing with so-called "empirically grounded typology" in qualitative research, Susanna Kluge [13] stresses that, in order to achieve a suitable interpretation of "typical", properties and dimensions should be elaborated upon and "dimensionalized" during the process of analysis by means of both data and theoretical knowledge. She also points out that, on the one hand, empirical investigations require theoretical knowledge, because they cannot be carried out purely inductively, and, on the other hand, the analysis should be based on empirical data, if meaningful statements about social reality are to be more than empirically remote constructs [13]. It is only when empirical analysis is combined with theoretical knowledge that "empirically grounded types" can be constructed. Only then, in her opinion, does the "model of empirically grounded type construction" show great openness and flexibility [13]. Furthermore, Kluge proposes the several-step construction model, where each stage of analysis can be realized through different analytical methods and techniques, and then the model complies with a variety of research questions and with different types of data. So, according to her, the most important steps are: (1) the combination of theoretical concepts and empirical subjective data, (2) the research strategy, which should combine different analytical methods and techniques and different types of data, (3) flexibility and openness for further development. She did not, however, elaborate on what constitutes the "ground" or the initial moments for developing such a combination.

Tom Wengraf proposed such an initial moment in constructing his biography-based typology as a one-case segment textual interpretation (sequence analysis) [32]. He proposed to start from one biographical text and then test first hypothesis by comparison with other cases. That strictly follows the principle of grounded theory developed by Strauss [8] and, in the case of biography research, this position of starting from text analysis seems very promising and helpful. It is mostly, however, the "blind" method of searching "special points" of research interest.

When the researcher begins with the definite aim of constructing typology from "below," then it is logical to start with the textual interpretation, bearing in mind some line of first text-structure division based on a definite theoretical conceptualization. Following Kenneth Bailey, such a "taxonomic" division of texts can be understood as a tool "for a classification of empirical entities" [1: 6] and such a taxonomical division could not only be strictly empirical but also verbal or conceptual [1: 6]. In sum, the methodological basis for an empirical grounded typology proposed here is based on the following principles:

1. It should be an empirically grounded typology oriented primarily towards subjectivity, with the aim of constructing wider attitudes or lifestyles
2. Subjective data should be supplemented with objective information where possible
3. Empirical data should be combined with a theoretical scheme
4. Different methods and types of data should be used at different steps of construction
5. It should start from a structuralized textual data analysis
6. The first taxonomical division of whole texts as an individual type of narration should be based on a theoretical conceptualization of the research topic.

2 Analytical Model of Construction Empirically Grounded Typology

The starting point for analysis and elaboration of the typology in any biographical research is a socio-biographical approach, where the focus is the tension between agency/structure [11, 24]. Out of this tension, there could be two possible configurations of analysis: "social imbedded in biography" and "biography embedded in social" [24]. While interested in the last analytical position—the actor's subjective perspective on his/her social activities—we are more interested in the types of biographical narrations as a "told story": how respondents present their life stories and describe their actions in terms of "acting," "experiencing," or enduring one's life. At the same time, how they mark their individual choices and the intercorrelation between internal and external conditions that influence their actions are objective and subjective factors [11]. Objective biographical facts as "lived life" are also presented in "told story" form as narrated episodes: episodes of education, job, mobility, marriage etc. Therefore, biographically oriented textual data provides an opportunity to begin a subjective/objective generalization from the form of textuality that corresponds with the primary principle of constructing typology considered earlier.

A biographical story usually includes two parts: strict narrative, as an uninterrupted story about own life and a part relating to questioning, which contains questions on the issues that are of most interest in research. The phase of narration itself could be considered a textual form of self-presentation, an individual "genre" of telling about oneself. This textual information could be used as a starting point for constructing a grounded typology and a first **conceptual and contextual taxonomy** of the entire narrative texts. The helpful instrument here, according to the interpretive biographical method [22], is the conceptual framework for further relevant data structuration and interpretation [34: 143]. For example, while analyzing individual cases from the viewpoint of "actor strategy," we could choose the conceptual dichotomy (taxonomy) of passive and active strategy, or stable and mobile strategy; the "story about success" versus the "story about failure," the "story about a chronic illness" versus the "story about rehabilitation," as well as any other socially meaningful dichotomy that fits the research discourse.

For the text analysis itself, we use the concept of "formal-textualism" introduced by Tom Wengraf [32]. It is based on the idea that the model of subjectivity could manifest itself through the "language of subjectivity," and in order to analyze narrated subjectivity one should concentrate on "genre-mediated speech," or the genre of storytelling that could be the basic conceptual division for further typological construction [32]. This concept seems very fruitful for the first step of grounded typology, as an initial reasoning that combines social context and subjectivity into a complex instrument useful in developing further steps, especially when we deal with an unknown field with unknown further dimensions. For example, we do not know how the subjective perception of "actor strategy" (adoptive or creative) coincides with real facts or other aspects of life strategy. Therefore, after the first division of text types according to the genre of storytelling, we could turn to a more flexible and open-ended search for other subjective/objective dimensions of the typological construction.

Out of that, we consider the type of narration as individual positioning vis-à-vis other actors and the social world in general [23, 24]. We must find the textual forms for presenting such actors' positions, to analyze how these actors, in social interaction with others, create and/or recreate such positions, moment by moment. In adopting and telling about his or her position, the individual reflexively rearranges their whole world. We understand how other individuals, objects, and situations are categorized, which of their properties are of importance, and how individuals, objects, and situations relate to one another [18, 23]. So, **positioning** considered through the narrative genre discloses a discursive relationship with the world and so could be considered as a broad linguistic construction of the position of the self, seen through verbal forms about regular actions (everyday acts) and as "value" position in general. Given that, we expected that our typology could be based first on structuring open narratives according to genre-mediated speech as a reflection about her or his experience in the form of verbal constructions. As the relationship between presented storytelling and "others," it could be considered in terms of discourse analysis. Thus, at the first step, we have some case groupings differentiated by the genre of "storytelling" or positioning considered in the wider frames of conceptual framework. At this stage, we will receive two contrast groupings, which could be interpreted as opposite positioning in frames of social contextualized reality [1, 32].

At the next step, while working with respondents' answers to additional questions (the "additional questioning" phase), where they specify some aspects and episodes of their lives, we could move to the stage of grouping comparison in search for other differences between these groups on the basis of their answers to additional questions from the researcher. This step is critical for the construction of grounded typology as a moment of flexible and open search and as the stage of "constant comparative procedure" [8]. As we did not assume beforehand what constituted the difference between these two groupings of subjective positioning; here we should explore each respondent's answers in order to find the most significant dimensions by which one grouping is in contrast to another. We search for the text fragments which could be relevant to the definite phenomena and, at the same time, as statements they could be compared along different interviews. We are trying to answer the questions: what constitutes the difference between these groupings of storytelling? Is this difference of any logically significance? What dimensions could be used in order to classify this difference? The form of a statement may contribute to answer the main research question: how do the values or episodes described differ verbally in each grouping and what is the meaning of this difference in a comparative perspective [9]? While using the tactics of qualitative content analysis, the researcher fixes subjective attitudes in the form of value statements about definite biographical dimension (for example, attitude towards career change, stability/mobility, priorities in job change, and so on). It doesn't mean that within one grouping all respondents answered in the same words, but, according to content qualitative analysis, we fix most contrasting attitudes in these two groupings [9]. It's also important to note what was said, in what order and in which context, which makes every statement in the text extremely important and gives opportunity to compare the same statements from different interviews. Through such "constant comparison" [8] we first formulate "grounded" multiplied hypotheses about further dimensions for comparison.

Two previous steps were mainly oriented on subjectivity. The next-step analyses are another type of biographical data. It is information about real facts and behavior. It supplies us with an additional information on how subjectivity shows itself in reality. Therefore, on this stage, the researcher moves from models of subjectivity to models of behavior, and this dimension is used in order to construct more general models or types. This stage also needs some preliminary contextual conceptualization about the focus of activity in order to choose a definite scale for measuring this dimension [7]: for example, the number of marriages/divorces, the number of job changes or migration episodes, and so on.

Finally, on the stage of combination of subjective/objective dimensions of constructed typology, when the construction process is complete, the researcher could also check its validity and common-sense credibility by expanding the dimensions of typification, attracting additional forms of individual activity in other segments of social space. As a result, the whole procedure of constructing an empirically grounded typology may be generated as a four-step "opened' strategy, combining both subjective and objective individual data. The main steps of constructing typology can be illustrated as the following set of steps:

1. Discourse analysis of narrated texts as genre of storytelling. Aim: to construct initial subjective groupings reflecting individual positioning.
2. Thematic content analysis of verbal answers as texts. Aim: to find and compare cases with definite opposition in attitudes within chosen groupings.
3. Measured dimensions of real individual actions. Aim: to compare types according to objective criteria.
4. Combination of subjective and objective dimensions in order to construct conceptual types as models of social behavior.

The whole research strategy of constructing the models of social behavior from biographical data lies on theoretical concepts in the definite sphere. It starts from analyzing the style of storytelling in the frames of the biographical scheme, goes through several stages in search of the difference between groupings in order to understand and interpret the processes and conditions and develop relevant properties into dimensions, and, finally, combines subjective attitudes with objective data. As a result, we can construct different types of socially oriented individual behavior as motivation-action strategy. The main challenges of the proposed strategy are:

- Maximum approximation to subjectivity grounded in firsthand textual information coming directly from respondents with minimal interference from the research team;
- Combining specific (individual) common-sense knowledge and general (wide social context) social knowledge;
- Combination of conceptual and empirical knowledge about the definite social sphere;
- Combination of different research techniques that improves the validity of the result.

Based on these rules of constructing a qualitative typology, we can conclude that our "empirically grounded typology" would be "grounded" on narration as well as on a theoretical concept of "agency" and "actor subjectivity."

3 Empirical Procedure of Elaborating Typology from Biographical Data Step by Step

As an example of such a methodological approach towards biographical data, for further analysis I will present the example of a set of biographical interviews focused on professional behavior and the configuration of the careers of professionals in Russia. Eighty-four in-depth interviews about professional life histories were conducted in 2016 in Russia. For more detailed analysis, I will use here information from 24 biographical interviews of successful professionals with those who started their job career as professionals and who now are in positions as middle managers in private and state enterprises and so they could be considered successful in their careers. They are 45–50 years old, women and men, who live in different regions of Russia, were educated as high-level professionals, and have already gone through the main stages of their careers. All respondents, as based on their current professional status, hold nearly equal positions in terms of their vertical mobility.

The objective of these in-depth interviews was an "internal perspective" on careers within the context of a wider biographical project. The focus on subjectivity was stimulated by a larger interest in the theoretical concept of subjective social mobility, which is focused on the "latent aspect" of career transition [16, 19, 29, 31] and on the concept of modern change in subjective attitudes towards professional career. According to Baumann, the changing social context and modern situation of flexible modernity on the labor market demands a new type of personality with modified human resources. Under these conditions, only the most flexible and adaptive personality is suitable; those who psychologically and physically are open to a new professional situation [2]. From that, our interest in professional careers turns more to its symbolical subjective meanings from the point of view of our respondents [12, 30]. Therefore, the main question was: to what degree are individual job trajectories stimulated by human resources or by external structural change in the labor market in the situation of drastic social change in modern Russia [25]? The interview itself was constructed as a detailed retrospective life story focused on transitions in work and life histories, with more detailed interest in critical points of the biographical project (biographical cross points) and individual attitudes towards subjective experience. The interview had three components:

- First – an uninterrupted narrative as storytelling about individual experiences and professional dynamics starting from the parent's family status to the respondent's current status.
- Second – a semi-structured interview with a formal guide to a detailed description of education, job, and family biographical experience.
- Final – open-ended questions about individual perspectives and values about professional and life success. The length of each interview was 1.5–2 h.

In an analytical sense, the information and its interpretation were oriented towards constructing a grounded typology of individual strategies as a combination of subjective and objective aspects of occupation career in times of social change. The whole plan of constructing typology is illustrated as a set of steps specified by content, aims, and used research resources (Table 1).

Table 1. The empirical scheme of constructing typology from biographical data

1st step – subjective criteria Grouping of active/reactive respondents out of their positioning towards job and career	Case, narrative, discourse analysis: to fix textual forms of positioning towards career and job
2nd step – subjective criteria Dimensions for comparison and contrasting of attitudes Coding of properties	Cases comparison by groupings, thematic content analysis: to compare and search for differences in subjective attitudes
3rd step – objective criteria Dimension: career mobility/stability	Cases comparison, real episodes of job change: to combine subjective and objective criteria
4th step – combination of subjective and objective criteria	Inductive interpretation of constructed types of occupation strategy

3.1 First Step: Grouping

First, following the identified principle of theoretical conceptualization and contextualization as the basis for typological construction, we analyzed general theoretical questions fit to the data. Our aim was to understand how professionals planned and realized their careers in terms of active/reactive positioning while describing their life and work experience in time of drastic social change in the form of told story in narrative form [5, 10].

According to the previously named principle of empirically grounded typology, we started from searching relevant textual forms as an individual "genre" of storytelling that could depict active/reactive positioning towards one's career. We focused on three episodes relevant to positioning towards career: the experiences of finding one's first job, job change (the story "as it was"), and broader narration about career success (the "my understanding of career success" story). Therefore, discourse analysis of storytelling (block of narration) is taken as a communicative process between the respondent and interviewer consisting of three verbal constructions of positioning that could form definite groupings:

- Positioning towards career in the situation of first job;
- Positioning towards job changes;
- More general positioning towards career success.

At this first step of grouping on the basis of positioning we were oriented by a wide preliminary hypothesis: was the professional strategy stimulated by the "self" and individual human resources (for example: "I decided that this position was not for me," "I addressed the agency") or by the external situation (for example, "Suddenly I was offered a new position," "It was only by chance that my relative offered me a job")? The grouping process was oriented on active and reactive subjective positioning towards career. We tried to find the general genre of storytelling that revealed itself through some grammatical or verbal equivalents of active or passive self-positioning and found it in the grammar opposition of verbs: "to do" and "to be." They not only indicate the passive/active positioning of individual narration, but also contain a volitional component as the willpower to realize one's plan. For example, the story of one group was organized by active verb constructions: "to make a career," "to think over further career," "to search for a new position," "to make a decision about a career," and so on. On the contrary, the other grouping used passive constructions for describing such episodes: "they put me in a position," "suddenly they offered me a job," "that was a matter of chance," "they told me to wait and see." The overview of the result of such opposition we can see in the following table (Table 2).

Table 2. Textual forms of self-presentation as the basis for first-step division and grouping of storytelling about professional career.

Grouping A. Storytelling about one's career as a set of individual choices and decisions Group of active positioning towards career	**Verb constructions:** "to do a career," "to think over the career plan," "to make a decision and begin to search" - *"...this is the result of my conscious choice, serious meditation over my life and as a result of making the decision" (Irina, CMR)* - *"Naturally, I welcomed such additional working hours, as this, I hope, will help in building my further career" (Alex, EKB)*
Grouping B. Storytelling about one's career as a set of accidents and luck. Group of "influenced by," "guided by the situation," "waiting for some chance"	**Verb constructions:** "climbing the stairs," "they put me," "suddenly they offered me a job," "that was a matter of chance," "I was lucky, at last they offered me a job." *"So, in fact, all these changes of my positions that was somehow a matter of chance" (Oleg, CMR),* *"... Maybe it was only some luck that they noticed me..." (Natalia, CMR)*

Out of these two possible types of positioning towards one's career based on textual forms of self-presentation appeared:

A. **"active" positioning**, where one's career experience is presented as a set of individual choices and personal decisions about professional movements; self-control of individual trajectory;
B. **"reactive" positioning**, where career experience is presented as a result of external influence of the labor market situation without any influence of the "self."

So, on the first step we focused on narratives about career and made groupings according to them. Moving further to the next step of analysis, we tried to find differences between these first identified groups.

3.2 Comparison and Difference

The **second step** had the aim of comparison: based on the comparison of the two groupings, we searched for their internal differences. As a result, we selected from different dimensions those that showed contrast for these two groupings. In what dimensions do they demonstrate extreme contrast in their attitudes? What are the properties of these differences? How can we classify and codify them? We made a circle of constant comparison of the text fragments revealed in the answers on open questions (the second block of the interview) in order to find most contrasting answers and as a result codify them, choosing those that fit broader theoretical concepts and background context of job situation. From that, we identified the most important dimensions for comparison on the level of subjectivity: 1. Attitudes towards autonomy. 2. Attitudes towards innovation. 3. Attitudes towards mobility. 4. Attitudes towards middle professional managerial status.

Moving from individual cases and language constructions on occupational career according to different statements we found several dimensions and codified them. Here, in table form (Table 3), we demonstrate those differences sometime by direct abstracts from definite interviews that are more vivid illustrations of extreme positions and sometimes only from short phrases taken from different interviews. This intermediate result could be interpreted in the conceptual frames as reflecting the changing situation on labor market during the last decades [21]. Subjective attitudes towards mobility, short-term employment, and the expansion of autonomy in the workplace seem to be in demand in a time of flexible modernity. We would not explore this aspect further, although we are reminded of Zygmunt Bauman's statement that flexible modernity requests an adequate, new type of personality. Under these conditions, only the most flexible and adaptive personality feels comfortable, those who psychologically and physically are ready to be mobile and adapt quickly to changes in their professional situation [2]. On the contrary, the group of "reactive" professionals, in this sense coincides with the traditional model of career in tough industrial time with strict work discipline and orientation on one-line life-long career. Nevertheless, for construction typology procedure it is an argument confirming that our typology grounded first on subjectivity and subjective-based criteria makes sense and sounds theoretically reasonable.

3.3 Changes in Professional Position – Actual Mobility

While our main research tool in the first two steps was text analysis and interpretation, we further shifted the focus to the analysis of more objective dimensions that demonstrate real acts of individual job strategy. Did general "storylines" correspond with real actions? Were subjective attitudes towards one's career the drivers for real career strategy during the course of one's life? We chose the dimension of mobility/stability in the workplace as the main criterion and compared the episodes of job change in these two contrast groupings. The aim of this was to find a meaningful relationship between subjectivity and real behavior in order to construct on this stage complex behavior patterns in professional life as a combination of subjective and objective dimensions. From storytelling analysis, we learned that, subjectively, the first grouping was more

Table 3. Selected attitudes of two groupings towards job career and their codes

	A. Active	B. Reactive
Code: autonomy in the working place	**Attitudes:** *control over one's self, control over one's time, overtime work, responsibility before, making individual decisions*	**Attitudes:** *the place in hierarchy, strict office time, responsibility, professional skills, experience, control of the chief, control over subordinates*
Code: innovation and novelty in workplace	**Attitudes:** *"important to learn something new" to acquire new knowledge, "use open opportunities," "to choose a new path," "to decide to move apart," "to begin from zero, to take a chance"*	**Attitudes:** *"I'm not sure that I can do the same at another place," "I don't know how I will adapt to new work partners, to a new job situation," "I was used to my old colleague," "I'm always scared in new situations"*
Code: job change, occupation mobility	**Attitudes:** *"It's a plus compared with my father. I'm perhaps more willing to change jobs. ... They (my parents) I know were panicked about changing jobs. They used to work in the same specialization until the last day of their career. Once they finished at school or institute, once they entered one firm and then... for their whole life. I'm scared of that." (Sergey, SMR)*	**Attitudes:** *"I'm not sure that I can do the same at another place," "I don't know how I will adapt to new work partners, to a new job situation," "I was used to my old colleagues," "I'm always scared in a new situation," "I prefer not to throw away already established businesses, not to exceed the level of my expectations" (Oleg, SMR)* *"My colleagues consider me a dinosaur, nobody nowadays stays for so long at one place, but I prefer it." (Alex, SMR)* *"I like when everything is stable, I'm afraid of losses." (Larisa, EKB)*
Code: professional status of administrator	**Attitudes:** *"... If I began to understand that I'm tired of this job, that I do not have opportunity to personal growth, I began to search for some new opportunities. I was never scared to search for something new. Yes, I have some financial freedom now, and do not look carefully at the food prices in the supermarket" (Vladimir, SMR)*	**Attitudes:** *"I'm more comfortable if I am in the position of some 'vice' commerce director, in any case I need somebody who leads me, some 'head' over me, some 'general' director." (Vladimir, EKB)* *"Now those people who worked with me are in my responsibility, in my submission. I'm not their direct boss, but I can, yes, I can command them, can influence them. Well, at the beginning that was funny, but now I am used to it." (Oleg, SMR)*

oriented on mobility while the second was oriented on stability. What were their linear moves in the time of drastic social change of the previous decades?

This dimension was presented as a set of movements from one position to another: from the starting point of the first job after receiving higher education to their actual position as a middle-level manager. We got a digital indicator. It appeared that, on average, the active type made 5–7 moves during their previous professional life, while the reactive type made only 1–3 moves. See the examples below.

Active Type:

Boris, CMR (railway engineer by degree):

- manual supervisor, railway construction company → railway engineer at state company → middle manager in power, engineering private firm → manager at international company in power engineering, state enterprise → financial director at nightclub → manager at international company in power engine trade → middle manager in private publishing trade company → middle manager, head of regional department in plumbing industry.

Andrey, EKB (economist by degree):

- supervisor manager at international company → manager at international tobacco trade company → local branch high administrator at "Pepsi" trade company → middle manager at international trade company → middle manager, head of local branch of Chinese trade company.

Reactive Type:

Svetlana, EKB (construction engineer by degree):

- engineer at large state plant → engineer at the same plant, which changed from state to private status → middle manager, head of the advertising department at the same plant.

Oleg, CMR: (railway engineer in training):

- manager in railway industry → manager in auto industry, private firm → middle manager, head of the transport department in a logistics private company.

Alex, CMR (engineer in gas industry by degree):

- lower controller in gas industry → entrepreneur → manager at trade private firm → financial director at small private trade company.

While constructing the typology, we started from subjective positioning towards career strategy and a complex of attitudes towards job. In this step, we changed the strategy, implying a more accurate indicator of real actions. It appeared that the first division on the basis of textuality—subjective positioning—coincides with real acts of mobility, and so our further developed typology of "active"/"reactive" makes sense. The first type demonstrated subjective control over their professional strategy and ability to structure their professional career according to individual "will" and as realization of her/his work/life plan for career development (self-motivation). The second type of professional career does not assume any individual plan, but was guided mainly by external factors, happening as a reaction of social change.

At this stage, however, there could appear a third, intermediate type of respondent, whose trajectory does not coincide with any single model of movement. For example, their real trajectories could include long episodes of unemployment or downward mobility. Here in the interests of a further analytical scheme of constructing typology, we excluded this type and will further consider only opposite extreme types of career patterns. Nevertheless, the intermediate type could be considered the result of negative social change influence on biography that led to a complete loss of individual control over one's career strategy.

This stage of reconstruction of patterns of real job mobility also has some contextual background connected with the specifics of Russian cases. The social and cultural change and economic restructuring over the previous decades, mainly at the beginning of the reforms (during the 1990s), the economic and cultural shock after the destruction of the "Soviet" type of industrial economy, the emergence and development of private business and the service industry transformed the situation in the labor market. It was worth losing the need for certain professions and for our respondents resulted in the need to change jobs or professionalization in order to use additional human resources or convert the old ones [26, 27]. That is, the specifics of the social context were reflected in the differences in the professional careers between the two types.

That remark shows that the construction of an empirically grounded typology needs an additional analytical resource for its construction: contextuality. In order to understand and interpret the differentiated subjectivities and patterns of behavior, the researcher should take into account the relevant social milieu, all relationships of our cases with micro- and macro-social reality [32]. Moreover, our short overview of the Russian labor market situation related to our respondents' positioning helps to understand them within the context of the social situation. Additionally, it serves as an argument for the credibility of our typology from the point of its contextualization and compliance with reality.

Therefore, as a result, we have two extreme types as patterns of professional strategy that stand in contrast according to the basis of their typology: to subjectivity, as individual positioning towards career strategy. Empirical textual material combined with theoretical concepts and contextualization provided the possibility of constructing an empirically grounded typology. With this help, we constructed two plausible types. The first is an "actor," where the individual is oriented on active and mobile occupational strategy, makes his/her own decisions, and tries to turn her/his own plan into reality. This type could be identified as an active agent of his/her professional biography. The contrasting "reactor" type could be seen as a passive object, waiting for a change of external situation and lacking self-control over his/her position in the labor market, whose decisions result only from external social demands. A reactive strategy makes someone more of an object of his/her transition rather than its subject: decisions in his/her (professional) biography just "happen."

3.4 Additional Dimension: Lifestyle

The lifestyle dimension is used here to demonstrate the abductive possibilities of proposed typology. We did not plan to investigate any other dimensions than professional because our primary aim was to construct types of professional career. Lifestyle was an additional dimension that gave us the opportunity to test the hypothesis of whether subjective and objective components of career strategies are caused by individual phenomena or are of a socio-cultural nature. This dimension provides an opportunity to check whether such a typology is connected with some cultural and social structural change within the strata of professionals: as Myers and Gutman named it, "lifestyle is the essence of social class", in their article with the same name "Lifestyle: the Essence of Social Class" [20]. Developing their concept further, we tried to answer whether it meant that our respondents who have the same social position subjectively perceive themselves as belonging to different class segments. We did not go deeper into the problems of lifestyle and social stratification here [for more see 28]; for our typology, it was more interesting to investigate whether they were oriented on different cultural milieu, in terms of both local and wider lifestyle patterns. As we noted earlier, their formal social position on the vertical social ladder is almost the same—professionals with the status of middle managers—but previously they demonstrated different behavior patterns of career strategy.

In the previous stages, respondents described themselves in their positioning towards their job and job transitions. Here we analyzed their discussion of lifestyle and, more widely, their surroundings—family and community life, comparing their quality of life with "others." We analyzed lifestyle orientations in the context of talk of long-life achievements and perspective: "what do you consider the most meaningful achievements (and perspectives) in your life?" We coded this dimension as future life perspectives, consumer priorities, and type of money spending.

Active type: lifestyle orientation on self-development and further prosperity

1. **Life perspectives:**
   - professional progress (*"further professional career," "improvement of my experience," "wish to be a vice-president of this company"*)
   - individual progress (*"I plan to go ahead with," "I like to jump into something unknown to me," "more books, more education–without it there's no way to go higher," "I visited more than ten or fifteen countries, gained a lot of life experience from that, and this alone I consider as a big achievement"*)

2. **Consumer priorities:**
   - traveling (*"I wish to travel more"*)
   - Prestigious housing (*private house, to have a nice house, to move to a more comfortable and warm climate*)

3. **Type of money spending:**
   - "more earning, more spending" (*"improve financial status," "I hope I would make more money at my company"*)

Reactive type: lifestyle more oriented on "stability" and saving

1. **Life perspectives:**
   - improvement of status in power (*"In fact, that's a small power, for sure, not a big one, but I have a little hope. Nevertheless, I hope that's all I will achieve"*)
   - individual perspective "not to lose previous achievements" (*"I have a stable position and hope not to lose it, not to fall lower," "Don't want to guess, but I would like to stay on the same level'*)
2. **Consumer priorities:**
   - housing (*to buy a small village house, or dacha*)
3. **Type of spending money:**
   - saving "just in case" (*"I would like to have a financial cushion first"*)

This additional dimension reflected the socio-cultural difference between the active and reactive types. The predominant orientation of the active type on further individual and professional development, recreation mobility (traveling), prestigious consumption, and further financial well-being coincides with the lifestyle model of the modern middle class in its western sense. On the other hand, the reactive type had as lifestyle priorities stability, a modest standard of living so as "to be as others" and "not to lose previous achievements" in status and financial sense demonstrated a more traditional "Soviet" type career strategy. That signifies the socio-cultural difference within the strata of professionals is explained by the Russian context of the changing situation moving between two socio-historical models. Not going deeper into such an interpretation, we could briefly note here that for the typology construction process this enriched our approach by embedding it into a wider biographical scheme, serving as support for the idea that definite types resulted from individual orientation to a definite surrounding and socio-cultural milieu. That creates the space for further research development of the typology as a hypothesis about socio-cultural differences between these types.

On the whole (see Fig. 1), the active type, by subjective and objective criteria, reflects the modern social trend of a professional career as a "reflective project of the self." According to the correlation between professional strategy and lifestyle, we can propose that in a cultural sense these types are moving in different directions. The first type is moving towards the model of self-control over its biography, as the other type reflects the fading trend of the predominance of social control. Such a position couldn't be considered as a position resulting from a lack of motivation or some kind of "irrational" choice, but simply as a choice of another typical pattern of an administrative system with a long-term professional career with one employer, which was common in industrial times. Social context as an additional indicator for empirically grounded typology reflects the socio-historical movement from past to present and into the future [32].

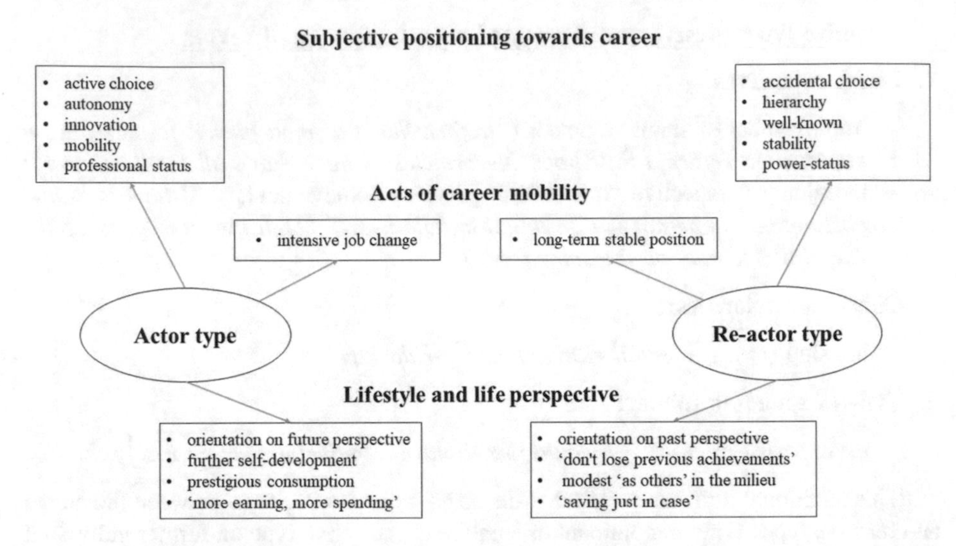

Fig. 1. Visual representation of "active" and "reactive" professional career strategies as types constructed from empirical biographical data.

4 Interpretation and Some Conclusions on Typology-Construction Oriented Biographical Research

In qualitative research, typology is mostly desired as an abstract goal, though in reality we stop at the interpretation and description of cases. Grounded theory methodology gives the possibility of constructing typology starting with one case description [13, 18, 32]. I proposed here another approach that starts from a grouping of similar cases and testing them based on the textual "genre" of self-presentation. Such a strategy has its advantages and restrictions.

As a methodological note, we should remark here that in order to start from textuality and trace a subjective-alternative grouping, we had to align carefully all the personal data of our respondents. I should note here that we chose from the whole set of biographical interviews those respondents who had the same background as professionals by education, moved up in their careers to the position of middle manager, and were about the same age. This similar background of the respondents provided the possibility of forming a comparison from their subjectivity and later turning to action-alternatives. That ensured the validity of further steps of typification and provided a bridge between subjectivity and objectivity in these alternatives. So, while starting a grounded typology from a grouping principle, the researcher should carefully choose the sample, both according to a theoretical concept and to definite objective indicators in order to ensure the basis for their subjective difference that improves its reliability and the validity of the results.

The model of typology is always organized in several stages as steps of moving between theoretical concepts and empirical data. Even if this is "grounded" analytical strategy, we first work with theoretical concepts that are the first way of analyzing unstructured data. From this, the researcher can understand the data and maintain an approach for further analysis and building of a typology. In our case of a proposed grounded typology, we demonstrated what could be the analytical opportunities of biographical research, grounded on socio-linguistic text-analysis and the first division of the textual data. The procedure started from such "subjective" properties as the grammatical form of textual self-presentation through the verbal form of "to be" or "to do" as mental models and incorporated the resulting socially sensitive typology of active/reactive dichotomy connected with different social settings in a changing social context, although I should note here a definite limitation of such a "contrast" typification; as I pointed out earlier, in reality there could very well be a continuum or several intermediate types. There may, however, be in social science this kind of simplest abstract distinction, moving in the direction of a Max Weberian "ideal type," which could be most promising, as it offers the possibility of tracing social dynamics along a socio-historical continuum. Tom Wengraf, who critically noted that sociology had a tendency to deal with relatively unhistorical pictures of the present, stressed the meaning of socio-historical context for biographical research and typology. In order to understand the present, we need to know both personal and interpersonal history, the history of context, then our typologies as a final generalization could lead us to relatively diachronic types as a demonstration of the historical development of systems or structures [34: 143, 160]. Here in this example I traced the two contrasting types of behavior patterns that reflect the socio-historical continuum and change in personal attributes as well as the change in wider social contexts from the "passing away type" to the "emerging type" with modified human resources that coincide with new tendencies in the labor market.

The advantages of our empirically grounded typology as an open and flexible research strategy permits us to move from the subjective/objective configuration of a career strategy to the so-called typology of structures [14, 32, 33]. During the process of construction, the proposed typology demonstrated its abductive character. In the final step, there emerged a new construction that was not anticipated: the combination of alternative career strategies with differentiated orientations in life values and lifestyles. That provided an opportunity to broaden the scope of individual-strategy types to the differentiation between socio-cultural milieus within the same segment of middle managers. As Tom Wengraf, following Layder, stresses, qualitative typologies shouldn't restrict themselves to the typology of action as typification of individual lifestyles and meanings, but should move further, to "typologies of typologies," or typologies of systems that could depict the settings and contexts of behavior patterns; then they should provide more explanations of social life [14, 32]. Our example typology of career strategies and its conceptualization showed such possibilities. It suits the modern context of social change in Russia.

It seems that such a research strategy of constructing a "grounded" typology, with all its advantages and restrictions, could serve as a useful tool in both biographical and wider methods of qualitative research, wherever a researcher aims to analyze textual data in terms of subjectivity/objectivity argumentation and interpretation in order to construct models of action in specific social contexts.

References

1. Bailey, K.: Typology and Taxonomies: An Introduction to Classification Techniques. SAGE, Thousand Oaks (1994)
2. Bauman, Z.: Liquid Modernity. Piter, St.-Petersburg, Russia (2008). (in Russian: Tekuchaya sovremennost)
3. Becker, H.: Through Values to Social Interpretation. Essays on Social Contexts, Actions, Types, and Prospects. Greenwood Press, New York (1968/1950)
4. Breckner, R., Rupp, S.: Discovering biographies in changing social worlds. The biographic interpretive method. In: Chamberlayne, P., Rustin, M., Wengraf, T. (eds.) Experiences of Social Exclusion: Biography and Social Policy in Europe, pp. 287–306. Policy Press, London (2002)
5. Diewald, M., Goedicke, A., Mayer, K.: After the Fall of the Wall: Life Courses in the Transformation of East Germany. Stanford University Press, Palo Alto (2006)
6. Flick, U.: Challenges for qualitative inquiry as a global endeavor. Qual. Inq. **20**(9), 1059–1063 (2014)
7. Gaupp, N.: School-to-work transitions—Findings from quantitative and qualitative approaches in youth transition research. FQS **14**(2), art 12 (2013)
8. Glaser, B., Strauss, A.: The Discovery of Grounded Theory: Strategies for Qualitative Research. Aldine Transactions, New Brunswick, NJ (1967). (Reprinted 2006)
9. Glaser, J., Laudel, G.: Life with and without coding: two methods for early-stage data analysis in qualitative research aiming at causal explanations. FQS **14**(2) (2013). http://qualitative-research.net/index.php/fqs/article/view/1886. Accessed 8 Dec 2016
10. Helemjae, E., Veerman, R.: Perception of Social Stratification in Estonia. Sotsiologicheskie issledovanija **7**, 38–48 (2010) (in Russian: Vospriyatie socialnoi stratifikatcii v Estonii)
11. Kazmierska, K.: Narrative interview as a method of biographical analysis. In: Fikfak, J., Adam, F., Garz, D. (eds.) Qualitative Research: Different Perspectives, Emerging Trends. Institut za slovensko Narodopisje, pp. 156–172. ZRC SAZU, Ljubljana (2004)
12. Kelley, C., Kelley, S.: Subjective social mobility: data from 30 nations. In: Haller, M., Jowell, R., Smith, T. (eds.) Charting the Globe: The International Social Survey Programme 1984–2009, pp. 106–125. Routledge, London (2009)
13. Kluge, S.: Empirically grounded construction of types and typologies in qualitative social research. FQS **1**(1) (2000). http://www.qualitative-research.net/index.php/fqs/article/view/1124. Accessed 21 Nov 2017
14. Layder, D.: Sociological Practice: Theory and Social Research. SAGE, London (1998)
15. Lazarsfeld, P., Barton, A.: Qualitative measurement in the social sciences. Classification, typologies, and indices. In: Lerner, D., Lasswell, H. (eds.) The Policy Sciences, pp. 155–192. Stanford University Press, Palo Alto (1951)
16. Marshall, V., Heinz, W., Kruger, H., Verma, A.: Restructuring Work and the Life Course. Toronto University Press, Toronto (2001)
17. McKinney, J.: Constructive Typology and Social Theory. Appleton Century Crofts, New York (1966)
18. Merlino, A.: From the analysis of argumentation to the generation of typologies: a model of qualitative data analysis. Qual. Rep. **19**(17), 1–24 (2014). https://nsuworks.nova.edu/tqr/vol19/iss17/1. Accessed 8 Dec 2017
19. Miles, A., Savage, M., Bühlmann, F.: Telling a modest story: accounts of men's upward mobility from the National Child Development Study. Br. J. Sociol. **62**(3), 418–441 (2011)
20. Myers, J., Gutman, J.: Life style: the essence of social class. In: Wells, W. (ed.) Life Style and Psychographics, pp. 243–267. Marketing Classics Press, Decatur, GA (2011)

21. Ranson, G.: Engineers and the Western Canadian Oil Industry: work and life changes in a boom-and-bust Decade. In: Marshall, V., Heinz, W., Kruger, H., Verma, A. (eds.) Restructuring work and the life course, pp. 462–473. University of Toronto Press, Toronto (2001)
22. Rosenthal, G., Fisher-Rosenthal, W.: The analysis of narrative-biographical interview. In: Flick, U., von Kardoff, E., Steinke, I. (eds.) A Companion to Qualitative Research, pp. 259–265. SAGE, London (2004)
23. Schreier, M.: Qualitative Content Analysis in Practice. SAGE, London (2012)
24. Schutze, F.: Biography analysis on the empirical base of autobiographical narratives: how to analyze autobiographical narrative interviews. Eur. Stud. Inequalities Soc. Cohes. **1**(2), 153–242 (2004). http://www.profit.uni.lodz.pl/pub/dok/6ca34cbaf07ece58cbd1b4f24371c8c8/European_Studies_2008_vol_1.pdf. Accessed 9 Dec 2017
25. Sellerberg, A., Leppänen, V.: A typology of narratives of social inclusion and exclusion: the case of bankrupt entrepreneurs. FQS **13**(1), art 26 (2012)
26. Semenova, V.: Patterns of vertical social mobility of professionals in times of social transformation: sociobiographical approach. In: Semenova, V., Chernysh, M., Vanke, A. (eds.) Social Mobility in Russia: Generational Aspect. Institute Sociology RAN, Moscow (2017). (in Russian: Patternyu Vertikalnoi Socialnoi Mobilnosti Professionalov v Period Sozialnych Transformatsyi: Sociobiographicheski Podhod. V kn: Socialnaya Mobilnost v Rossii: Pokolencheski Aspect)
27. Shkaratan, O.: Sociology of Inequality. Theory and Reality. National Research University Higher School of Economics, Moscow (2012). (in Russian: Sociologya neravenstva. Teoriya i Realnost)
28. Sobel, M.: Lifestyle and Social Structure: Concepts, Definitions, Analyses. Quantitative Studies in Social Relations. Academic Press, New York (1981)
29. Tampubolon, G., Savage, M.: Intergenerational and intragenerational social mobility in Britain. In: Connelly, R., Gayle, V., Lambert, P. (eds.) Social Stratification: Trends and Processes. Routledge, London (2016)
30. Travers, M.: Qualitative sociology and social class. Sociol. Res. Online **4**(1), 1–13 (1999). http://www.socresonline.org.uk/4/1/travers.html. Accessed 23 Sept 2016
31. van den Berg, M.: Subjective social mobility: definitions and expectations of 'moving up' of poor Moroccan women in Netherlands. Int. Sociol. **26**(4), 503–523 (2011)
32. Walker, C.: 'I don't really like tedious, monotonous work': working-class young women, service sector employment and social mobility in contemporary Russia. Sociology **49**(1), 106–122 (2015)
33. Weiss, A.: Comparative research on highly skilled migrants. Can qualitative interviews be used in order to reconstruct a class position? FQS **7**(3), art 2 (2006)
34. Wengraf, T.: Uncovering the general from within the particular: from contingencies to typologies in the understanding cases. In: Chamberlayne, P., Bornat, J., Wengraf, T. (eds.) The Turn to Biographical Methods in Social Science, pp. 140–164. Routledge, London (2000)

Capturing Nordic Identifications Through Participatory Photography

Solveig Cornér[1(✉)], Maria Forsius[1], Gunilla Holm[1], Harriet Zilliacus[1], and Elisabet Öhrn[2]

[1] Faculty of Education, University of Helsinki, Helsinki, Finland
solveig.corner@helsinki.fi
[2] Department of Education and Special Education, University of Gothenburg, Gothenburg, Sweden

Abstract. This study explores how participatory photography can be used in researching upper secondary students' identifications with what it means to live in one of four Nordic countries. The study draws on students' constructions and interpretations of photographs. For this article the data analyzed consisted of 571 photographs taken during spring 2018 by a total of 104 students in the metropolitan areas in Helsinki, Stockholm, Oslo and Copenhagen. The analysis of the photographs and their captions show that students associated themselves mostly in a positive way with the Nordic region, though also some critical attitudes were identified. Visual ethnography in education as a method enhanced the upper secondary students' way of giving meaning to what living in the Nordic countries means to them. Moreover, the method enables the students to become co-researchers together with the research team in both an aesthetic and narrative way. The study offers insights into how participatory photography can be as a useful and activating method in both local and cross-national research.

Keywords: Participatory photography · Identification
Upper secondary students · Nordic region

1 Introduction

Photography as a research method in qualitative research is growing and evokes a lot of interest. It is argued that children and young people are seldom heard in the school environment and therefore photographs offer possibilities for young peoples' engagement in research (Lodge 2009). In addition, Lodge (2009) points out, that photography as a method is especially useful in the engagement of marginalized and shy students in school.

Our study is conducted in four countries in the Nordic region. The Nordic countries are geographically and culturally close and often referred to as the Nordic region. The Nordic region consists of Finland, Sweden, Norway, Denmark and Iceland, as well as Greenland, the Faroe Islands and the Åland islands. Education has been vital to the development of the Nordic welfare model and might even, as Antikainen (2006) suggests, be seen as one of its main preconditions. The educational systems in the Nordic countries emphasize the equality of opportunity for all of its citizens

A. P. Costa et al. (Eds.): WCQR 2018, AISC 861, pp. 136–145, 2019.
https://doi.org/10.1007/978-3-030-01406-3_12

(Isopahkala-Bouret et al. 2018). According to a recent study conducted by the Nordic council of ministers (Andreasson and Stende 2017), collaboration in different areas between the Nordic countries is highly valued by the Nordic citizens. However, considering the seemingly high interest in the shared opportunities by the Nordic citizens, we know surprisingly little about how young people experience the meaning of living in a country belonging to the Nordic region. Given the importance of future Nordic societal development and collaboration, there is a need to better understand the perspectives of young people and what shared perceptions in the Nordic region means to them. At the moment, the research literature on what the Nordic is from young peoples' perspective, is scarce.

The overall study focused on how young people in Finland, Sweden, Denmark and Norway perceive what it means to live in a Nordic country and what being Nordic means. The study was conducted through participatory photography and by this, our aim was to produce both robust information on the foci and different kinds of understanding and information (Pink 2012; Rose 2016). We draw on a recent empirical study from four Nordic countries to analyze this, and also on the authors' previous work related to the topics and empirical work on gender, class and ethnicity in urban schools (Öhrn 2009, 2011, 2012, Öhrn and Weiner 2017) and justice through education, bilingual school space, intercultural education, arts-based visual research and ethnographic analyses (Holm 2014, 2018; Holm et al. 2015, 2018). The focus of this paper is though specifically on how participatory photography can be used in exploring young people's identifications with, in this case, the Nordic.

2 What Does It Mean to Be Nordic?

Youth identifications with the Nordic is seen in this study as being citizens in one of the Nordic countries from their perspective e.g. what it means to be "Nordic". Rather than seeing identifications as fixed, the term "identification", more strongly than "identity", expresses a process, which includes relating to other individuals, communities, situations or categories within a particular context. From a non-essential view, identifications are perceived as fluid and changing tied to specific cultures rather than being fixed and stable (Anthias 2011; Hall 1996).

The identifications of youth today reflect the increasingly globalising and postmodern society, which is under constant and rapid change. Youth form their identities in the context of multiple traditions through a mix of local and global traditions, via both first hand and virtual media reality, which transcend the borders of nations and separate 'cultures' (Verschueren 2008). There is today more space to negotiate and chose our identities. However, there are notably also inequalities in the distribution of opportunities among youth (Gewirtz and Cribb 2008). Identifications include a number of different aspects, such as, gender, home culture, country or dominant culture, ethnicity or appearance, religion, language, sexuality, subcultures/hobbies, profession, and social class. Included are also individual qualities, which can for instance refer to individual abilities or disabilities. These different identifications intersect, that is, the different aspects inter-relate and crosscut within people's lives and in social relations (Anthias 2011).

Identifications are closely connected with a sense of belonging as they express membership in a social group, community or nationhood. They can create forms of solidarity bringing people and groups together. Students' identities are importantly developed within the context of the school and with peers. The process of identification, necessarily also involves construction of boundaries and exclusion, separating ourselves from what we are not (Anthias 2011). The school is a primary arena for dealing with a whole array of differences, and students construct their identifications as much through notions of how others see "us" and "who we are not" as through conceptions of who "we" are (Reay 2009, 277–278). A complex view of identifications is reflected onto the view of citizenship, and that youth today have plural and diverse citizenships. The national level and the nation state are still relevant levels, but globalization challenges static and one-dimensional models of citizenships with new transnational levels such as the Nordic, the global or the cosmopolitan. Classic civic belonging is substantially and continually being transformed in the landscape of social media, consumerism, and popular culture. Therefore, diverse forms of belonging and citizenship, which transcend national boarders, need to be considered when studying youth (Banks 2008; Hoikkala 2009; Mansikka and Holm 2011).

3 Aim

The aim of this study is to explore how upper secondary students in the Nordic capitals identify themselves with the Nordic region by using participatory photography. The following research question was addressed: How can participatory photography as a research method be used to study upper secondary school students' identifications with the Nordic and the Nordic region? What does it mean to them to live in a Nordic country?

4 The Study

The principal aim of this study was to investigate and understand how participatory photography can be used to study upper secondary students' identifications with the Nordic and the Nordic region. As Holm (2018) states, "participatory photography means that the research participants take the photographs, and in this way become a kind of co-researchers/co-ethnographers". While the participants take the photos and are engaged from the beginning of the project, it enables the researchers to get a sense of what the participants want to show as important or interesting from their own worlds (Holm 2018). Further, participatory photography can be described as a way to study identities and identifications, since it allows the participants, in our case upper secondary students, to control how they portray their perceptions of who they are (Holm et al. 2015).

The data consist of 571 photographs taken in spring 2018 and are read in the context of the upper secondary students' urban environment, with decent housing, good facilities for cultural and leisure activities, but also visible segregation, conflicts and inequalities (Öhrn and Weiner 2017). A total of 104 upper secondary students from four upper secondary schools in metropolitan Helsinki (n = 17), Stockholm (n = 24), Oslo (n = 30), and Copenhagen (n = 33) participated in the study. The research team collectively decided what schools would be invited to participate in the project based on their location in the capital regions. In the second stage, the principals of each school were contacted by one of the researchers and the principal explored the teachers' interest in participating in the research. The upper secondary schools varied from each other. However, the student population is well represented in terms of age and area, since the mean age in each country was nearly the same and the schools were all located in the urban metropolitan areas in each country. A majority (f = 63/60.5%) of the participants were female. Some schools were more culturally diverse than others. The demographics of the participants and the number of photographs are shown in Table 1.

Table 1. Demographics (gender and age) for participants in Finland, Sweden, Norway and Denmark, number of photographs

Country	Participants f/(%)	Gender m/f	Age (mean)	Photographs taken/country
Finland	17 (16)	1/16	17	65
Sweden	24 (23)	13/11	18	132
Norway	30 (29)	10/20	17	133
Demark	33 (32)	17/16	16.5	241
Total	104 (100)	41/63	17	571

The data were collected, by two of the researchers in the four countries between February and May 2018. The students were asked to photograph, with their mobile phones, themes that they associated with the Nordic region and that in some way captured how they experienced living in a Nordic country. In order to make the photographs more understandable to the viewer, the students were asked to write a short caption to every photo. Students could take as many photographs as they liked but could upload only the ten best photographs per student in order to keep the number of photographs manageable. The photographs were then uploaded to a shared project platform. Prior to conducting their task, the students were introduced to the aim of the project and additionally took part in a brainstorming session about the Nordic region, where the Nordic was further discussed. The students had 1-3 weeks to do photographs depending on the schools' schedules. All students participating in the project gave their written consent.

4.1 Analysis

The analysis started with a careful overview of the photographs (Keats 2009). When analyzing the photographs, the captions helped the researchers to interpret the students' photographs as intended by the students (Holm et al. 2015). This was especially crucial when interpreting photographs that were more symbolic or metaphorical. The research team categorized the students' photographs according to the emerging themes, but also into informative photographs and symbolic photographs (Holm et al. 2015; Elliot et al. 2017). A preliminary descriptive coding into a theme was made during the first phase of analyzing the photographs. The decision of coding a certain photograph as a theme was made when the photograph captured something important about the data associated to the research question (Braun and Clarke 2006; Rose 2016). The main themes identified in the analysis process of the photographs were belonging to the welfare state, youth culture, individualism versus collectivism, the Nordic nature, Nordic traditions, architecture, healthy living, and technology. Three of the authors conducted the analysis process as a team. The authors conducted the categorization of the photographs into themes collectively.

5 Results

The preliminary results show that the students' reflected upon several themes in relation to their identification of the Nordic region in the photographs. The students perceived their privileges of living in the Nordic welfare states (such as free education, health care, social security as well as safety and equality) as one of the most important themes. The youth culture was another theme frequently portrayed and discussed. Other common Nordic themes were youth culture in relation to alcohol, and in contrast, healthy living (healthy food, sport, outdoor activities). The students also associated the frequent use of technology and social media in everyday life as a way of being a Nordic citizen. The results illustrate further that students constructed both informative and symbolic photographs in relation to the various themes.

5.1 Informative Photographs

As in Elliot et al. (2017), informative photographs tended to focus on situations that directly related to experiences significant for the students. For example, a frequent theme captured by the students in all Nordic countries, were their association with equality. The students pointed out in their pictures that, due to the Nordic welfare states, people have equal possibilities to live a good life and to choose the kind of life they want to live. They brought up equality in several ways: between genders, between adults and children and as equal rights for people with disabilities as well as for sexual minorities. As examples, a Norwegian student illustrated the rights for all people in the Nordic countries to get married (Fig. 1) whereas a Danish student illustrated gender equality by capturing a man on parental leave (Fig. 2). The students associated, in addition, the strong well-fare society with the opportunity for free education for everyone (Fig. 3).

Fig. 1. "Same-sex marriage is legal in all Nordic countries" (Norwegian student)

Fig. 2. "Men use their right to parental leave, and take care of the children, more frequently than in other countries" (Danish student)

Fig. 3. "Children in the Nordic countries are entitled to free education and books" (Swedish student)

Further, in connection to the youth culture, the students frequently chose to illustrate the active life that is typical for the Nordic region, since, for example, bicycling and skiing (Figs. 4 and 5) are common activities among the youth.

The students' also systematically showed their appreciation for the four seasons and the beautiful nature in the Nordic countries (Fig. 6). However, in opposite to the emphasis on healthy living, the students shared many photographs describing a youth culture associated with heavy alcohol use (Fig. 7).

Fig. 4. "It's Nordic to bike and think about the environment" (Danish student)

Fig. 5. "Skiing is a Nordic tradition" (Swedish student)

Fig. 6. "Our nature is unique and wonderful – especially in wintertime" (Finnish student)

Fig. 7. "To party is a part of life" (Danish student)

5.2 Symbolic Photographs

Symbolic photographs are more complex and more difficult to interpret without understanding the photographers' habitus and the context (Holm et al. 2015). However, in most cases they also deal with deeper, more serious issues that might be difficult for the students to express in words. A Finnish student captured, for instance, the Nordic welfare state as feeling safe in the Nordic countries in a symbolic photograph (Fig. 8), accompanied by a describing caption. A Swedish student additionally illustrated the perception that there is an individualistic culture in the Nordic countries with a photograph of a drawing where all the focus is on the individual (Fig. 9). The meaning of these symbolic photographs and other similar ones would be difficult to understand without the captions (Holm et al. 2015).

Students' perceptions of the possibilities and opportunities living in a Nordic country were often quite idealistic and uncritical. For example, students emphasized gender equality, which is better than in most other countries but by no means perfect.

However, some critical reflections were brought forth by several students through photographs about, for example, the alcohol centered youth culture.

Fig. 8. "I feel safe living in Finland. It is possible to walk alone on the street and express your opinion without having to be afraid." (Finnish student)

Fig. 9. "We have an individualistic culture in the Nordic countries" (Norwegian student)

6 Conclusion

This study focused on how young people through participatory photography can show what living in the Nordic countries means to them. The method worked well when exploring different aspects of the students' identifications with what it means to live in a Nordic country. In line with Holm (2018), the method enables the students to become co-researchers together with the research team in both an aesthetic and narrative way, since the students were in charge of constructing their photographs. Students had to reflect on their own lives and the societies they live in as well as the neighboring countries in order to photograph their thoughts. Further, consistent with Holm (2018), using participatory photography with captions enabled the participants to bring forth the reason why they have taken a photograph and what they want to emphasize. The captions pushed students to compose thoughtful photographs in order for the photographs to have an understandable meaning. In summary, this study brings forth insights into how participatory photography can serve as a useful and activating method with young people in both local and cross- national educational research.

Acknowledgements. This research was supported by the Nordic Centre of Excellence 'Justice through Education' financed by NordForsk grant # 57741.

References

Andreasson, U., Stende, T.: Ett värdefullt samarbete: Den nordiska befolkningens syn på Norden (A valuable co-operation: The view of the Nordic co-operation by the Nordic people). Nordic Council of Ministers (2017). Homepage. http://norden.diva-portal.org/smash/record.jsf?pid=diva2%3A1152251&dswid=-2647. Accessed 4 Dec 2017

Anthias, F.: Intersections and translocations: new paradigms for thinking about cultural diversity and social identities. Eur. Educ. Res. J. **10**(2), 204–216 (2011)

Antikainen, A.: In search of the Nordic model in education. Scand. J. Educ. Res. **50**(3), 229–243 (2006)

Banks, J.A.: Diversity, group identity, and citizenship education in a global age. Educ. Res. **37** (3), 129–139 (2008). https://doi.org/10.3102/0013189x08317501

Braun, V., Clarke, V.: Using thematic analysis in Psychology. Qual. Res. Psychol. **3**(2), 77−101 (2006)

Elliot, D.L., Reid, K., Baumfield, V.: Capturing visual metaphors and tales: innovative or elusive? Int. J. Res. Method Educ. **40**(5), 480–496 (2017). https://doi.org/10.1080/1743727x.2016.1181164

Gewirtz, S., Cribb, A.: Taking identity seriously: Dilemmas for education policy and practice. Eur. Educ. Res. J. **7**(1), 39–49 (2008)

Hall, S.: Who needs 'identity'? In: Hall, S., Gay, P. (eds.) Questions of Cultural Identity, pp. 1–17. Sage, London (1996)

Holm, G.: Photography as a research method. In: Leavy, P. (ed.) The Oxford Handbook of Qualitative Research. Oxford University Press, New York (2014)

Holm, G., Londen, M., Mansikka, J.-E.: Interpreting visual (and verbal) data: Teenagers' views on belonging to a language minority group. In: Smeyers, P., Bridges, D., Burbules, N.C., Griffiths, M. (eds.) International Handbook of Interpretation in Educational Research: Part One. Springer, Dordrecht (2015)

Holm, G.: Visual ethnography in education. In: Beach, D., Bagley, C., Marquez da Silva, S. (eds.) The Wiley Handbook of Ethnography of Education. Wiley, Blackwell (2018)

Holm, G., Sahlström, F., Zilliacus, H.: Arts-based visual research. In: Leavy, P. (ed.) Handbook of Arts-Based Research, pp. 311–335. Guilford Press, New York (2018)

Hoikkala, T.: The diversity of youth citizenships in the European Union. Nord. J. Youth Res. **17** (1), 5–24 (2009). https://doi.org/10.1177/110330880801700102

Isopahkala-Bouret, U., et al.: Access and stratification in Nordic higher education. A review of cross-cutting research. Educ. Inq. **9**(1), 142–154 (2018). https://doi.org/10.1080/20004508.2018.1429769

Keats, P.A.: Multiple text analysis in narrative research. Visual, written, and spoken stories of experience. Qual. Res. **9**, 181–195 (2009)

Lodge, C.: About face: Visual research involving children. Education **37**(4), 3–13 (2009)

Mansikka, J.-E., Holm, G.: On reflexivity and suspension: Perspectives on a cosmopolitan attitude in education. Nord. Stud. Educ. **31**, 76–85 (2011)

Verschueren, J.: Intercultural communication and the challenges of migration. Lang. Intercult. Commun. **8**(1), 21–35 (2008). https://doi.org/10.2167/laic298.0

Pink, S.: Doing Visual Ethnography. Sage, London (2012)

Reay, D.: Identity making in the classrooms. In: Wetherell, M., Mohanty, C.T. (eds.) The Sage Handbook of Identities, pp. 277–294. Sage, London (2009)

Rose, G.: Visual Methodologies. Sage, London (2016)

Öhrn, E.: Challenging sexism? Gender and ethnicity in the secondary school. Scand. J. Educ. Res. **53**(6), 579–590 (2009)

Öhrn, E.: Class and ethnicity at work. Segregation and conflict in a Swedish secondary school. Educ. Inq. **2**(2), 345–355 (2011)

Öhrn, E.: Urban education and segregation: The responses from young people. Eur. Educ. Res. J. **11**(1), 45–57 (2012)

Öhrn, E., Weiner, G.: Urban education in the Nordic countries: Section editors' introduction. In: Pink, W., Noblit, G. (eds.) Second International Handbook of Urban Education, pp. 649–669. Springer, Cham (2017)

Biomechanics of Nurse Midwives in the Delivery: Contribution of Qualitative Research

Armando David Sousa[1], Cristina Lavareda Baixinho[2,3], Fátima Mendes Marques[2], Mário Cardoso[2], and Maria Helena Presado[2(✉)]

[1] Hospital Center of Funchal, Funchal, Madeira, Portugal
armandodav@gmail.com
[2] Nursing School of Lisbon, Nursing Research & Development Unit, Lisbon, Portugal
mhpresado@esel.pt
[3] Center for Innovative Care and Health Technology – CiTheCare, Lisbon, Portugal

Abstract. The prevalence of musculoskeletal injuries is high in maternal and obstetrical nurses. In this study, we used film recording and analysis to develop a qualitative and descriptive way of analysing the postures of the experts during labour and understand how intervention during labour influences the postures and identify strategies for the prevention of musculoskeletal injuries. Ten maternal and obstetrical specialist nurses participated in the films, and the results evidenced changes in posture when changing from a static position to dynamics and allow to make suggestions, in the work contexts, for the prevention of musculoskeletal injuries in the midwives and to improve the quality of life of the professionals.

Keywords: Nurse midwives · Musculoskeletal injuries · Biomechanics

1 Introduction

Musculoskeletal injuries related to work (MIRW) have been diagnosed for a long time as a serious problem for the working population [1]. Interestingly, it is in the health field that they take overwhelming proportions with high prevalence associated to lesions and symptoms such as back pain, muscular fatigue and pathologies, such as hernias, tendinitis and repetitive work lesions [1–4].

This public health quasi epidemic can reach a prevalence of 65.4% in health professionals and it is the cause, directly or indirectly, for absenteeism, depicting 56% of temporary service absence causes [5], with associated high costs to the lesions treatment and reduction of the professionals' productivity [3].

Some authors [1] classify it as the main health problem among registered nurses (RNs). However, they highlight that the information available about its impact is still limited. Different studies, performed with nurses, consider that its origins reside in

A. P. Costa et al. (Eds.): WCQR 2018, AISC 861, pp. 146–155, 2019.
https://doi.org/10.1007/978-3-030-01406-3_13

problems related with the movement of the hospitalized persons. However, this is not an etiological factor in maternal health care and obstetrics [6], where parturients are not mobilization and transference dependent. The manipulated loads (i.e. newborns), have lower weights then adults.

From the foregoing, the authors consider that research in midwives has not been as productive as in other health professionals who work in emergency and inpatient services for adults [6]. Investigations performed in this specific context highlight the nature of the professional activity, with the adoption of painful postures in the period of delivery. The prone-supination movements, the low collaboration of the parturient that is going to be moved and/or mobilized and the inability to predict her behaviour during the task in hands, are considered a risk factor [4, 6]. We cannot forget, obviously, the individual and organizational risk factors, such as space configuration, which can make it harder for Nurses Midwives (NM) to position correctly in order to perform specific tasks during the period of delivery [4, 6] and the insufficient dimensions of the rooms, working rooms and delivery rooms, which will condition the adoption of adequate postures [4, 6].

As aforementioned, the NM maintain incorrect postures for long periods, disregarding corporal alignment, choosing an orthostatic posture with accentuated neck and lower back flexion during cervical dilation assessment (tocological exam), labour (second stage in the period of delivery) and placenta delivery (third stage in the period of delivery), keeping at the same time repetitive movements, applying force and elevation of the members to 90º [6, 7].

A recent study performed in the simulated high-fidelity practice environment, whose goal was to analyse the students' postures of the master course in maternal health and obstetrics, fully characterises these specific risks. Results evidence that the posture alterations and corporal dynamics in professionals remain. During labour, when there is an alternation of static positions for Dynamics, a body misalignment is observed evidenced by shoulder unevenness, lateral head tilting, as well as anterior body tilting and arms and forearms flexion, diminishing stability, and in the adoption of the static position there is no alternation of the body weight through the lower members [6].

However, in the richness of the findings of this investigation we can not fail to mention that a high-fidelity mannequin was used for the delivery simulation, which by itself can influence and benefit results, since it does not comprise the labour realism in a natural situation of clinical practice, it being a controlled situation, the stress is less and it is an important factor to allow a broad comprehension about MIRW in NM.

Some authors observe that in situations with higher levels of stress, when the work is performed in higher levels of anxiety and lesser probability of control, the lesion risks are significant, since these are serious aggravation factors [8].

2 Method

This investigation, with a qualitative and descriptive approach nature has as objectives: to analyse the NM postures during the delivery period; to understand the influence of nursing interventions during labour on body alignment, posture and balance.

Filming the clinical practice in eutocic deliveries allows for direct and detailed observation, as well as the analysis of the postures and conditioning factors of the principles of biomechanics.

This type of research establishes a relationship between the practices and behaviours of professionals in compliance with the principles of biomechanics. The phases of labour and the motivations for biomechanical decision making allow a comprehensive understanding of the phenomenon. The use of films allows emotional detachment for reflective analysis. The possibility of reviewing and stopping the image [9] allows capturing the surrounding environment.

The participants were 10 NM who participated in eutocic deliveries in a Portuguese hospital, who authorized the study.

Since there are no specific interpretation methods for image analysis that allow the systematization of all procedures [10], it should be noted that careful planning was done taking into account the time available for research, cost, experience of the researchers, skills of the filming technician, ethical-legal issues. Authorization was requested from the Board of Directors and the Ethics Committee.

Prior to filming, all the participants in the film (doctors, nurses, operative assistants and parturient/family) received information about the purpose of the study, the methodology of analysis of the images and signed the informed consent term.

Anonymity and confidentiality of the recordings was guaranteed; they were informed that in the use of "frames" of the filming, in order to divulge the investigation, faces would be covered so as not to allow identification.

Two fixed GoPro Hero 4 V2 cameras were selected and placed in the recording of 10 low-risk eutocic deliveries in a horizontal position, allowing the analysis of the postures adopted by the NM during the second and third stages of labour. One of the researchers present in the room performed the control of the filming.

The videos were visualized and analyzed based on the grid elaborated in another investigation [4] and following to the categories in accordance with the principles of biomechanics: (1) body movement; (2) body alignment; (3) balance; (4) strength; (5) friction. Five researchers were involved in the analysis of the findings and two in the validation of the categories.

After consensus of selected excerpts, they were coded. The codification, categorization and interpretation of the films was made directly as advocated by Flick [11].

The images were worked on by assigning a colour code that facilitated the coding and the definition of the categories, in a structured and interconnected way, using webQDA® software. In the definition of the categories, representativeness, completeness, homogeneity and relevance of the subject were guaranteed.

This study was approved by the Administrative Council of the Hospital Center of Funchal and the Ethics Committee of SESARAM (Serviço de Saúde da Região Autónoma da Madeira) with the reference no. 44/2017 (nov. 14). Throughout the entire research process, anonymity and confidentiality of the findings was guaranteed.

3 Results

The participants in this study are 9 women and 1 man, graduated in nursing, with training in NM. The average age is 47 years, with work experience of an average of 23 years of service and 13 years as NM. The weekly workload of 5 of the participants is over 42 h, 4 participants between 35 and 40 h and one participant with 35 h a week.

In the analysis of the videos, the risk factors related to the activity and the psychosocial risk factors were taken into account. Four categories emerged in relation to activity-related risk factors (Fig. 1) and a category related to psychosocial risk factors (autonomy in guiding labour).

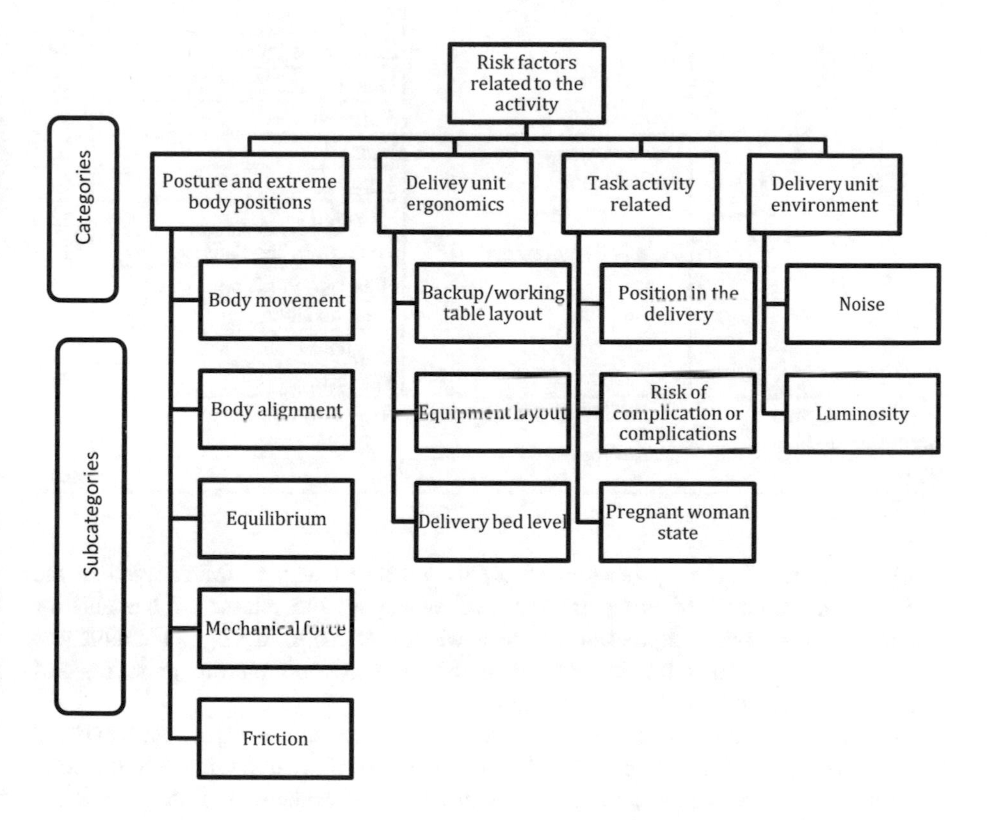

Fig. 1. Categories and subcategories of activity-related risk factors

From the subcategories it was necessary to create other subcategories with the purpose of synthesizing the action. The use of the videos made it possible to observe in detail the NM approach (Table 1) during the second and third stages of labor, providing the analysis of the postures and conditioning factors of the principles of biomechanics. Three biomechanical evaluation moments were defined in the second stage and two in the third stage, according to the analysis of the videos.

Table 1. NM biomechanical risk factors during labour: observation units (FI) by category and subcategory

Risk factors	Category	Subcategory	FI
Activity related	Posture and extreme body positions	Body movement	48
		Body alignment	6
		Balance	147
		Mechanical force	34
		Friction	36
		Subtotal	**248**
	Delivery unit ergonomics	Backup/working table layout	15
		Equipment layout	9
		Delivery bed level	10
		Subtotal	**34**
	Delivery unit environment	Noise	2
		Luminosity	10
		Subtotal	**12**
	Task activity related	Position in the delivery	10
		Risk of complications or complications	4
		Pregnant woman state	10
		Subtotal	**24**
Psychosocial and organizational	Autonomy in delivery orientation		11
Total			**325**

At the time of delivery (second stage) the posture that the NM adopted in the protection or removal of the perineum was evaluated; the release of the anterior shoulder; and the facial expression of the newborn. At the third stage, attention was given to the controlled traction of the umbilical cord and uterine pressure, and inspection of the integrity of the placenta.

Regarding the activity-related risk factors, in the posture category and extreme body positions, with 248 references, it was possible to verify that in relation to the body movement there are several activities in which the NM remains in a static position, being the protection of the perineum at the moment of the output of the presentation and the traction of the umbilical cord and uterine massage, examples of activities in which the NM remains a few minutes in the same position. It should be noted that in one of the deliveries blood was drawn from the umbilical cord (stem cells) and that the NM remained static for 4 min.

Concerning the balance, with 147 recording units, we observed that the deliveries performed in the sitting position maintained the balance (support base, center of gravity and alignment) in all activities, while the deliveries performed in the orthostatic position (Fig. 2) presented a greater imbalance in the maintenance of the line and center of gravity in relation to the protection activities of the perineum and release of the anterior shoulder. In the change of the static position to dynamics, body alignment is not guaranteed in most cases, and the passage is made in imbalance.

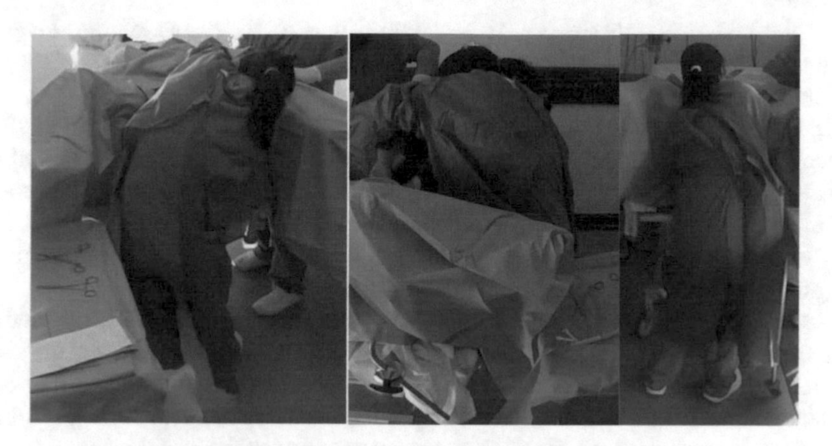

Fig. 2. Frames of the videos that reveal positions adopted by NM during the birth

It has been found that, in the occurrence of possible complications such as tight cervical circulars, and hypotonic newborns requiring resuscitation, NMs adopt attitudes where body alignment and balance have not been taken into account.

In a more careful observation of the videos, it was verified that there was only body misalignment of the NM that performs the seated delivery, in the orientation of another professional, evidenced by the unevenness of the shoulders caused by the lateral approach to the perineum, a consequence of the existence of two professionals in the field of action.

Regarding the deliveries performed in the orthostatic position, we verified the occurrence of a corporal misalignment at the moment of the activities of the perineum, release of the anterior shoulder and inspection of the placenta. It was verified that placenta inspection was not performed on the support table, but close to the field of action (near the egg-elimination container), providing 20º to 60º torso flexion and static position for a few minutes.

The moments where there was greater mechanical force exerted by the NM with 34 references were, at the moment of the protection of the perineum, having been realized a movement of pressure on the perineum, the liberation of the previous shoulder of the newborn, being the application of force in the sense of gravity to enable posterior shoulder release and controlled traction of the umbilical cord.

Regarding the subcategory Friction, we found 36 references that emphasize the facial expression performed to the newborn for release of secretions, protection of the perineum and uterine massage at the time of placenta delivery.

Regarding the Ergonomics category of the delivery unit, we observed 34 references, relative to the positioning of the support table/work in the field of action, there being reference in five deliveries to the position after the field of action of the same, forcing lateral rotation of the torso, flexion from 45º to 90º of the arm and <60º flexion of the forearm, to allow access to the equipment/material. The arrangement of the material on the worktable according to its priority was taken into account in nine of the deliveries.

In the bed-level subcategory, it was found that in six deliveries, the level was adequate and in four it was at a lower height, forcing the torso flexion at the time of the NM approach.

In relation to the environment category of the birthing unit that obtained 12 references, we found that noise was one of them with two references, agitation and cries of pain of the pregnant woman (without epidural analgesia by fast labour), and another due to the number of professionals in the room. With regard to luminosity, all deliveries had the pantof (light projection equipment specific for indirect light emission) designed for the field of action and natural luminosity.

Regarding the task activity category that obtained 24 references, we found that three of the deliveries were performed in the sitting position and seven in the orthostatic position. Reference was made to four situations of risk of complications that corresponded to cervical circulars and to the hypotonic newborn that had to be resuscitated. Regarding the parturient, we found that seven were cooperative and, in three situations, the women were agitated and non-cooperative.

With regard to the psychosocial and organizational risk factor, in the category of autonomy in the orientation of delivery, through video analysis we can verify that there are six references to interferences of other professionals during delivery, both in the delivery and in the execution of procedures - "(…) the doctor who is at the head of the pregnant woman, gives directions at the same time, taking opposite directions from the NM (now push/now breath)" (O7).

4 Discussion

The MIRW phenomenon in health professionals has been studied in the international sphere, but we have verified that in the production of evidence, the biomedical model focuses on the risk factors and the quantification of risk through the positivist research paradigm. Although this approach made it possible to unveil the problem and much has been gained from it, it should be noted that it does not cover the risk complexity nor the dynamic decision-making that professionals take about their appreciation and control.

Researches note that studies among midwifery practitioners are limited in the literature [12]. The literature review showed timid results emerging from other studies that used a qualitative approach to the study of the problem in focus [4, 6, 13, 14].

These research findings are in line with other studies that conclude that the nature of the activity requires that the professional spend long periods of time standing. Most of the time is spent with "body misalignment" that is evidenced by the flexion of the torso, anterior inclination to the head, upper limbs with anterior position to the torso and a base of very little support, which contributes to the decrease of balance and misalignment of the shoulders with the rest of the body [4, 6, 7].

This study highlights that the moment when the professional protects the perineum is in a static position that is maintained in time, and can reach 4 min, and despite not being in a loading position, it changes the principles of biomechanics, contributing to misalignment and incorrect posture in the standing position. Static positions held for a long time are per se a risk factor for injury that can be increased by other risks [12, 15].

Some authors consider balance the major biomechanical principle without which alignment, safety movement and proper posture can not be guaranteed [6, 16]. Body balance represents the state of firmness of position, maximizing function, with minimum effort and muscular work, in which there is a stability of forces that oppose each other [16].

The analysis of the videos and the options of the different professionals to perform the activity during the expulsive period leads to the observation that the professionals who performed it in the seated position maintained the balance (support base, centre of gravity and alignment) in all activities. However, the deliveries performed in an orthostatic position presented a greater imbalance in the maintenance of the line and centre of gravity in relation to the protection activities of the perineum and release of the anterior shoulder.

It should be noted that the expulsive period has specific risks associated with the task itself, such as sudden movements of pronosupination, to facilitate the newborn extraction, which are performed with the hands raised to the shoulders level, and with the shoulders in tension and elevated. These findings are in agreement with other studies [6, 17]. Even when in a sitting position, the professional is subject to postures and extreme positions associated with body movement, mechanical strength and friction.

This finding should be taken into account in the postgraduate and professional continued training, since it is a good preventive practice of MIRW and may even justify the use of simulated high-fidelity practice for the training of these professionals. Subsequently, the training associated with Ergonomic modifications have been associated with a decrease in the prevalence of MIRW [18].

The force applied at the moment of the expulsive period, which can also be called the load, is necessary for any movement, although there are many types of force that affect the biomechanical structure of movement from an ergonomic perspective. A biomechanical imbalance can occur when the internal force required is higher than the person's strength and this imbalance has a high potential to result in injury [12].

The same authors consider that in the genesis of the lesions there may be forced activities during labour, such as pushing, pulling, and moving parturients or heavy, harmful objects and awkward work postures, such as repetitive tasks, working in extreme and stressful body postures [12].

Nine out of ten participants have an effective workload of more than 35 h. Some studies report that there is an increased risk of injury with the number of working hours and the decrease in rest periods [15]. Therefore, one recommendation for prevention is the definition of break periods and implementation of work practice [18]. These two activities are not only beneficial to work ergonomics, but also allow for a better quality of work and productivity [18].

The results show that the non-organization of the workspace and the non-anticipation of the movements by the professional constitute a risk that increases in emergencies. This aspect corroborates the previous results, which indicate that in moments when there is a need to go to the support table to get material, the manual handling of the load is carried out with the object not being carried near the body, and the arms being distant decrease the strength to carry it. The strength in the mobilization of loads is supported essentially by the handgrip [6], which may justify the high prevalence of lesions of the upper limbs of the specialists [12].

We can not end the discussion without mentioning the potential of using the footage to understand the phenomenon under study. We share the vision of Garcez, Duarte and Eisenberg [9] by valuing the possibility of emotional detachment for the reflexive analysis of the material, with the possibility of reviewing or freezing the image, but if this aspect is important for the study, it is also for the professionals' training. The value of autoscopy in NM behaviour change, possibly using research-action methodologies, should be the subject of research.

5 Conclusions

The specificity of skilled nursing interventions during labour influences the posture adopted by the NM. The use of video footage enabled a detailed observation of the NMs approach during the second and third stages of labour, facilitating the analysis of the principles of biomechanics and other conditioning factors.

Body misalignment is evident when alternations of static to dynamic positions arise. The flexion of the torso, anterior inclination of the head, upper limbs with anterior position to the torso and a base of little support, contribute to body balance decrease. The choice of the professional to sit for the delivery is a safe practice to maintain body alignment, ensuring balance in all of its activities. Misalignment is only observed when the nurse performs the delivery sitting in the guidance of another professional, because the pedagogical orientation of the professional in training implies that the expert has simultaneously two focuses of attention.

In the orthostatic position adoption, there is no alteration of the body weight by the lower limbs, the centre of balance is outside the base of support and back flexion is more pronounced. The analysis of the videos allowed the identification of risk factors, but also revealed possibilities for prevention.

This study allows a better understanding of the risk of musculoskeletal injury and the use of biomechanical principles in midwives. The results also allow us to recommend the adequacy of spaces and equipment in the workplace. It is important to make specific training on MIRW prevention, included not only in school curricula but also in clinical work contexts, using simulated high-fidelity practice.

The limitations of the study relate to the intentionality of the choice of contexts and participants. We consider that in this research process, the methodology used with the use of video recording evidences their fundamental contribution for qualitative research, which we intend to develop and apply to improve the quality of life of professionals.

References

1. Thinkhamrop, W., Sawaengdee, K., Tangcharoensathien, V., Theerawit, T., Laohasiriwong, W., Saengsuwan, J., Hurst, C.P.: Burden of musculoskeletal disorders among registered nurses: Evidence from the Thai nurse cohort study. BMC Nurs. **21**(16), 68 (2017)
2. Carneiro, P., Braga, A.C., Barroso, M.: Enfermagem em contexto domiciliário- influência das condições de trabalho. Int. J. Work. Cond. **7**, 1–16 (2014)
3. Serranheira, F., Uva, A.S.: Frequência de lesões músculo-esqueléticas relacionadas com o trabalho e das lombalgias em enfermeiro(a)s: estudo. Autoridade para as Condições de Trabalho, Lisboa (2015)
4. Baixinho, C.L., Presado, M.H., Marques, F.M., Cardoso, M.: Biomechanical safety in the clinical practice of nurses specializing in maternal health and obstetrics. Rev Bras Promoç Saúde **29**(Supl), 36–43 (2016)
5. Jellad, A., Lajili, H., Boudokhane, S., Migaou, H., Maatallah, S., Frih, J.B.S.: Musculoskeletal disorders among Tunisian hospital staff: Prevalence and risk factors. Egypt. Rheumatol. **35**, 59–63 (2013)
6. Presado, M.H., Cardoso, M., Marques, F.M., Baixinho, C.L.: Posturas dos estudantes durante o trabalho de parto: análise de filmes de prática simulada. In: Atas 6º Congresso Ibero-Americano em Investigação Qualitativa, 2, pp. 488–497. CIAIQ, Portugal (2017)
7. Ganer, N.: Work related Musculoskeletal disorders among healthcare professional and their preventive measure: a report. IJSRSET **2**(4), 693–698 (2016)
8. Abeldu, J.K., Offei, E.B.: Musculoskeletal disorders among first-year Ghanaian students in a nursing college. Afr. Health Sci. **15**(2), 444–449 (2015)
9. Garcez, A., Duarte, R., Eisenberg, Z.: Produção e análise de vídeogravações em pesquisas qualitativas. Educação e Pesquisa **37**(2), 249–262 (2011)
10. Uchoa A.G.F., Godoi C.K., Mastella, A.S.: Análise Qualitativa de Material Texto-audiovisual: por uma metodologia integradora. In: Atas CIAIQ 2016, vol. 3, pp. 417–422. CIAIQ, Portugal (2016)
11. Flick, U.: Introducción a la investigación cualitativa. Morata, Madrid (2004)
12. Okuyucu, K.A., Jeve, Y., Doshani, A.: Work-related musculoskeletal injuries amongst obstetrics and gynaecology trainees in East Midland region of the UK. Arch. Gynecol. Obstet. **296**, 489–494 (2017)
13. Long, M.H., Bogossian, F.E., Johnston, V.: The prevalence of work - related neck, shoulder, and upper back musculoskeletal disorders among midwives, nurses, and physicians: a systematic review. Work. Health Saf. **61**, 223–229 (2013)
14. Long, M.H., Johnston, V., Bogossian, F.E.: Helping women but hurting ourselves? Neck and upper back musculoskeletal symptoms in a cohort of Australian Midwives. Midwifery **29**, 359–367 (2013)
15. Ladd, D., Henry, R.A.: Helping coworkers and helping the organization: the role of support perceptions, exchange ideology, and conscientiousness. J. Appl. Soc. Psychol. **30**(10), 2028–2049 (2000)
16. Potter, P.A., Perry, A.G., Storkert, P.A., Hall, A.M.: Fundamentos de Enfermagem, 8th edn. Elsevier Editora, Rio de Janeiro (2013)
17. Taghinejad, H., Azadi, A., Suhrabi, Z., Sayedinia, M.: Musculoskeletal disorders and their related risk factors among Iranian nurses. Biotech. Health Sci. **3**(1), e34473 (2016)
18. Nkhata, L.A., Louw, Q., Brink, Y., Mweshi, M.M.: Review on effects of ergonomic interventions for nurses on function, neuromuscular pain and quality of life. J. Prev. Rehabil. Med. **1**(2), 53–60 (2016)

Teacher Narratives on the Practice of Conflict Mediation

Elisabete Pinto da Costa[1(✉)] and Susana Sá[2]

[1] Researcher at the Research Interdisciplinary Research Centre for Education and Development (CeiD), Porto Lusófona University, Porto, Portugal
elisabete.pinto.costa@gmail.com
[2] Researcher at the Research Centre of Child Studies (CIEC), University of Minho, Braga, Portugal
susanaemiliasa@gmail.com

Abstract. In education policies in Portugal, teacher training must provide a response to the objective of "The Promotion of Schooling Success". In general, the teachers have revealed difficulties in the domains of interpersonal relations. The study is set in the context of the training of 140 teachers, from various education cycles. It deals with the theme of conflict mediation and was undertaken at 9 Portuguese schools. The purpose of our research was to gain a greater understanding of teachers' perceptions regarding the *praxis* of conflict mediation in the school context. A qualitative research study of several cases was carried out. The data was gathered from written narratives and critical reports drafted by teachers at the end of the training programme. Data analysis derived from the content analysis and was supported by webQDA® software. The results observed indicated that mediation is applicable and worthwhile; it is considered to generate positive effects even when it constitutes a challenge for the consolidation of acquired competencies.

Keywords: School mediation · Teacher training · Qualitative research

1 Introduction

The contemporary school has been confronted with countless social problems, which have urged the institution to reflect upon itself. Demands have been made on schools to perform different tasks and play new roles in order to address the plurality of social and educational responsibilities attributed to it, designated by author functions [18]. The complex challenge posed is that of a search for new and effective management formulae to be implemented in the school's social, relational and cultural space. This should be set within a logic of socialization, citizenship, inclusion and quality, in both the social and educational aspects.

However, the existence of dissension and the breakdown of relationships is revealed by the rates of indiscipline, conflicts and violence. These have led to the fragmentation of normal school procedures, so that the culture of discipline, which is supposed to exist in this context, has been questioned [22].

A. P. Costa et al. (Eds.): WCQR 2018, AISC 861, pp. 156–169, 2019.
https://doi.org/10.1007/978-3-030-01406-3_14

Therefore, the need to foster coexistence, through the positive management of relations and conflicts, has become a priority for schools. This social dimension has also become an educational, pedagogical and organizational issue [20]. This was highlighted in the Report of the International Commission on Education for the Twenty-First Century [8], which presents the learning of coexistence as one of the greatest challenges in Education for the contemporary world. Thus, "learning to live together" - namely by debating, disagreeing, placing oneself in the position of the other, generating consensus, defining and following common objectives, within a rationale of mutual respect, tolerance and cooperation – has been acknowledged as being one of the principal concerns of Education and, consequently, of Schools. In order to "learn to live together" we must learn and practice competencies associated to relationships and civic behaviour.

We must reflect on how schools, teachers and students are ready to work on the issue of coexistence, namely the aspects that are related to situations of conflict - current or imminent - which jeopardize the traditional models and procedures adopted. Peaceful coexistence occurs when conflicts are not repressed or ignored; yet, it is rather a question of dealing with situations of this nature by resorting to suitable skills, strategies and procedures, which are grounded on dialogue, collaboration and accountability.

1.1 Teacher Training and Conflict Mediation

Besides the upheavals generated in socialization, which can occur in the school area itself, the negative effects of this issue have been detected mostly in the context of teaching-learning relations. The problem has thus affected the performance of teacher and student roles, undermining the effectiveness of the classroom and, more generally, of the school itself [2].

The school thus plays an extremely important role in the fostering of conscious, critical and creative beings, who participate and are committed to the process of their own development. It is generally acknowledged that in addition to teaching, schools should also educate students holistically. The roles of social and relationship skills have become crucial in the education and training of youngsters. This perspective emerged in the changes which occurred in the curricular reorganization of Primary and Secondary Education (Decree-Law no 139/2012, dated 5th July). It is in this context that we conclude that a "smart curriculum school" must consider the constant changes which society is subjected to, as well as the adjustments required, when confronted with a heterogeneous population [12, 14].

The adaptation and enhancement of educational practice has become a necessity, if not a requirement, for the teacher who is an integral part of an ever-changing global society. In this sense, teacher training consists of a means of competency renewal, which will enable the teacher to work with all students according to their real potential. We will thus be able to address the objectives of inclusive education and the management of diversity in an improved manner.

When continuous training focuses on the school's problematic issues and its projects, which are able to value students and teachers' experiences, and articulates these with the life quality of those involved, then it has the potential to improve interpersonal relationships, as well as the teaching and learning process itself. This can be reflected both in better academic performance and also in greater satisfaction experienced by teachers.

Continuous training in this context is considered to constitute a strategy which links scientific knowledge to practice [26]. This is of particular relevance, since it enables a reflection on the different types of knowledge, techniques and attitudes of teachers [11], who are thereby willing to participate in a process that will allow for the creation of a new educational praxis.

Within the context of training, the life stories presented seek to reassert the importance of human reflexivity [5]. They also provide a greater understanding of the process through which people signify/justify their personal and training trajectories, or how they construct these meanings within the complex dynamics that occur between the individual and the social [7].

Continuous training in conflict mediation, within the school context, is set within a constructivist vision (personalistic, contractual, interactive and reflexive). This begins with a contextualized reflection in order to assemble the training features in context, placed in a framework of regulation and incremented work practices and processes [17].

One has seen an increase in the interest shown by the training centres of Portuguese schools with regard to continuous training programmes, designed by universities in order to empower teachers to become agents of change, as well as to promote social and educational quality. The training programmes operates in line with the methodology of practical or procedural know-how.

Teacher training in this area can be integrated into *broad scope* school mediation projects [20], or can be provided as a single training course, although one might then have to determine which structures and entities will be involved in mediation. Research has primarily focused on the effects and results of mediation programmes on students, and the impact of mediation projects on the school's social climate. The development of research on the training of teachers as mediators implies breaking new ground, although studies of a quantitative or mixed nature have been identified. These have collated teachers' perspectives and have primarily addressed the previously mentioned areas of importance. Examples of these are: [6, 13, 15, 19, 21, 23].

2 Methodology

The research objective consisted of a scrutiny of 9 different school contexts, 140 narratives and 140 critical reports drafted by teachers. As such, we opted for an analysis methodology which would reveal specific features, both of the communities themselves as well as of the opinions expressed by the respective teachers at the different schools studied. To this end, we chose to proceed with research of a qualitative and essentially exploratory nature, which would be applied to the study of multiple cases. In this particular study, one of the authors was also one of the researchers, as well as being the coordinator of the training programme and a member of the training team.

A substantial amount of qualitative research was dedicated to the natural environment of the study in question. One attempted to establish the closest contact possible with the participants [16], thus assuming that "the researcher's subjectivity and that of the subjects studied is inherent to the process of investigation [10], with the analysed data tending towards "an inductive approach" [4].

The general objective of this study consisted of identifying the learning curve and practice of the conflict mediation skills acquired and experienced by teachers. In order to proceed with the investigation, one established the following specific objectives: gaining an insight as to whether mediation can be applied to the school context; analysing the extent to which mediation can be articulated with the reality of schools; determining how teachers practice mediation; understanding the most evident aspects of mediation practice; establishing the effects and results of mediation practice and identifying the positive and negative features of mediation practice.

A problematic issue in our research study might be expressed as follows: after specialized training, what are the teachers' perceptions of mediation practice in the school context?

In order to address the specific nature of the objectives and of the subject of this research, we concluded that the most suitable approach would be a qualitative one, since "qualitative methods cannot be considered to be dissociated from the research process and the theme studied" [10, p. 19].

In the analysis of multiple cases, such as those presented here, we divided them into sub-cases of the holistic case [9]. These often helped to reinforce the discoveries made in the entire study, since the multiple cases in question were selected due to the fact that they replicated each case, contained deliberate and contrasting comparisons or presented variations based on hypotheses [28].

One will use the term *reference units* for units which comprise a sentence or group of words that make sense and are meaningful [1].

Our study is of a uniform design, and was implemented in 9 school contexts, with 140 teachers. The variables included are those of Activity Narratives (henceforth designated as N) and Reports of Critical Reflection (henceforth designated as R). We considered the following features pertaining to the participating subjects (teacher-mediator): Gender, Education Level taught and in which Region of the country. We also considered the following features for the subjects involved (mediated student): Age and Education Cycle (it would have been interesting to add the Statute feature but all of the mediated subjects were students, who were never peers or other participants of the education community).

Finally, it is pertinent to highlight the procedures considered to respect the ethical issues involved. Teachers were requested to provide two documents: (a) an Activity Narrative of a mediation experience in a real context (presented orally in the last training session and later sent as a written document to the coordinator of training/researcher); (b) a Report of Critical Reflection about the training course (submitted thirty days after the end of training). Both of these documents were sent online. We obtained the approval of the training centres of the schools, to which the teachers belonged, in order to use the data in scientific studies. The confidentiality of the information provided was also ensured. Thus, with the purpose of assuring the secrecy of the contexts studied and the anonymity of all of the participants, we proceeded with a coding system. The Narratives were designated N1 to N140, the Reports were identified as R1 to R140, the mediated students were asked to provide a fictitious name, and the schools were designated A1 to A9.

Training took place between the months of October and April, in the academic year 2017–2018, in the context of a Training Course. The workload ranged from 25 to 30 h of classroom training, and 25 to 30 h of self and individual work. The training courses were certified by the competent body. They were developed in partnership with centres dedicated to the continuous training of teachers and requested by schools. These training activities were set within a framework of school projects, which were developed with the purpose of pursuing the directives of the public policy established by the National Programme to Promote Schooling Success, presented in the Council of Ministers [24], Resolution no 23/2016, dated 24th March and included in Axis 4: Quality and innovation of the education and training system for the Operational Programme Human Capital (POCH). The 9 Training courses, which were undertaken in the central and northern regions of Portugal, were carried out by four trainers from the team of the Mediation Institute at *Universidade Lusófona do Porto*, who are experts in mediation and possess a great deal of experience in providing training for the different education cycles as well as for adults.

3 Data Treatment

We chose to favour the Narrative feature in the data collection technique. The methodology consisted, initially, of a separate analysis of the teachers' results. With the purpose of ensuring the comparison and triangulation of data, one subsequently and whenever possible, proceeded with a comparative analysis of the results obtained from the schools.

Bearing the study objectives, we sought to undertake a detailed and thorough description so as to ensure the validation and credibility of the qualitative study [1]. We considered the "need to establish some strategies. The most noteworthy of these is the triangulation of the various sources collected, namely, that of viewing the same phenomenon from different angles" [25, p. 9]. Furthermore, we opted to favour data triangulation, a method which ascertains whether the information gathered can be confirmed by another source (theoretical). The transparency inherent to the entire method was harnessed to ensure the merit, credibility and reliability of the investigation [27, p. 151].

3.1 The Analysis Matrix

By using webQDA® software, we were able to include the data (from the "Activity Narratives" and from the "Reports of Critical Reflection") in the internal sources. We examined the 20 most frequently used words, which were conditioned to a minimum of 7 letters. Those mentioned most often were: Training (880), Context (874), Mediation (861), Conflict (743), followed by Mediator (684) and Mediated (626), Colleagues (501) and Improvement (482). One then confirmed the dimension of the analysis: Training in Context, as well as the three vertexes on which the Training programme and objectives were based: Mediation, as a method of social and educational intervention; the transforming action of the Mediator, from the perspective of the Improvement of the subjects (Mediated) and of the Context, with the purpose of fostering an awareness of the dissemination by Colleagues-Teachers of training in the required skills, as well as the need for a greater involvement of Colleagues-students.

We were conscious of the characteristics inherent to each community, and of the reduced number of studies on the specific theme. It was thus important to create a matrix of internal coherence, which would allow for a homogeneous analysis and would be directed towards three objectives: (a) not lose sight of the research issues at hand; (b) enable the triangulation and comparison of the various data *corpora*; (c) allow for the comparison of contexts. The matrix which was then created proved to facilitate these objectives and is presented in Table 1.

Table 1. Internal coherence of investigation for the dimension "training in context"

Investigation issue	Investigation objectives	Data *Corpus*	Analysis types	Observations and expectations
What are the teachers' perceptions regarding mediation practice, after specialized training, in a school context?	Understand whether mediation can be applied to the school context; Verify the extent to which mediation is articulated with the school's reality; Establish how teachers practice mediation; Understand which are the most evident aspects of mediation practice; Identify the effects/results of mediation practice; Acknowledge the positive and negative aspects of mediation practice	Activity narratives Reports of Critical Reflection	Content analysis	One expects to be able to compare the data from the teachers' Narratives and to correlate these with the data from the Reports of Critical Reflection and the results achieved

One began with the "free-floating reading" [3], of the documents, in order to establish initial contact with the material. This was followed by further reading due to the wealth and extension of the analysed *corpus*. After this stage, the larger categories began to emerge inductively, in accordance with the pre-established objectives, as well as the theoretical framework and the results which ensued from the reading of the Narratives and Reports. We found 9 categories: Types of Conflicts; Mediator's Characteristics; Mediator's Skills; Dynamics of the Mediation Process; Results; Positive Aspects; Negative Aspects, Expectations; Proposals.

We then proceeded with the coding process, for which a category and subcategory tree was built. This is presented in Table 2.

Table 2. Categories and subcategories of the dimension of the training in context in conflict mediation

Dimension	Categories	Subcategories		
FORMATION IN CONTEXT IN CONFLICT MEDIATION	Types of Conflict	How they contacted with the conflict		
		Who proposed the resolution of the conflict		
		Which conflicts were mediated	Relationship	Physical attacks
				Verbal attacks
				Humiliation, offenses, defamation…
			Interests and needs	
			Normatives	
			Activities	
			Cultural	
	Mediator's Skills	Use of techniques	Active listening – (show interest, clarify, paraphrase, doing echo, summarize)	
			Questioning – (open, closed, circular, crossed, reflexive questions)	
			Mild language	
			Assertiveness	
			Communication regulator	
			Solution provider – (assist in the creation of options and choice of a mutually satisfactory solution)	
	Mediator's Characteristics	Neutrality		
		Impartiality		
		Competence		
		Confidentiality		
	Dynamics of the mediation process	Development of the pre-mediation	Create trust in mediation	
			Predisposition for collaboration	
			Willingness	
			Information on mediation	
			Information on the mediator's role	
			Rules:	
			confidentiality	
			impartiality	
			Consent	
			Attention to pre-mediation procedures	
			Attention to mediation procedures	
		Development of the mediation (investigate and manage the conflict)	Main Strategies	Positions, interests and needs
				Active listening
				Communication regulation
				Promoting thinking
				Promoting responsibility
			Techniques	FFNS method (facts, feelings, needs and solutions)
				Questions: Open, closed, crossed, circular
				Negotiation (proposing solutions to be analysed, Evaluation of the solutions)
				Acceptance of a solution (mutually satisfactory)
				Compromise (agreement)
				Follow up (assessment of the compromise fulfilment)
			Mediation models: resolutive / transformative	
	Results	Resolution / transformation of the conflict		
		Learning of relational skills		
		Improving coexistence		
	Positive Aspects	The most satisfaction as mediator and in mediation		
		The most satisfaction about those involved		
		satisfaction about the resolution of the conflict		
	Negative Aspects	Fragilities		
		Constraints		
	Expectations	Purposes of mediation (as school culture).		
	Proposals	Proposals for a better functioning of mediation		

3.2 Analysis and Discussion of Results

Once the previous steps were completed, we started to question the data to answer the research question, obtaining the respective matrix of questioning facilitated by the webQDA® software. We question the data regarding the number of references by category and by schools, whose results are presented in Table 3.

Table 3. References by category and school

	Type of Conflict	Mediator's Skills	Mediator's Characteristics	Dynamics of the mediation process	Results	Positive Aspects	Negative Aspects	Expectations	Proposals
A1	75	28	789	10	235	67	10	16	15
A2	21	5	432	9	90	9	10	17	14
A3	35	11	637	8	207	19	12	19	18
A4	12	7	235	16	73	7	6	18	19
A5	28	6	534	16	186	11	12	20	18
A6	18	7	371	10	71	10	9	14	14
A7	87	29	836	9	203	56	12	13	11
A8	92	26	943	15	306	71	15	20	18
A9	85	23	831	14	301	73	9	20	17

From the analysis of Table 3, we conclude that the category that was most referenced, by schools, both Narratives and in the Reports, was the Mediator's Characteristics, followed by the categories Results, Conflicts, Positive Aspects and Negative Aspects. These will be the categories that we will analyse in depth. This shows that teachers have focused on the new role in the management of interpersonal relationships and conflicts.

It should be noted that the Expectations and Proposals categories have only one reference (if any) for each number of trainees enrolled in the training by each school. We obtained a total of 140 trainees, 137 references in the Expectations category and 127 references in the Proposals category. In this way, it can be seen that many of the teachers were not very reflective about their future mission as mediators.

After analysing the results by category and schools, we proposed to carry out a more advanced analysis, since we chose to present the results of the questioning between categories considering all teachers and schools as a single group.

Questioning the category of Type of Conflicts according to the category Results (see Fig. 1), we verified that the mediator-teacher had direct contact with the conflict situation and it was the one who proposed the mediation process (100 reference units), attempting its resolution or transformation, as it can be seen from the following narratives: "I saw that the two students insulted each other in the classroom" (N67); "both students were fighting in the playground" (N3). It should be noted that when it came to conflicts in activities, teachers pointed out only 5 references on the resolution, for example: "they gave a handshake and went to the playground to play" (N16); "I was admired and suddenly they were already talking directly to each other, not needing me" (N42).

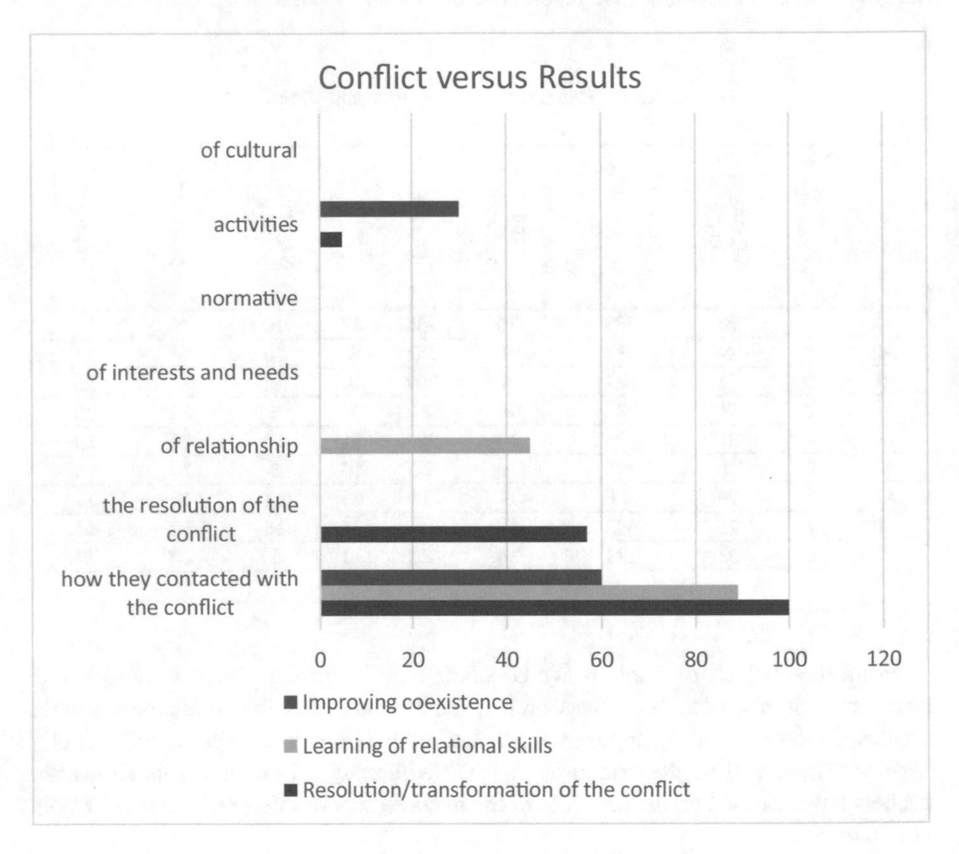

Fig. 1. Conflicts versus results

We also verified that the mediation had an effect in the improvement of the coexistence between: (a) mediated (60 reference units): "After appointing a new interview, they don't miss and with much more at ease between them" (N3); (b) mediators (30 reference units): "I feel I am more tolerant" (R57). From the data analysis it also arose a high interest in the learning of mediator skills: (a) from the mediated (89 reference units), "they were the first to ask to belong to the conflict mediation team at the school" (R34); (b) from mediators (45 reference units), "I am available to act as a mediator, yes…" (R20).

We analysed in detail the Type of Conflicts and we verified that most of them were Physical attacks (90 reference units), followed by Verbal attacks (31 reference units) and, finally, Humiliation (2 reference units), 17 referenced conflict situations were not identified by the teachers.

Analysing the situations of conflicts in all education cycles, the first cycle stands out (92 occurrences), where the teachers refer that they do not yet denote this type of episodes, generally confusing with violence and indiscipline. We also verified, in this case, the complete success in the resolution. Secondary education has less incidence of conflicts (7 occurrences) and are less successful in improving coexistence: "They kept talking to each other, but without that intimacy that had always united them for a long time. Now it's a cordial relationship." (R59).

When we questioned the data about the Mediator's Characteristics and Positive or Negative Aspects, we obtained Fig. 2:

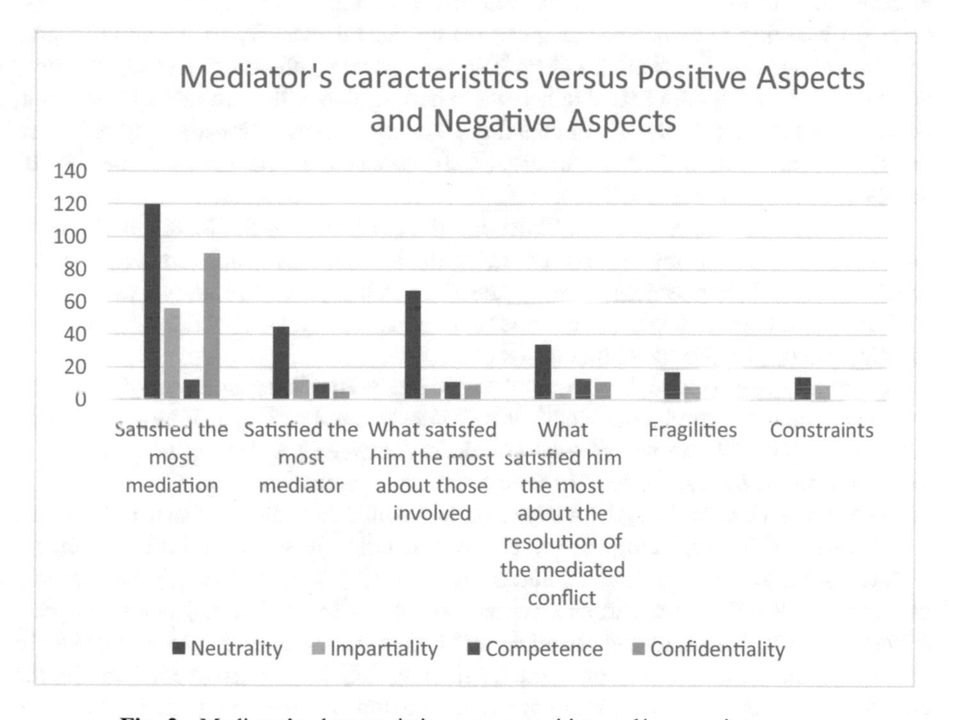

Fig. 2. Mediator's characteristics versus positive and/or negative aspects

Regarding the Mediator's Characteristics by Positive Aspects, the teachers were una-nimous in referring to Neutrality in the following areas: (a) satisfaction in mediation (120 reference units); (b) satisfaction as mediator (45 reference units); (c) satisfaction regarding the participation of those involved (67 reference units) and (d) satisfaction regarding the conflict resolution (34 reference units); we quote three illustrative examples: "being neutral helps the dialogue" (N15); "The use of neutrality facilitates mediation" (N102); "I think that being neutral in the process, the students trust us more" (R20). Confidentiality stands out as a Positive Aspect in Satisfaction in the Mediation (90 reference units), as can be seen from the following examples: "students feel more comfortable in mediation if they know that the mediator is a confidant" (N56), "my students felt more relieved when they knew that the subject would remain within those walls" (R31).

As for the Mediator's Characteristics for Negative Aspects, "Neutrality" stands out, again, in terms of Fragilities (24 reference units) and Constraints (12 reference units). Examples of fragilities as for: (a) separation of roles; (b) to follow the strategies of the mediation process, can be observed in the following narratives: "What I found most difficult in the mediation process was being class leader of my students and having to put myself in another role" (R83); "I was so obsessed with the stages of mediation that I forgot to draw up the contract of agreement" (R71); "Will I remember to be impartial in the future?" (R92). Impartiality was also pointed out among the Fragilities (8 reference units), to the extent that the action of the mediator-teacher can be under pressure: (a) knowing the students' realities; (b) by the inevitability of the adult to give solutions: "But don't you think you're being a recurrent offender?" (N137); "I know you both…" (N1); "I admit I had to help them find a solution that would suit them both, because I know them" (R104). Impartiality also emerged as a Constraint (9 reference units), for example, with this narrative: "My concern is that I might one day be involved in the problem" (R35).

With regard to the Mediator's Skills and the Dynamics of the Mediation Process, we have verified that the teachers fit the mediation strategies and techniques in the resolution model. For example: "After attending training and learning to use conflict mediation strategies in context, I became more attentive and available, at school and in the classroom, to solve the conflicts" (R17).

Other results of our questioning among the analysed categories revealed that there was no correlation between the Mediator's Gender in the Results; Positive Aspects; and Negative Aspects. There was also no correlation between the Age of the Mediated in relation to the Mediation's Results, within the same education cycle.

In terms of Expectations, the analysis of data collected from 140 teachers showed a unanimous position regarding the practice of mediation in school, in order to: (a) facilitate teacher-student and student-student dialogue; (b) improve teaching and learning in a more harmonious way, in an environment of understanding and mutual respect; (c) contribute to reduce school abandonment and underachievement, as it is shown in this particular narrative: "When there is no conflict, the environment both in the classroom and outside the classroom would be much more harmonious providing openness for the implementation of other activities in the teaching-learning context." (R69). However, we have identified narratives that demonstrate that cultural and professional change is a structuring challenge: "There are variables that we have to remove from the school environment, such as indiscipline, so that we can work on other variables, such as underachievement, abandonment, and compliance with programs, because that was what we prepared ourselves for pedagogically" (R82). That is, the idea remains that conflict must be repressed or avoided, when conflicts are inviting to interpersonal relations and it is important to take advantage of them, as opportunities, to learn relational skills and growth.

The proposals pointed out by the teachers in relation to mediation followed two directions: at the macro level: (i) by improving the functioning of mediation at the school; (ii) by raising the awareness of the educational community towards the creation of mediation offices; (iii) by specialized training in mediation by all teachers; and at meso level: by adopting educational policies consistent with mediation. In this regard, we refer to the following narratives: "It would be vital to have colleagues, with specific training in conflict mediation, exempt from our classes, members of a school office for this purpose, in order to provide neutral and effective mediation, supporting us in the pedagogical background." (R47); "A school mediation office, derived from government policies, would be an added value to support teachers in their day-to-day teaching-learning process" (R90).

4 Conclusions

From the results analysis, and to answer our research question: "What is the teachers' perception about the practice of mediation, after a specialized training, in a school context?", we registered a positive impact of the mediation training process in the personal and professional development.

We verified that mediation has applicability in the school context, from the first cycle up to secondary education, allowing the teacher to assume a central role in the coexistence. In fact, this was a unanimous reference in the teachers' proposals in the critical reports, in which they expressed two main needs: (1) all teachers having training in conflict mediation; (2) the existence of a mediation office in the school that allows the management of interpersonal relations and conflicts, in order to foster a harmonious environment in the school premises. In this way, it is confirmed that mediation can have articulation with the reality of teachers and students in the school context.

Teachers' learning of the principles, strategies and techniques for conflict mediation was revealed in their application in the experiential context. Realizing this allows the valuation of training in context, where the practical part is essential. We believe that formations with only theoretical part will not be as successful in the teachers' personal and professional development. This is also supported by the way in which teachers focus on the *modus operandis* (indicated in the Mediator's Skills category), with special emphasis on the management of feelings of the mediator-teacher indicated in the Mediator's Characteristics Category - Neutrality/Impartiality). We recognize that teachers intend to find solutions to the situations of rupture and disruption, maintaining a remedial perspective of the mediation. It is important to consolidate learning, providing training with more workload, in order to consolidate the transforming perspective of mediation, which calls for the empowerment among individuals.

The most evident element of the practice of mediation was, undoubtedly, Neutrality. This characteristic/technique was positively highlighted, since it was one that: (a) satisfied most the teacher as mediator; (b) provided a satisfaction for the participation of the mediated; and (c) was more valued in the results. Here lies the potential of education, training and self-moralization of the methodology of mediation, replacing the classic strategies, still in force and generalized, of hetero-moralization, focused on the teacher figure, who feels he is not heard and effective with this last type of approach. Therefore, the new strategy has also generated more Constraints and more Fragilities among teachers, because, despite the training, they recognized that this strategy is difficult to apply due to old habits, the tasks that increase consecutively, the traditional pedagogical role, still attached to the instruction paradigm, and to a distant relation to social issues that take a lot of time and disturb the teacher-student relationships.

In general, we can conclude that the formation in mediation in context has been successful, although it is challenging to consolidate the learning of the conflict mediation praxis, which will sustain the culture of mediation within the educational community.

References

1. Amado, J., Costa, A.P., Crusoé, N.: Manual de Investigação Qualitativa em Educação. Imprensa da Universidade de Coimbra, Coimbra (2013)
2. Amado, J., Freire, I.: A(s) Indisciplina(s) na Escola - Compreender para Prevenir. Almedina, Coimbra (2009)
3. Bardin, L.: Análise de Conteúdo, 70. Edições, Lisboa (2015)
4. Bogdan, R., Biklen, S.: Investigação Qualitativa em Educação. Uma introdução à teoria e aos métodos. Porto Editora, Porto (1994)
5. Bolívar, A.: "De nobis ipsis silemus?": Epistemología de la investigación biográfico-narrativa en educación. Revista Electrónica de Investigación Educativa **4**(1) (2002). http://redie.uabc.uabc.mx/vol4no1/contenido-bolivar.html. Accessed 18 May 2018
6. Boqué, M.C., García, L.: Evaluación diferida de la formación del profesorado en convivencia y mediación. REIFOP **13**(3), 87–94 (2010)
7. Delory-Momberger, C.: A condição biográfica: ensaios sobre a narrativa de si na modernidade avançada. EDUFRN, Natal-RN (2012)
8. Delors, J. (Coord.): Educação. Um tesouro a descobrir. Relatório da Comissão Internacional sobre Educação para o século XXI. Asa, Porto (1996)
9. Duarte, J.B.: Estudos de caso em educação. Revista Lusófona de Educação **11**, 113–132 (2008)
10. Flick, U.: Métodos Qualitativos na Investigação Científica. Monitor, Lisboa (2005)
11. Formosinho, J.: Modelos Organizacionais de Formação Contínua de Professores. In: Tavares, J., Moreira, J., Oliveira, R. (eds.) Formação Contínua de Professores: realidades e perspectivas, pp. 237–257. Universidade de Aveiro, Aveiro (1991)
12. Leite, C.: Para uma escola curricularmente inteligente. Edições ASA, Porto (2003)
13. Ibarrola-García, S., Iriarte Redín, C.: La influencia positiva de la mediación escolar en la mejora de la calidad docente e institucional: percepciones del profesor mediador. Profesorado. Revista de Currículum y Formación de Profesorado **17**(1), 367–384 (2013)
14. Martins, F., Leite, C.: O Currículo Nacional do Ensino Básico e as Orientações Curriculares de Geografia: representações dos autores e (re)interpretações dos professores. Indagatio Didactica **3**(1), 80–94 (2011)
15. Martins, A., Machado, C., Furlanetto, E.: Mediação de Conflitos em Escolas: Entre Normas e Percepções docentes. Cadernos de Pesquisa **46**(161), 566–592 (2016)
16. Merriam, S.B.: Qualitative Research and Case Study Applications in Education. Jossey-Bass, San Francisco (1998)
17. Nóvoa, A.: Concepções e práticas de formação contínua de professores. In: J. Tavares, J. Moreira R. Oliveira (eds.) Formação Contínua de Professores: realidades e perspectivas, pp. 15–38. Universidade de Aveiro, Aveiro (1991)
18. Nóvoa. A.: Evidentemente: Histórias da educação. Edições Asa, Porto (2005)

19. Paula, A., Durante, V., Fantacini, R.: A importância do papel do professor mediador diante dos conflitos no cotidiano escolar. Educação, Batatais **6**(1), 53–68 (2016)
20. Pinto da Costa, E.: Mediação de Conflitos: Construção de um Projeto de Melhoria de Escola. PHD, ULHT, Lisboa, Portugal (2016)
21. Pulido, R., Fajardo, T., Pleguezuelos, L., Gregorio, R.: La mediación escolar en la comunidad de Madrid: Análisis del impacto de la formación en el profesorado y alumnado en el IES "Las Américas" de Parla. Revista de Mediación **3**(6), 32–43 (2010)
22. Quaresma, L.: Interação e indisciplina na escola. In: Abrantes, P. (org.) Tendências e controvérsias em sociologia da educação, pp. 159–171. Mundos Sociais, Lisboa (2010)
23. Quinquiolo, N.: O papel do professor como mediador de conflitos entre crianças da educação infantil. Revista Ciências Humanas - Educação e Desenvolvimento Humano - UNITAU **10**(1, 18), 116–125 (2017)
24. Resolução do Conselho de Ministros n.º 23/2016, de 24 de março
25. Sá, S.O., Costa, A.P.: Critérios de Qualidade de um Estudo Qualitativo (Carta Editorial). Revista Eixo **5**(3), 9–12 (2016)
26. Silva, A.: A formação contínua de professores: uma reflexão sobre as práticas e as práticas de reflexão em formação. Educação & Sociedade **21**(72), 89–109 (2000)
27. Souza, N., de Costa, A.P., Souza, F. de: Desafio e inovação do estudo de caso com apoio das tecnologias. In: de Souza, F., de Souza, D., Costa, A.P. (orgs.) Investigação Qualitativa: Inovação, Dilemas e Desafios, vol. 2, pp. 143–162. Oliveira de Azeméis, Aveiro, Ludomedia (2015)
28. Yin, R.K. Case Studies Research: Design and Methods. 3ª ed. Sage, Thousands Oaks (2003)

Millennials' Representations Regarding Cohabitation: A Qualitative Exploratory Study

Mariana Silva, Gonçalo Reis, and Catarina Brandão(✉)

Faculty of Psychology and Education Sciences of the University of Porto, Porto, Portugal
mariana96dtms@gmail.com, goncalomanuelreis@gmail.com, catarina@fpce.up.pt

Abstract. This study focus on the representations of Millennials regarding cohabitation, the challenges associated to the process of transitioning to live with a partner and the strategies to use in order to manage cohabitation successfully. A qualitative exploratory study was conducted using an online questionnaire and a semi-structured interview with a couple cohabitating. Data was analyzed using qualitative thematic analysis with the support of NVivo11. Results suggest the importance of communication and planning before transitioning to cohabitation, as well as ensure space for oneself. Workload, financial management, coping with the routine and housework are anticipated as challenges. The couple's experience is coherent with many of the representations regarding cohabitation. Curiously enough, social support was absent in the questionnaire data, but is experienced as fundamental by the couple. Knowing what Millennials think regarding cohabitation allows understanding how they approach this new chapter in life.

Keywords: Cohabitation · Transition · Millennials · Exploratory research

1 Introduction

1.1 From Dating to Living Together

Since the second half of the 19th century, social and institutional norms have faded, allowing individuals to redefine how to proceed from dating to permanent commitments [1]. Living together without marrying has become a common trend [2, 3] and the rate of cohabitation is increasing [4], which indicates that it is viewed as a normative step in relationships. This union is short-lived, lasting, usually, two years [5] and, when the third-year mark is reached, it tends to culminate in either dissolution or marriage.

Several studies (e.g., [6–8]) mention a behavior called stayover (also known as visiting or part-time cohabitation), during which couples spend time together at the house of one of the partners, while maintaining separate residences. Stayover is seen as precursor to cohabitation, since individuals often cannot recall the precise moment they stopped overstaying and started cohabiting. The couple stays together on and off, and therefore feels as if they are living together anyway, and does not go through a decision-making process; on the opposite, cohabitation "just happens" [9].

A. P. Costa et al. (Eds.): WCQR 2018, AISC 861, pp. 170–185, 2019.
https://doi.org/10.1007/978-3-030-01406-3_15

Motivations to cohabit have a relevant impact on the type of relationship that will develop [9]. Cohabitors emphasize emotional («'love', 'caring', 'companionship' and 'sharing'», p. 204) and pragmatic motives, such as economic advantages (sharing a house and household items or expenses, therefore saving money) and experimenting marriage (as a trial or to test the relationship), to deal with some insecurities, and to try a joint bank account [9].

Cohabitation, being complex and diverse, can be viewed negatively. Ambiguity can be a consequence of lack of agreement between partners regarding their role or relationship [10–13]. That can also lead to an increased difficulty in developing clear and shared meanings about the relationship [12]. The consequence is poor relational quality and adjustment, and high relational instability [4, 12, 14, 15].

According to Rhoades, Stanley and Markman [16], couples who live together to test the relationship have more negative relationships and communication; low adjustment, confidence and dedication; high insecurity; and a great risk of depression and anxiety. Couples who live together for external motives: economic – polled finances – or for convenience – having all items in one house [17] and to test the relationship – have poor relationship quality; low commitment and satisfaction; and high ambivalence and conflict [18]. Internally based commitment is related to dedication, positive attributes about the partner and the relationship, and living together to spend time together and experience intimacy, which leads to positive relationship quality [13, 17]. Satisfaction with relational sacrifices affects the response of partners to conflict [18]. When a relational sacrifice for the other partner or the relationship is made, that brings benefits for the relationship [19, 20]. Relational sacrifices lead to couple satisfaction, functioning and commitment [21–23]. Couples who cohabit to spend more time together, may be more willing to sacrifice for the relationship [13, 17, 22].

Housework is one of the challenges of cohabitation. Women traditionally do routine tasks, less pleasant and more time-consuming (e.g., cooking) and men are typically in charge of occasional household tasks, more flexible and enjoyable (e.g., caring for the yard) [24]. Yet, married couples that have cohabited, show an equivalent division of housework and more egalitarian expectations and experiences [24].

Financial affairs are a key factor in young adults' relationships [25]. Cohabitation reflects a lower investment on the part of couples, when compared to getting married. This lower investment may explain an also lower financial commitment [26]. For example, cohabitors are less likely to share finances, have a joint bank account or a residence in both names [27–29]. Financial affairs are a predictor of conflict and might reduce intimacy [25] and lead to dissolution [30–34]. Routines are also a concern. The higher levels of freedom in cohabitors' lifestyles (concerning the role and what is expected from each partner) may explain the ambiguity they end up feeling in their routine [35]. Workload and the conciliation between personal life, work and the routine, is also a challenge. Nonstandard work schedules are more disruptive in the early that in later year of marriage, since it is important for the couple to establish a strong bond, before being separate during the days [36]. The physical demands of working nonstandard shifts might exacerbate the psychological stress caused by those schedules and increase chances of separation [37].

Although the tendency is to stress the negative aspects of cohabitation [12], this way of living is increasingly common [4, 15]. Besides, since during emerging adulthood individuals cohabit multiple times [38], they learn more about relationships and that persists throughout their life [25]. Hence, couples should be encouraged to evaluate their motivations to cohabit. If the answer is 'to spend more time together', that will benefit their relationships [18].

1.2 Millennials and Relationships

Today's emerging adults belong to the millennial generation, to which it is crucial to pay attention. The Millennial cohort includes all individuals born between 1981 and 2000, approximately [39]. Most Millennials think it does not make a difference whether a person is married or single: for example, 67% of Millennials say that happiness is not related to whether you are single or married, and 61% say that social status is irrelevant to marital status [40]. Moreover, Millennials are more likely than older generations to say that it is easier for a single person to get ahead in a career [40]. According to Shields-Dutton [41], Millennials are the commanders of an "ever changing society", where marriage is no longer the only suitable step to form a family and cohabitation is an increasingly common option.

Millennials have very specific opinions regarding cohabitation or love. From 1976 to 2008, the proportion of adolescents who agreed that premarital cohabitation was a good testing ground for marriage increased by 75%, and data collected from 2011 to 2013, point that 64% of men and women agreed that living together before marriage could help to prevent divorce [42, 43]. One in four Millennials, who had ever cohabited, cohabited more than once: the instability of cohabiting unions increased, as 30% of Baby Boomers in contrast to 46% of Millennials dissolved their first cohabiting union. Overall, one quarter of Millennials will probably cohabitate, in contrast to only 16% of the Baby Boomer cohort [44].

It appears that relationships have become far more unstable for Millennials than they were for Baby Boomers or Generation X [43]. Even though young adults pursue romantic relationships that are more serious and longer lasting than those during adolescence, they are not particularly committing to marriage [45]. They are keenly aware of the values reflected by their parents, but have their own ideas and practices [46]. Millennials are a "Me Generation", since they avoid giving up their freedom and independence by dictating their lives to a romantic partner. They seek "openness and adventure and exploration and passion" and avoid "a relationship that in any way limits [them]", as stated by Riley, a Millennial who was interviewed by McGuire [46]. Young adulthood paths have become less clear and more diverse, and while these changes in roles and pathways may seem a benefit to young adults attempting to find their way, it ultimately may induce uncertainty and vulnerability [47].

1.3 Goal of the Study

Considering, on the one hand, the lack of literature regarding the process of transitioning to cohabitation and, on the other hand, the fact that it seems that this will be increasingly a trend, the goal of this study is to explore and identify Millennials' representations regarding cohabitation with a romantic partner.

2 Method

We developed a qualitative exploratory research in order to explore (1) the challenges that Millennials associate to cohabitation; (2) how Millennials would prepare the process of transitioning to cohabitation; and (3) the strategies they believe to have a positive impact in transitioning to cohabitation.

2.1 Participants

Data was gathered from 30 individuals who had a romantic partner but did not cohabit, 18 being women (60%) and 12 men (40%), with a mean age of 22.56, ranging between 20 and 27 years (individuals born from 1991 to 1998). The majority were students (73.3%), one was an intern (3.3%) and the rest were workers (23.3%). Participants were in a relationship for an average of 3.71 years, ranging from 10 months to 12 years. Seventy-five percent of the participants (n = 25) have considered the possibility of living with a partner in the future, 25% might consider that possibility (n = 5) and no one said they had not considered it. Regarding having shared a residence with someone other than relatives for at least six months, 66.6% (n = 20) said no, 33.3% (n = 10) responded affirmatively, all with classmates and two of whom with a romantic partner also. P (number) represents participants in the questionnaire.

In order to deepen some of the results obtained, semi-structured interviews were conducted with a heterosexual couple selected by convenience. At the time of the interview the couple cohabitate for eight months, after dating for four years. Both are 22 years old, working students, and taking a master - the woman is a senior; the man is in the fourth year. None of them has shared before a home except with relatives, and their residence did not belong to neither of them, previously. The man is represented by R; the woman by A.

2.2 Data Gathering Instruments and Procedures

We developed a questionnaire, which presented demographic questions and three open questions (e.g., what challenges do you think you might encounter while living with a romantic partner?). The questionnaire was subject to a pilot study with an individual belonging to the study's population - a man with 22 years old, presently in a romantic relationship. The questionnaire was then made available at Google Forms® and the link was sent to individuals belonging to the Millennial generation. It was available during two weeks.

A semi-structured interview was also developed. The script presented three open questions (e.g., Tell me about the challenges you experienced during the transition to cohabitation) and was developed considering the study's research objectives. The interviews were held individually with each member of the couple; both at the participant's house. They were audio recorded and participants signed an informed consent.

2.3 Data Analysis

Our corpus of analysis is the questionnaires' open-ended answers and the interviews transcripts, which was subject to qualitative thematic analysis according to Braun and Clarke [48], to whom thematic analysis "is a method for identifying, analyzing and reporting patterns (themes) within data." (p. 79), thus been adequate to our research objectives. We conducted an inductive analysis and identified themes at the semantic or explicit level.

Regarding the process itself, in a first phase we familiarize yourselves with the all data, transcribing the interviews, and validating them with the interviewees. The data from the questionnaires was imported to Excel and then to NVivo11 (QSR), as well as the interviews' transcripts. Once the data was in NVivo, we read and reread it, noting down initial ideas and organizing these ideas in to memos. This allowed to generate our initial codes (second phase), using *Code* and *In Vivo Coding* features. We then collated these codes into potential themes, gathering all data relevant to each (third phase). Next, the themes were reviewed (fourth phase) using the *Merge* feature, and we produced thematic maps (*Project map* in NVivo) of the analysis in each of our research objectives (see Fig. 1 for an example). The codding was conducted by the senior researcher, and discussed with the two other researchers. In the fifth phase of the process we defined and named the themes, making them more robust and meaningful for each research objective. Then, in the sixth and final phase of the analysis, we produced the report of our analysis, identifying meaningful extracts and relating the analysis with the literature on the topic.

The process of data analysis was guided by the study's goals and we look at identifying themes in each of the questions presented to participants. In addition, we gave a definition to each code, so the researchers could return to the data at any given moment and understand the meaning of all codes.

3 Results Presentation and Discussion

Next, the results regarding each research objective are presented and discussed. We first present the data coming from the questionnaires; then we consider the experience of the interviewed couple, in a process where we move from description to interpretation.

3.1 Challenges That Millennials Associate to Cohabitation

The main challenges anticipated by the Millennials in our study are - how to manage the dynamic as a couple; how to reconcile each other's resources; and ways of being (see Fig. 1).

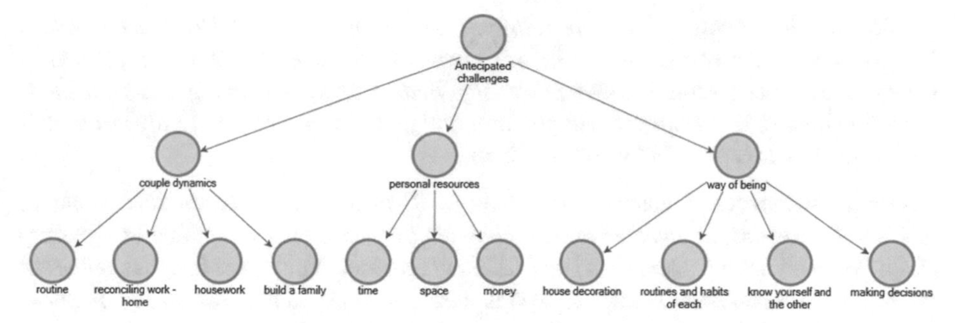

Fig. 1. Themes regarding challenges of cohabitation

Couple's Dynamics. The challenges related to the dynamic of the couple include dividing the housework («*Discussion on division of household tasks*», P8), and the couple avoiding falling into routine («*Do not fall into the rut*», P29). There is also the theme of work-life balance («*I think the main challenge is to reconcile personal life and work life in a harmonious way*», P20), which seems close to starting a family («*to raise a family*», P16).

Managing Personal Resources. Regarding the theme – managing the resources of each member of the couple, three sub-themes were identified. The first regards finances («*match each one's way of managing expenses*», P19). The second subtheme is time management («*spend time with the families of origin*», P9) and the third refers to space, specifically how to manage sharing with someone the same physical space («*we share the same space*», P28).

Way of Being. Another theme in the data is the challenge of being with someone that is different. Specifically, the idea that getting to know each other better, due to the cohabitation, will be challenging, as well as reconciling routines and habits (e.g., regarding food) and each other's personalities («*live with the differences between us*», P18). The need to make decisions together also emerged, enhancing the importance of good communication in the couple.

Regarding the interview data, two themes were identified. The interviewees struggled (and sometimes still do) with their dynamic as a couple (e.g., routines, housework, reconciling work and home), and personal resources (finances).

Couple's Dynamics. With regard to Routines, R. states that they are «*still trying to implement a (...) fixed routine*». They were dating for four years, so they already «*had a certain routine*», that had to be broken «*to start a new one*». A. considers that it is hard «*until you assimilate all the routines*». Even though they had one, they «*had never lived under the same roof*», and «*it's always challenging (...) to find our own space*».

Reconciling work and home - R. is concerned with time conciliation, «*I am either at college or working and A. is also either at college or working, so we often meet (...) at the end of the day or (...) to have dinner*» and «*try to divide tasks and reconcile with our workload*».

For R. Housework is a «*new reality*», which «*many times [lead to] conflicts because I hadn't cleaned, or she hadn't cleaned*». A. says that the idea of "help", always caused her discomfort «*When you say 'help', you are assuming that someone is actually doing it, and someone will go there and just give a hand, and I always told R. 'When we live together 'help' won't exist'*».

Personal Resources. Finances - according to R. now «*it's harder to save money... like we did [before]... whether for holidays or for weekends away*», due to «*grocery shopping*» and having «*lunch (...) and [dinner] outside*». For A. the fact that neither of them has «*a comfortable wage*», leaves her «*anxious (...) if something happens unexpectedly*».

Time management is hampered by Millennials workweeks, which include evenings and weekends and take much of Millennials' energy. According to Claps [49] this leaves Millennials little energy to focus on their relationships. Dedicating time to starting a family might indeed be a challenge. Dating and maintaining a romantic relationship requires focus and effort, which can be challenging particularly for this population [49], since they are trying to establish their career.

Dividing the housework is also a challenge in cohabitation, which our participants acknowledge and the couple experienced. Nonetheless, couples who cohabit do have a more balanced division of it, in comparison with married couples that never cohabited [24].

We identified in the data from the questionnaires that having to share the same physical space with a partner is seen as a challenge. This may be related to the ambiguity that cohabitors frequently experience in relation to their specific role in the relation in their everyday life and routines [35]. Therefore, if the role of each is not well defined, all the dynamics involved in sharing a space may be compromised. Also, this result may be related with the fact that Millennials are described as being impatient, subject to various pressures and privileged [50], which may turn more difficult sharing a common space. This is coherent with the fact that spending more time together and sharing a space was associated by some of the participants in the questionnaire with the possibility of experiencing more conflicts («*Becoming saturated with being with this person, because we spent more time together than before*», P27), as if distance played a protective role regarding conflicts in the couple.

Let us now consider the challenge of reconciling each other's financial resources and how these resources are used. This was, as seen, a challenged faced by the interviewed couple. These results emphasize the importance of educators supporting college students regarding how to manage their finances in order to avoid debt or financial instability [51]. The discomfort felt in A.'s discourse may also be explained by the discouragement that Millennials felt during their first attempts at financial independence, which happened during global recession [52].

The couple associated the experienced challenges to conflicts having emotional and cognitive impact: «*there's always some conflicts*» which leads to being «*a bit more depressed*». According to R. this sometimes makes them question whether their decision was the right one («*many times even think if (...) it was too soon (...) maybe later would have been better? We would have been more mature to assume a residence together*»). A. also mentioned «*periods in which you feel very angry, very frustrated*» and «*moments of tension*». To note that, according to A., this did not always happen – «*in the beginning (...) it was such a positive thing, that we forgot about negative aspects*». This suggests that it is important to prepare romantic couples before transitioning to cohabitation, in order for them to develop competencies, which will allow them to handle the specificities of sharing a house with a partner.

3.2 How Millennials Would Prepare the Process of Transitioning to Cohabitation

Here we want to understand how Millennials would prepare the process of transitioning to cohabitation with a romantic partner. The themes identified regarding this are - Planning; Securing several aspects of the relationship; and Communicating. Stayover is also mentioned as a way of preparing the process (in the form of a training), as well as being empathic regarding the partner along the process, making an effort to understand each other's difficulties and fears (see Fig. 2).

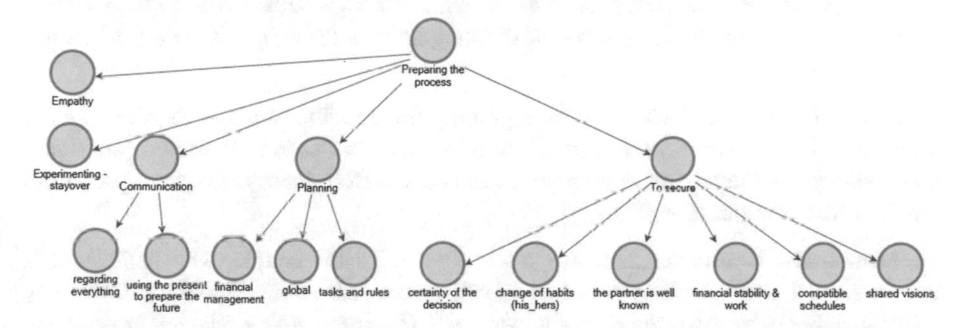

Fig. 2. Themes regarding how Millennials would prepare the transition to cohabitation

Planning. Planning seems to be understood as something important, even if some participants do not specify exactly what they would plan, talking globally. Others are specific – they would plan financial issues («*plan what each will pay and leave a margin for financial contingencies*» P9) and the tasks and rules to follow («*Make a plan of action, distribution of tasks*», P10).

Securing. Participants would try to secure several dimensions before transitioning. These include feeling secure regarding the step they are going to take («*to make sure it was the best thing for both*», P24), that they really know the person with whom they are going to share a house with («*understand habits and opinions in relation to routines, use of money, chores and couple dynamics*», P7), that each member of the couple has his own stability («*there is no financially dependent member*», P20). There is also reference to assuring that their schedules are compatible and own habits are changed, in favor of the couple («*change some more irregular routines*», P4), and that the members of the couple share visions regarding life and the future («*agreement on the main values and expectations*», P20).

Communicate. The theme Communication emerges autonomously in the data and also associated with Planning and Securing (presented above), in the sense that talking with the partner would allow for a better planning and for securing dimensions such as knowing each other better («*Debate on more sensitive issues*», P13). When focused autonomously, participants referred to communicating regarding things in a global and general way, but also using the present to prepare the future. This means talking about how they behave and manage things in the present moment of their relationship, as a way to prepare living together, discussing how events should be dealt with once the couple starts cohabiting («*Talk about events that happen now and that can serve as examples in the future on this issue*», P21).

The couple interviewed presented several suggestions they believe would be important for those who are planning transition to cohabitation - Planning; Securing some aspects; Communicating; Being empathic; and to Relativize.

Planning. The couple considers planning an important strategy if people think on moving to cohabitation. They should talk on how things will be when they start living together («*and we often talked of "Oh, when..." and we talked about it*», A.). This suggests, once again, the importance of thinking on how the couple will do things once cohabitating.

Securing. Here we find being secure regarding the decision to move together («*to be sure... that's what you want to do*», R) and that the partner is well known («*and that you really know the person next to you... ok?*», R.). Both these ideas were present in the data from the questionnaires.

Communicate. Communication was also a theme for the couple, who believes it is important that couples be prepared and available to talk about everything once they move together («*how was the day, if it went well, if... if something relevant happened*», R.), in a sincere and open way («*say "Look, I did not like this, I did not like that*"», A.).

Empathy. It is important to be able to put oneself in the other's shoes («*we must also know how to accept*», A.) and learn how to **Relativize**. The couple considers it is important to develop the ability to step aside from situations or things that may upset you at a given moment («*and look with a cooler look at things*», A.), avoiding impulsive reactions that can harm the couple.

According to William Andrews Tipper [53], the financials decisions made by Millennials have a high degree of caution and are based on a strong association between finance and risk, which explains the emergence of planning finances as a theme in the data from the questionnaires. Despite the absence of this theme in the couple's interviews, we did find the reference to the importance of planning how things will be once the couple cohabits.

Security emerged in our data, as found in Carmichael's and Whittaker's study [9], where some participants stressed their need to feel safe regarding the transition and, therefore, started cohabitating with the goal of testing how a marriage would be - they call it "marriage trial". They believed marriage trial could help to know better the person who they are living with and, consequently, avoid a divorce. It also helps to understand the routines and habits of the other member of the couple, which are both important at a practical and emotional level.

Assuring financial security before transitioning was also a strategy for our participants (despite not being suggested by the couple as something to consider before moving together). This is also coherent with the study by Carmichael and Whittaker, who found that sometimes cohabitors deferred living together for pragmatic reasons, for instance, if the couple or one of the members was not yet financially stable to cohabit.

3.3 Strategies with a Positive Impact in the Transition to Cohabitation

We now focus on the strategies that once cohabitating should be adopted, in order to guarantee the cohabitation success. Millennials refer themselves to several strategies they believe would have a positive impact when starting cohabiting with someone (see Fig. 3). These were organized into six main themes: Adapting to the partner; Communication; Financial management; Focus on the self; Sharing; and Spending time together.

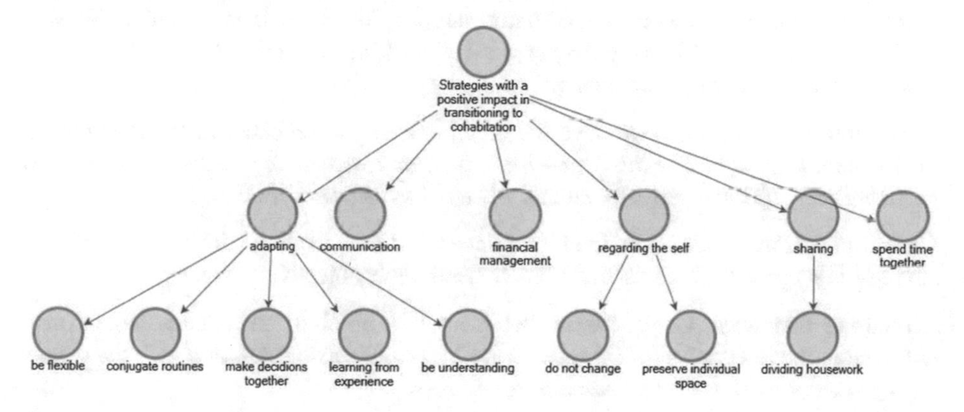

Fig. 3. Themes regarding positive strategies in cohabitation

Adapting. Millennials see adapting as a way to ensure that cohabitation is a success. This implies being flexible («*I would try to be more open to try things*», P26); and understanding («*Greater acceptance and availability*», P4). It also implies to conjugate the couple's routines («*greater synchronization of personal lives*», P17); making joint decisions («*Trying to compromise on some things and try to adapt myself to the fact that decisions are not just mine*», P3); and learning from the events in the relationship («*learning from mistakes and not using them to build conflicts*», P7).

Regarding the Self. At the same time as the above theme underlies the importance of adjusting to the other, "Regarding the self" underlies the importance of protecting one's self. The individual must ensure that being in a relationship, in the form of cohabitation, does not annul or cause change in what he is («*without ever forgetting the personal plans (like being with friends) of each one*», P14), even if one has to adjust to routines and habits.

Sharing. To ensure there is an equilibrium regarding the housework is also seen as a strategy, being curious to note that six women refer to this strategy, against three men («*I would have to help with the housework*», P6).

Communication. Once again, communication is seen as important to guarantee that cohabitation goes well («*To meet the challenges I would try to talk to him*», P5).

Managing Finances. Having a good financial management is also a theme in the data. It is important to considerer the money available for the couple and how to use it («*decide what each one pays based on the income of each*», P9).

Spending Time Together. Finally, to guarantee that the couple spends time together emerges as something that is seen as important, assuring that the couple would protect the relationship and escapes routine, keeping their interest in being together («*activities that allow the couple to approach each other*», P20).

Let us now consider the couple's experience regarding the strategies that they saw having a positive impact once they were cohabitating.

Adapting. This was also a theme to the couple, to whom it is important to stay attentive to the other («*knowing the person well and knowing when he is well, when he is not... a look is enough, its enough*", R.).

Regarding the Self. Here we have the theme - trying to be better in the relationship («*"I cannot always be like this", and there it is, as I was saying, learning to be calm, realizing that the same strategy cannot always be adopted*», A.).

Communication. Once again, the importance of communicating with the partner. This strategy is present in the data since our first research objective.

Managing Finances. Guaranteeing that there is a good financial management also emerged in the interviews («*"Okay, we have these responsibilities, and with the money I get from my work I have to manage for it"*», A.).

Getting Social Support. Here we have a theme until now absent from the data - having access to social support. According to R. their parents play an important role regarding the success of the couple's cohabitation («*either at lunches or dinners... washing and drying clothes (...). We have dinner at A. parents' house and at my parents' house, we can save some money*»).

Communication and Managing finances are always present in the data, suggesting that these are, indeed, important themes for our participants concerning cohabitation. They seem to associate it not only to a good transition to cohabitation as well as to succeed in it. In addition, communication seemed a fundamental strategy for the couple, before and during cohabitation. Despite referring to communication as something important, we are aware that this not implies that a person will use this strategy or that he/she will proficient at it. Nevertheless, we believe it suggest that participants in our study are sensitive regarding the role that this strategy plays in a relationship.

Getting social support was only present in the data from the couple. One may speculate if people who are not yet cohabiting are not sensitive to the importance of social support for couples. Particularly in the beginning, or when the couple has children or has a low income, due to a fragile working condition or economic crises.

In fact, Millennial parents remain intimately involved in their children's lives in college and into their careers [56] being an extremely important source of support in important decisions.

The theme "regarding the self" is worth underlining, since it reflects being available to look at oneself and make changes, in order to beneficiate the relationship. This openness may be fundamental, indeed, considering that nonstandard work schedules, which are common to Millennials, may prevent the couple from establishing and maintaining a strong bond [36].

4 Final Considerations

Little is known regarding the process of moving from dating to cohabitation, which turns this research both pertinent and timely. The participants in our study anticipate a number of possible issues when transitioning into cohabitation, namely, the conciliation of time spent at home and at work and the division of housework. That goes hand in hand with the finding that millennial workers prioritize work-life balance more than previous generations [54]. In addition, Millennials tend to prefer the egalitarian model of union (in which the husband and wife both have jobs and both take care of the household and children) over the traditional model [40]. Starting a family is something that seems to attract the concern of Millennials in this study. In fact, most Millennials look forward to marry and have children [40].

Kalmijn and Bernasco [55] suggested that cohabiting couples at the beginning of the century did not have strong desire for autonomy, privacy, and independence. However, the Millennials in our study showed much concern over each partner's personal time, space and money, suggesting the increased relevance of these aspects to the Millennial generation.

Millennials in our study believe that good practices to prepare the transition would include focusing on maintaining an open and healthy communication and planning how the couple would manage various issues, including financial matters. Communication is, indeed, one of the aspects that makes cohabitation last [9]. Planning and communicating about financial matters is also essential for Millennials who are cohabiting, since they approach their financial decisions with a high degree of caution [53]. Making sure that the couple has compatible schedules and visions about life and guaranteeing each member is able to adapt its own habits appear as essential [9]. Moreover, knowing the partner and establishing a culture of empathy emerges as essential to make the transition flow smoothly. In addition, the ability to experiment living together (practicing stayover) emerges as a useful strategy since couples can use it to wait for a more appropriate time in the relationship to move forward or end up opting for someone else that comes along, without having to deal with commitment problems and hurt feelings [7].

In fact, the couple implemented much of the strategies that were identified in the questionnaires, and consider them as positive practices. Namely, adaptation to the partner in form of conjugation of routines or creation of an understanding environment, where communication was key; regarding the self, Millennials in our study stated that it was vital to preserve the personal space, and to always try to improve for the sake of the relationship, while being careful not to annul themselves. However, a new idea came when the couple talked about the support they were receiving from their parents. Indeed, according to Alsop [56], Millennials rely on their parents for emotional and financial support.

This research has some limitations. Having interviewed only a couple implies that we focused a specific case. Both members of the couple are working students, which itself may imply specific constraints to cohabitation. Nonetheless, considering that Millennials continue to invest on their skills after having a degree, building their own academic profile and career, may turn this couple not very deviant from the typical Millennials' couple.

This study gave voice to individuals belonging to the Millennials' generation, trying to understand their personal expectations regarding a process that they will probably experience. Giving voice to people creates the space for exploring their own expectations, reflecting on how one thinks about himself, about possible future life events and their meaning. This space for self-knowledge is typically inherent to qualitative research, and we believe it was accomplished in this specific research. The individuals to whom we gave voice to (whether through the questionnaire or interview) had the space to think regarding their own perspectives and evaluate them. When we present data from different sources, we also create the opportunity for those who participated in the research, those who belong to the same population or those who deal with similar processes or populations, to develop their own understanding, which may influence their behaviors and/or competencies and representations.

The use of a Computer Assisted Qualitative Data Analysis software (CAQDAS) was also fundamental in our research. It allowed us to organize and store the research's design, the gathering data instruments, memos and the data itself, while ensuring an easy access to all of these elements. We managed data from different sources, analyzing it without taking the data from its context. It also allowed operationalizing qualitative thematic analysis, codding and representing ideas and themes, namely using thematic maps.

Future studies should consider enlarge and diversify the sample, to access more participants and test the representations we identified here (in the form of themes) using a deductive approach. It would be important to conduct a longitudinal research, in order to understand the phases of the transitioning process and the strategies mobilized by the couple to cope with them.

Knowing the perspectives of Millennials regarding cohabitation provides information relevant to consider in the design of developmental strategies, to guarantee that the couples transitioning adopt the practices associated to quality in a relationship and to the success in cohabitation.

References

1. Jamison, T., Prouxl, C.: Stayovers in emerging adulthood: who stays over and why? Pers. Relationsh. **20**(1), 155–169 (2013)
2. Carmichael, G.A.: Consensual partnering in the more developed countries. J. Aust. Popul. Assoc. **12**(1), 51–86 (1995)

3. Carmichael, G., Mason, C.: Consensual partnering in Australia: a review and 1991 census profile. J. Aust. Popul. Assoc. **15**(2), 131–154 (1998)
4. Cherlin, A.J.: Demographic trends in the United States: a review of research in the 2000s. J. Marriage Fam. **72**(3), 403–419 (2010)
5. Kennedy, S., Bumpass, L.: Cohabitation and children's living arrangements: new estimates from the United States. Demogr. Res. **19**, 1663–1692 (2008)
6. Arnett, J.J.: Emerging Adulthood: The Winding Road from the Late Teens Through the Early Twenties. Oxford University Press, New York (2004)
7. Jamison, T., Ganong, L.: We're not living together: stayover relationships among college-educated emerging adults. J. Soc. Pers. Relationsh. **28**(4), 536–557 (2011)
8. Knab, J.T.: Cohabitation: sharpening a fuzzy concept. Unpublished manuscript, Bendheim-Thoman Center for Research on Child Wellbeing (2005)
9. Carmichael, G., Whittaker, A.: Living in Australia: qualitative insights into a complex phenomenon. J. Fam. Stud. **13**(2), 202–223 (2007)
10. Manning, W.D., Smock, P.J.: Measuring and modeling cohabitation: new perspectives from qualitative data. J. Marriage Fam. **67**(4), 989–1002 (2005)
11. Sassler, S.: The process of entering into cohabiting unions. J. Marriage Fam. **66**(2), 491–505 (2004)
12. Stanley, S., Rhoades, G., Markman, H.: Sliding versus deciding: inertia and the premarital cohabitation effect. Fam. Relat. **55**(4), 499–509 (2006)
13. Willoughby, B., Carroll, J., Busby, D.: The different effects of "living together": determining and comparing types of cohabiting couples. J. Soc. Pers. Relationsh. **29**(3), 397–419 (2011)
14. Rhoades, G., Stanley, S., Markman, H.: A longitudinal investigation of commitment dynamics in cohabiting relationships. J. Fam. Issues **33**(3), 369–390 (2012)
15. Sassler, S.: Partnering across the life course: sex, relationships, and mate selection. J. Marriage Fam. **72**(3), 557–575 (2010)
16. Rhoades, G., Stanley, S., Markman, H.: Couples' reasons for cohabitation: associations with individual well-being and relationship quality. J. Fam. Issues **30**(2), 233–358 (2009)
17. Stanley, S., Markman, S.: Assessing commitment in personal relationships. J. Marriage Fam. **54**(3), 595–608 (1992)
18. Tang, C., Curran, M., Arroyo, A.: Cohabitors' reasons for living together, satisfaction with sacrifices, and relationship quality. Marriage Fam. Rev. **50**(7), 598–620 (2014)
19. Finkel, E.J., Rusbult, C.E.: Prorelationship motivation: an interdependence theory analysis of situations with conflicting interests. In: Handbook of Motivation Science, pp. 547–560. Guilford Press, New York (2008)
20. Kelley, H.H.: Personal Relationships: Their Structures and Processes. Erlbaum, Hillsdale (1979)
21. Rusbult, C.E., Van Lange, P.A.: Interdependence, interaction, and relationships. Annu. Rev. Psychol. **54**(1), 351–375 (2003)
22. Van Lange, P.A.M., Rusbult, C.E., Drigotas, S.M., Arriaga, X.B., Witcher, B.S., Cox, C.L.: Willingness to sacrifice in close relationships. J. Pers. Soc. Psychol. **72**(6), 1373–1395 (1997)
23. Wieselquist, J., Rusbult, C.E., Foster, C.A., Agnew, C.R.: Commitment, pro-relationship behavior, and trust in close relationships. J. Pers. Soc. Psychol. **77**(5), 942–966 (1999)
24. Batalova, J.A., Cohen, P.N.: Premarital cohabitation and housework: couples in cross-national perspective. J. Marriage Fam. **64**(3), 743–755 (2002)
25. Hardie, J.H., Lucas, A.: Economic factors and relationship quality among young couples: comparing cohabitation and marriage. J. Marriage Fam. **72**(5), 1141–1154 (2010)
26. Brines, J., Joyner, K.: The ties that bind: principles of cohesion in cohabitation and marriage. Am. Sociol. Rev. **64**(3), 333–355 (1999)

27. Blumstein, P., Schwartz, P.: American Couples. William Morrow and Company, New York (1983)
28. Rindfuss, R.R., VandenHeuvel, A.: Cohabitation: a precursor to marriage or an alternative to being single? Popul. Dev. Rev. **16**(4), 703–726 (1990)
29. Heimdal, K.R., Houseknecht, S.K.: Cohabiting and married couples' income organization: approaches in Sweden and the United States. J. Marriage Fam. **65**(3), 525–538 (2003)
30. Burstein, N.R.: Economic influences on marriage and divorce. J. Policy Anal. Manag. J. Assoc. Public Policy Anal. Manag. **26**(2), 387–429 (2007)
31. Hoffman, S.D., Duncan, G.J.: The effect of incomes, wages, and AFDC benefits on marital disruption. J. Human Resour. **30**(1), 19–41 (1995)
32. Kalmijn, M., Loeve, A., Manting, D.: Income dynamics in couples and the dissolution of marriage and cohabitation. Demography **44**(1), 159–179 (2007)
33. Lewin, A.C.: The effect of economic stability on family stability among welfare recipients. Eval. Rev. **29**(3), 223–240 (2005)
34. South, S.J.: Time-dependent effects of wives' employment on marital dissolution. Am. Sociol. Rev. **66**(2), 226–245 (2001)
35. Nock, S.L.: A comparison of marriages and cohabiting relationships. J. Fam. Issues **16**(1), 53–76 (1995)
36. Presser, H.B.: Employment schedules among dual-earner spouses and the division of household labor by gender. Am. Sociol. Rev. **59**(3), 348–364 (1994)
37. Presser, H.B.: Nonstandard work schedules and marital instability. J. Marriage Fam. **62**(1), 93–110 (2000)
38. Lichter, D.T., Qian, Z.: Serial cohabitation and the marital life course. J. Marriage Fam. **70** (4), 861–878 (2008)
39. Howe, N., Strauss, W.: Millennials rising: the next great generation. Vintage Books, New York (2009)
40. Wang, W., Taylor, P.: For Millennials, parenthood trumps marriage (2011). http://pewsocialtrends.org/2011/03/09/for-millennials-parenthood-trumps-marriage/
41. Shields-Duton, K.: Attitudes toward cohabitation: a cross sectional study. Master's thesis (2016). Stars. Accessed 26 May 2018
42. Bogle, R., Wu, H.: Thirty years of change in marriage and union formation attitudes, 1976–2008 (FP-10–03). National Center for Family and marriage Research (2010). http://ncfmr.bgsu.edu/pdf/family_profiles/file83691.pdf
43. Eickmeyer, K.: Even more 'premarital divorce': cohabitation and multiple union dissolutions during young adulthood. Electronic Thesis or Dissertation (2016). https://etd.ohiolink.edu/. Accessed 26 May 2018
44. Eickmeyer, K.J., Manning, W.D.: Serial cohabitation in young adulthood: baby boomers to millennials. J. Marriage Fam. **80**(4), 826–840 (2018)
45. Arnett, J.J.: Emerging adulthood: a theory of development from the late teens through the twenties. Am. Psychol. **55**(5), 469–480 (2000)
46. McGuire, K.: Millennials' perceptions of how their capacity for romantic love developed and manifests. Master's thesis (2015). Scholarworks. Accessed 26 May 2018
47. Settersten, R.A.: The contemporary context of young adulthood in the USA: from demography to development, from private troubles to public issues. In: Booth, A., Brown, S., Landale, N., Manning, W., McHale, S. (eds.) Early Adulthood in a Family Context 2010, National Symposium on Family Issues, vol. 2, pp. 3–26. Springer, New York (2012)
48. Braun, V., Clarke, V.: Using thematic analysis in psychology. Qual. Res. Psychol. **3**(2), 77–101 (2006). https://doi.org/10.1191/1478088706qp063oa
49. Claps, E.: The Millennial generation and the workplace. Master's thesis (2010). DigitalGeorgeTown. Accessed 5 Oct 2018

50. Kaplan, A.B., Darvil, K.: Think [And Practice] Like a Lawyer: Legal Research for the New Millennials. Legal Communication & Rhetoric: JALWD, 8 (2011). SSRN. https://ssrn.com/abstract=1933474
51. Allen, K., Kinchen, V.: Financial management practices of college students. Glob. J. Bus. Res. **3**(1), 105–116. http://www.theibfr.com/archive/gjbr-v3-n1-2009.pdf#page=107
52. Kemp, B.D.: The recession generation: an examination of Millennials' Money management habits. Master's thesis (2016). https://scholarworks.unr.edu/bitstream/handle/11714/3288/Kemp%2c%20Brienna%202016%20Recession%20Generation%20%20An%20Examination%20of%20Millennials%E2%80%99%20Money%20Management%20Habits.pdf?sequence=1&isAllowed=y. Accessed 5 Oct 2018
53. Tipper, W.A.: The future savings challenge: the implications of generation Y's attitude to finance and sustainability (2014). https://www.green-alliance.org.uk/thefuturesavingschallenge.php
54. Carless, S.A., Wintle, J.: Applicant attraction: the role of recruiter function, work-life balance policies and career salience. Int. J. Sel. Assess. **15**(4), 394–404 (2007)
55. Kalmijn, M., Bernasco, W.: Joint and separated lifestyles in couple relationships. J. Marriage Fam. **63**(3), 639–654 (2001)
56. Alsop, R.: The Trophy Kids Grow up: How the Millennial Generation is Shaking up the Workplace. Jossey-Bass, New York (2008)

A Conceptual Model for Action and Design Research

Telmo Antonio Henriques[1(✉)] and Henrique O'Neill[2]

[1] Department of Information Science and Technology, ISCTE-IUL, Lisbon, Portugal
telmo_antonio_henriques@iscte.pt
[2] ISCTE Business School, ISCTE-IUL, Lisbon, Portugal

Abstract. Organizational research has a pattern of special characteristics which make a clear distinction from other research paradigms. When using this kind of approaches – mainly those which are based on Action and Design – the Interpretivist, Constructivist, and Participatory perspectives dominate. They have already proven to have strong foundations – including ways of doing, data, and results – which turn these two paradigmatic approaches into effective ways for getting knowledge, doing things, and promoting change. The objective of the current article is to present a top-level conceptual model – under the form a tri-dimensional perspective – for Action and Design Research. It combines the traditional scientific, engineering, and organization development approaches – depicting how an organization can, simultaneously, solve problems, produce actionable knowledge, change, and artifacts. It has been developed using a Design Science Research approach, tested in a major organizational change program, and successfully used to teach research methods essentials to Master and DBA students.

Keywords: Action Research · Design Science Research · Conceptual model

1 Introduction

Organizational research, as a particular instantiation of "real world research", has a pattern of special characteristics. They set the scene for a clear distinction from other kind of research settings, including laboratory, empirical, or other participative approaches to community-based research.

Making a clear differentiation between practitioners and academics, Robson [1] highlights a set of distinct researcher's interests, work nature, and typical approaches, contrasting between "real world" and "traditional academic" researchers.

Particularly, concerning the typical approach of "real world" researchers, he clearly denotes their main working characteristics, including "their interest in solving problems and getting large effects, with concern for actionable factors, almost always working in the 'field', with strict time and budget constraints, being generalist researchers, with a need for familiarity with a wide range of methods and approaches, oriented to client needs, and with a strong need for well-developed social skills".

A. P. Costa et al. (Eds.): WCQR 2018, AISC 861, pp. 186–201, 2019.
https://doi.org/10.1007/978-3-030-01406-3_16

In strict line of congruence with this characterization, the most frequent approaches to organizational research – either from an Organization Development perspective or from an Engineering viewpoint – are often grounded on Action and Design paradigms, in opposition to the traditional non-invasive and observational Positivist approaches.

Action Research (AR) – with its ontological, epistemological, and methodological believes, aims, nature of knowledge, values, ethics, voice, and inquirer posture – fits clearly within this classification of Participatory Research.

On the other hand, considering the design perspective, and focusing the discussion on the "sciences of the artificial", Simon [2] distinguishes from the task of the science disciplines, "as being to teach about natural things and how they are and how they work", and the task of engineering schools, as consisting of "to teach about artificial things and how to design and make artifacts that have desired properties".

Asserting that "engineers are not the only professional designers" he argues that "the intellectual activity that produces material artifacts is not different from the one that prescribes remedies for a sick patient, or devises a new sales plan for a company, or establishes a social welfare policy for a state."

This perspective and assertions set a landmark argument for the relevance of Design Science Research (DSR) – as a valid approach to define problems, identify requirements, design, implement, and test artifacts which solve specific problems.

When using this two kind of approaches – either based on Action (AR) or Design (DSR) activities to solve problems and to produce knowledge – the patterns of Interpretivist and Constructivist perspectives are clearly in the field.

Considering its foundations, processes, data, and results, these paradigmatic approaches do not turn the research process onto a less rigorous activity. In fact, a lot of research work – with successful field applications and associated peer reviewed publications – has emerged to confirm the rigor and relevance attributes for a pragmatic use of these two paradigms, either isolated or in combination.

Nevertheless, within this area still persists a relevant gap, and opportunity for applied research, between Gregor and Hevner's [3] "well-developed design theory about embedded phenomena" and level 1 contribution types on "situation implementation of artifacts".

Particularly, some kind of Gregor and Hevner's [3] level-2 contributions – including what Vaishnavi and Kuechler [4] refer as frameworks ("real or conceptual guides to serve as support or guide") and methods ("sets of steps used to perform tasks") – have its own space for development.

Namely those specifically focused on perspectives about the main steps and data involved in AR and DSR approaches denote a clear research gap. More specifically, there is a clear opportunity to develop a global overview for the AR and DSR paradigms, evidencing its main data and process cornerstones, integrating the science, organization development, and engineering perspectives.

It combines these three main cornerstones of problem-solving, in order to produce emergent knowledge, effective change, and useful artifacts. It is the main target of this publication.

The need for such a conceptual model arises from the specific need expressed by IS Management master students, struggling with so many different Action and Design

Science research traditions and approaches, to have a pragmatic overview for these applied research paradigms.

It is currently presented on the basis of a systematic literature review, but it is founded on empirical research developed by the authors, and tested by specific organizational research practice [5, 6]. Its current version has been matured on the basis of a teaching empirical practice, and the associated Action Research Process Model's dimension has already been developed and published [7].

So, the current research is directly associated with the answer to some main questions, which are relevant - for graduate students and for practitioners - on their first steps using AR and DSR, namely:

- What are the essential steps of the research process which will allow me to introduce rigor on my knowledge-generation practice, and what are the pieces of information that must be used and produced at each step?
- What are the essential steps of the design process which allow me to introduce relevance on my engineering practice, progressing from problems to artifacts, and what are the pieces of information that must be used and produced at each step?
- What are the essential steps of the change process which allow me to effectively advance on my organization development practice, and what are the pieces of information that must be used and produced at each step?

In order to provide an appropriate answer to these essential questions, the model must achieve a main objective of producing a pragmatic view of the main data and processes necessary to solve a multidimensional problem within an organizational context and involving these three dimensions.

Considering Boonstra's [8] principle that "there is no one best way in organizing and changing" and Burnes' [9] argument that "the ability to manage change is now recognized as a core organizational competence, challenging the idea that there can be a one best way to do it", the model to be produced should be necessarily simple, clear, and pragmatic (in order to be well understood and easily adopted), but also sufficiently general and flexible (allowing it to be easily adapted to distinct problem areas, disciplines, application contexts, and circumstances).

These essential attributes of the model determine that, on its main orientations, it must not be prescriptive but mainly supportive (will answer to "what to do" rather than "how to do" questions).

Also, considering its basic purpose, the model ought to deliver a simple, pragmatic, and useful basis to teach Master, DBA, and PhD students on applied Organizational Research Methods – providing an overview of its main dimensions and helping them to discover their own way and main references for its contextualized application.

Furthermore, addressing its field application, it aims to support organizational professionals, while researching inside their own organizations – combining applied research with design and organization development, as a professional challenge.

For the purpose, and based on a literature review, focused on the process and data scopes of Research, Action, and Design, a tri-dimensional top-level view of the essential aspects which emerge from the main references on the field has been developed.

The model has been field-tested, with success, along a strategic transformational change intervention within an IS/IT Unit of a major Portuguese Bank - which has addressed, at a first program, the areas of Organizational Culture and Values, Leadership Development, and Employee Engagement. Subsequent structured projects have also been setup and developed, benefiting from action and behavioral learning developed during the first stage, and covering the areas of Training and Development, Internal Communication, ICT Process Maturity alignment, and Project & Portfolio Management Tools' alignment and implementation.

The associated research has been the main focus of a successful Doctoral Dissertation in Information Science and Technology [5], addressing the main areas of Information Systems, Quality, and Organization Development.

2 Literature Review

In order to clearly identify and define the design requirements for a conceptual model representing a pragmatic view of AR and DSR, it was crucial to review the most relevant literature aspects for both research paradigms, with a special focus on the main associated process and data.

Considering the combination of the two approaches, relevance must also be given to the literature on Action Design Research.

These main elements represent the Organization Development and the Engineering dimensions.

Furthermore, encompassing the need to obtain a whole perspective of the main requirements involved in research activities, a complete literature review should be conducted in order to make explicit their major process and data components – thus representing the Science dimension.

2.1 The Science Dimension – Research Process and Data Implications

Independently of the specific epistemological, ontological, and methodological characteristics associated with each inquiry paradigm, research activities exhibit a set of common patterns and requirements, which will allow us to recognize its main processes and data components.

As a first reference in this field concerning the research process Saunders, Lewis and Thornhill [10] denote that "most research textbooks represent research as a multi-stage process that you must follow in order to undertake and complete your research project".

For this research path, they specifically recommend a set of main activities: "(1) formulate and clarify your research topic, (2) critically review the literature, (3) understand your philosophy and approach, (4) formulate your research design, (5) negotiate access and address ethical issues, (6) plan your data collection and collect data, (7) analyze your data using qualitative and/or quantitative methods, (8) write your project report and prepare your presentation and (9) submit your project report and give your presentation."

This view offers us a first insight on the main research process steps, which the authors deeply develop on their book devoted to teach research methods to business students.

Yin [11], focusing on the Case Study Research approach, describes it as a "linear but iterative process", including a path starting with a "thorough literature review and the careful and thoughtful posing of research questions or objectives".

Particularly, his approach includes six major interrelated steps: "(1) plan, (2) design, (3) prepare to collect evidence, (4) collect evidence, (5) analyze evidence, and (6) share".

Concerning the area of Social Research, Bryman [12] summarizes the process in seven essential components: "(1) literature review, (2) concepts and theories, (3) research questions, (4) sampling cases, (5) data collection, (6) data analysis, and (7) writing up".

Similar patterns are recognizable in several reference publications within the domain of Research Methods and Research Design (e.g. [13, 14]).

Together, they highlight a set of common data elements, which integrate:

1. A literature review, covering the main aspects of the research disciplines and method;
2. A definition of the research approach;
3. The explicit and clear identification of the research targets;
4. The production of a research design;
5. The reporting on the effective development of research activities, including evidence;
6. The identification of the research results;
7. The evaluation of the research;
8. The production and publication of emergent knowledge.

Also, on a process perspective, and in strict accordance with the production of these main data results, the associated activities should include some essential steps:

1. A systematic review of relevant external (and internal) knowledge;
2. The definition of a research approach;
3. The formulation of research targets (questions, objectives and hypothesis);
4. The design of the research, in terms of process, data and tools to be used;
5. The effective development of the research activities;
6. The execution of the field processes associated with the research;
7. A formal evaluation of the research; and
8. The generation of relevant external knowledge.

These are the main data and process elements which have been considered, as strictly necessary, to integrate the Science cornerstone of our model (Fig. 1).

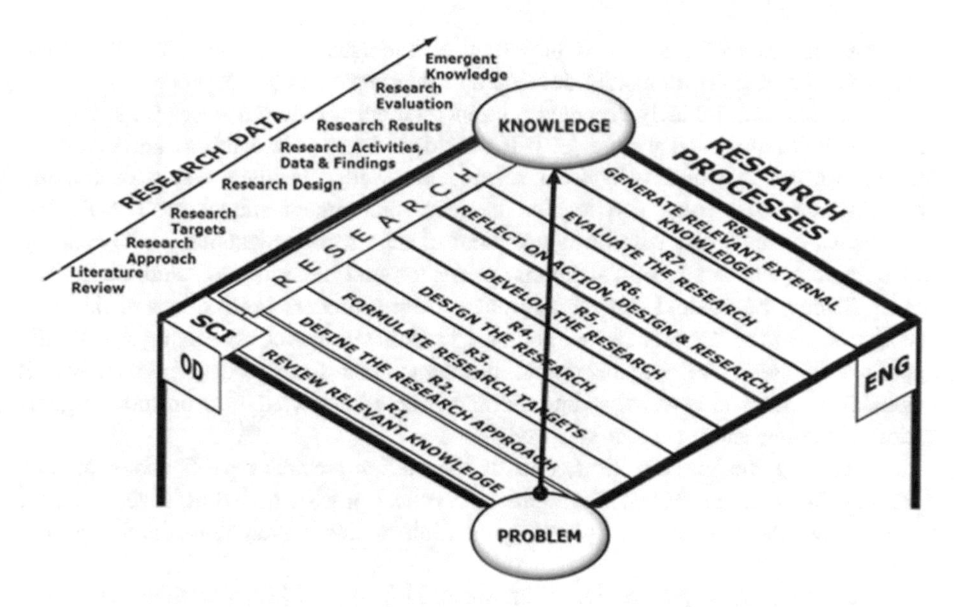

Fig. 1. ADR foundations: the science perspective

2.2 The Organization Development Dimension – AR Process and Data

Within social contexts Action Research has already proven to be an effective approach to promote development through engaging change interventions: solving problems, embodying peoples' aspirations, and meeting their dreams and deep expectations.

Literature is plenty of narratives detailing its successful application within distinct social contexts and using a multiplicity of approaches.

Academic publications have reflected this multiplicity of successful approaches, starting from the relevant works and vision of Kurt Lewin [15], and its further development within prominent institutions and research centers – including, among others, the Tavistock Institute for Human Relations, the Institute for Social Research in Industry, the Work Research Institute of Norway, and the Centre for Action Research at the University of Bath.

These fundamental developments have been reported by academic publications. One of those – the Action Research Journal – has, since 2003, played a major role on the area, as a major publication devoted to this research paradigm, and providing regular literature reviews [16–19] on the subject.

Also, major developments have been subject of specific dissemination via seminal and structural books, handbooks, and encyclopedias [20–25].

Furthermore, from a professional perspective, several cases of successful organizational change and development processes, associated with participative interventions targeting major organizational transformations [26] have been reported.

Being Action Research a process – with "the double burden of testing hypotheses and effecting some (putatively) desired change in the situation" [27], where "there are two action research cycles operating in parallel" [24], a "core action research cycle" and a "thesis action research cycle" [28], involving "two goals: solve a problem and contribute to science" [29] – it is important to understand how the "action" and the "research" dimensions intimately develop and interleave along this process.

In this context, a widely accepted reference model for Action Research is the one provided by Shani and Pasmore [30]. It considers Action Research as an "emergent inquiry process in which behavioral science knowledge is integrated with existing organizational knowledge and applied to solve real organizational problems", "simultaneously concerned with bringing about change in organizations, in developing self-help competencies in organizational members, and in adding to scientific knowledge". Setting the context and the approach, from a process point of view, this perspective shows the importance of promoting organizational change using a systemic approach to solve real organizational problems, and from a data perspective, it enhances the relevance of using internal and external knowledge to promote organizational learning and emergent knowledge.

Concerning the process itself, one of the earlier prevalent publications on this domain [31] considers Action Research as a cyclical process including several stages, namely "Diagnosing, Action planning, Action taking, Evaluating, and Specific learning."

Also, on a process perspective, Kemmis and McTaggart [32] – in their systematic and reflective model – consider AR as integrating four main phases: "Planning, Acting, Observing, and Reflecting."

This perspective, mainly adds to the previous views a dimension of observation and reflection, which connects the results of action with the evaluation of the research.

Further on, Coghlan and Brannick [24] have identified a set of initial typical questions to be addressed as part of a research proposal. They include main interrogations about the action dimension, namely: "what is the action?; what is the rationale for this action?; why is it worth doing?; what is the desired future?; what is the present situation?; what is the plan to move from here to there?; what is the time schedule?; with whom will you collaborate?; where do you, as the researcher, fit into the action?; what are the ethical challenges?"

For the research dimension these authors consider the main relevant questions: "What is the rationale for researching this action? What is the contribution to knowledge that this research intends to make? How do you intend to inquire into the action? How do you ensure quality and rigor in your action research?"

Subjacent to these questions is the need to ensure a set of processes in the Action dimension to diagnose the current situation and to get consensus on the desired situation, as well as, to plan the intervention, before action, and its evaluation. On the other hand – concerning the Research perspective of these questions – relevance is given to the need to provide a research approach rationale, to set research targets and to establish research evaluation criteria.

Evoking Mezirow's [33] forms of reflection – which cross the main territories of intentions, planning, action, and outcomes – they emphasize that "As an action researcher you try to understand your intentions, to develop appropriate plans and strategies, to be skilled at carrying them out, to reflect on how well you have carried out the plans, and to evaluate their results".

Referring to the "dual imperatives of action research", McKay and Marshall [34] enhance "its interest in and commitment to organizational problem solving, and its interest in and commitment to research, and the production of new insights and knowledge".

Altogether, these pivotal perspectives on Action Research provide us a wide perspective of its main requirements, which – on an action perspective – must include: (1) the traditional planned change processes (identify organizational objectives and the change approach, develop a diagnosis, design the intervention, do the intervention, evaluate the results) and (2) the associated information usually reported at a change intervention (change objectives and approach, organizational diagnosis, intervention design and plan, change results, change evaluation);

Joining together both perspectives – of Research and Action – an aggregated view, integrating the Science and Organization Development dimensions, can now be depicted (Fig. 2).

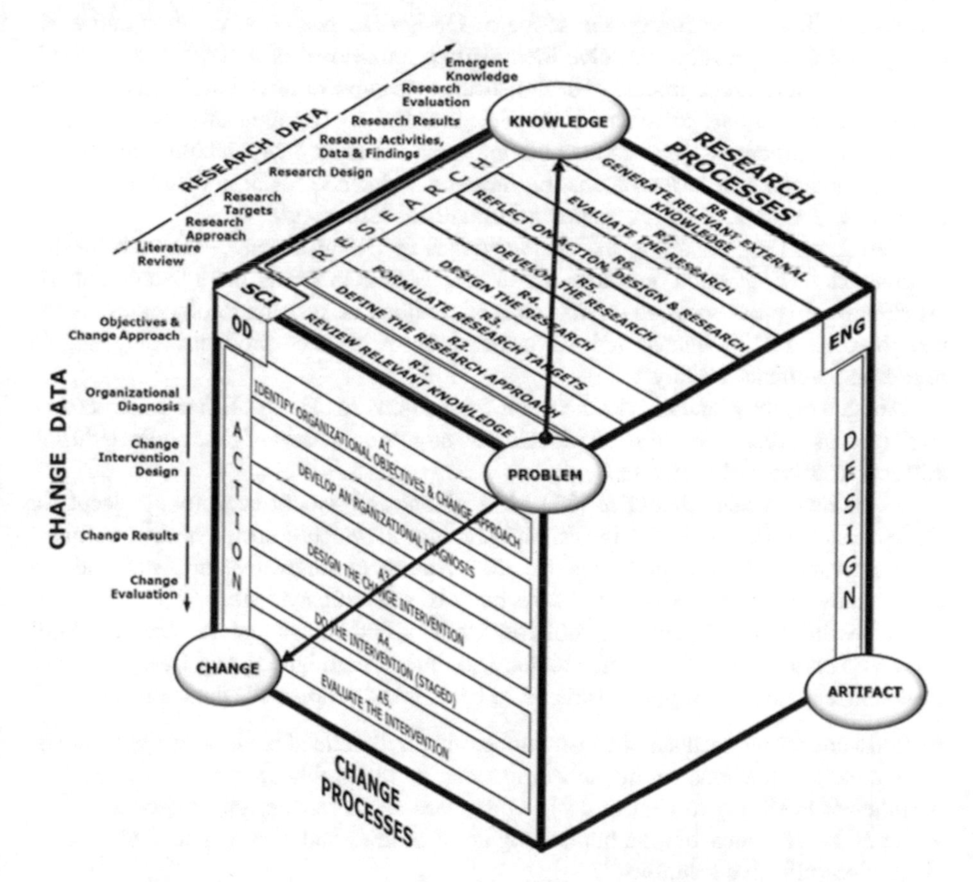

Fig. 2. ADR foundations: science and organization development perspectives

2.3 The Engineering Dimension – Design Science Research Process and Data

A third dimension for our 3-cornerstone perspective can now be introduced via the traditional Engineering approach to the Design of useful Artifacts which solves relevant organizational Problems. It will be further aggregated with the Science (Research) dimension to promote relevant Knowledge generation.

Concerning this subject, Simon [2], considering the task of engineering schools as teaching about artificial things ("how to design and how to make artifacts that have desired properties") asserts that "engineers are not the only professional designers, and that everyone designs who devises courses of action aimed at changing existing situations into preferred ones". He argues that "the intellectual activity that produces material artifacts is no different fundamentally from the one that prescribes remedies for a sick patient, devises a new sales plan for a company, or establishes a social welfare policy for a state."

Peffers [35] elaborating on the scope of Design Science – as "creating and evaluating IT artifacts intended to solve identified organizational problems" – describes it as involving "a rigorous process to design artifacts to solve observed problems, to make research contributions, to evaluate the designs, and to communicate the results to appropriate audiences". For these authors, such artifacts "may include constructs, models, methods, and instantiations, but might also include social innovations or new properties of technical, social, and/or informational resources".

Van Aken [36], regarding the characteristics of Design Science Research (DSR), emphasizes that "research questions are driven by field problem, there is an emphasis on solution-oriented knowledge, linking interventions or systems to outcomes, as the key to solve field problems, and the justification of research products being largely based on pragmatic validity".

Altogether, these authors set the main foundations for Design Science Research – as a rigorous research activity – evidencing its targets on a basis of relevance: to design artifacts to solve real problems.

Also, Hevner and Chatterjee [37] set a global and widely accepted, concept for Design Science Research as "a research paradigm in which a designer answers questions relevant to human problems via the creation of innovative artifacts, thereby contributing with new knowledge to the body of scientific evidence".

According to this definition, problems, artifacts, and knowledge are central to DSR and based on this groundings, these authors go further, highlighting that Design Science Research addresses what are considered to be "wicked problems" characterized by:

- "unstable requirements and constraints based on ill-defined environmental contexts;
- complex interactions among subcomponents of the problem;
- inherent flexibility to change design processes as well as design artifacts;
- a critical dependence upon human cognitive abilities and also on social abilities to produce effective solutions."

Reeves [38] – elaborating on the DSR path from problems to solutions – suggests a chain of main processes which include: "analysis of problems, development of solutions, test and refinement, reflection, and enhanced implementation".

Hevner [39], using an elaborated process, envisions Design Science Research as integrating a three-cycle approach and processes, including:

- "a Relevance Cycle (requirements; field testing) – bridging the contextual environment of the research project with the design science activities;
- a Rigor Cycle (build design artifacts and processes; evaluate) – connecting the design science activities with the knowledge base of scientific foundations, experience, and expertise that informs the research project; and
- a Design Cycle (grounding; additions to knowledge base) – iterating between the core activities of building and evaluating the design artifacts and processes of the research."

This model highlights the main objects and actors within the application domain (people, organizational systems, and technical systems, and its problems and opportunities), the associated knowledge base, including scientific theories and methods, experience and expertise, and meta-artifacts (either design products as design processes), and the DSR process itself.

Also, discussing the main DSR activities, Peffers [35] describes a Process Model which includes six main stages: (1) identify problem and motivate, (2) define objectives of a solution, (3) design and development, (4) demonstration, (5) evaluation and (6) communication.

Offerman [40], based on a comparison of DSR activities, propose an outline for the DSR process which includes: (1) problem identification, (2) solution design, and (3) evaluation.

Vaishnavi and Kuechler [4] propose a process for DSR, including: (1) Awareness of Problem, (2) Suggestion, (3) Development, (4) Evaluation and (5) Conclusion.

As a synthesis – emerging from these major approaches to DSR – we can recognize, as common elements, some main processes, including:

1. definition of the problem,
2. definition of the associated requirements,
3. design of an appropriate solution, usually under the form of an artifact,
4. development of the artifact, and
5. its test and evaluation.

Also, along the whole process, some major information components – which must be, progressively and congruently, produced and reported – can be recognized, including:

1. problem definition,
2. requirements' definition,
3. solution definition,
4. artifact development and testing, and
5. design evaluation.

Being DSR a research process, it must, naturally, aggregate the data and process requirements previously described as applicable to the Research perspective.

So, a two-dimensional perspective for DSR can be depicted as follows (Fig. 3).

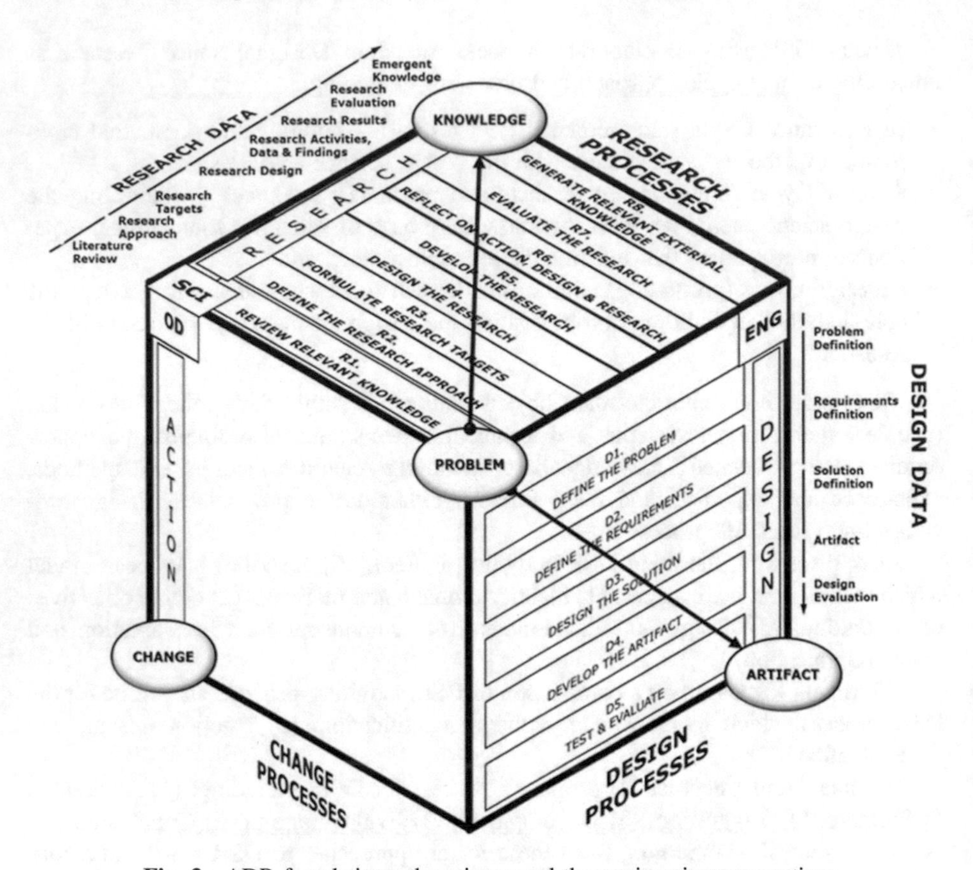

Fig. 3. ADR foundations: the science and the engineering perspectives

2.4 Combining the Two Approaches: Organization Development and Engineering

Sein [41], combining the Action and the Design components subjacent to the AR and DSR paradigms, has developed an Action Design Research (ADR) approach as "a research method for generating prescriptive design knowledge through building and evaluating ensemble IT artifacts in an organizational setting".

It deals with two major interrelated challenges: (1) "addressing a problem situation encountered in a specific organizational setting by intervening and evaluating, and (2) constructing and evaluating an IT artifact that addresses the class of problems typified by the encountered situation."

The method focuses on the building, intervention, and evaluation of an artifact, which reflects not only the theoretical precursors and intent of the researchers but also the influence of users and its ongoing use in context.

According to these authors the whole process involves four major steps: "(1) problem formulation, (2) building, intervention, and evaluation, (3) reflection and Learning, and (4) formalization of learning."

This confirms, from an integrated point of view, the double-dimension perspectives previously developed and presented for Action Research and Design Science Research.

3 Summary: A Global Perspective from the Literature

A systematic literature review has provided us with an overview of the main data and process requirements for each perspective (dimension) of our model.

Integrating the three perspectives – from the Science, the Organization Development and the Engineering approaches – a final view of our Model's foundation can be depicted (Fig. 4).

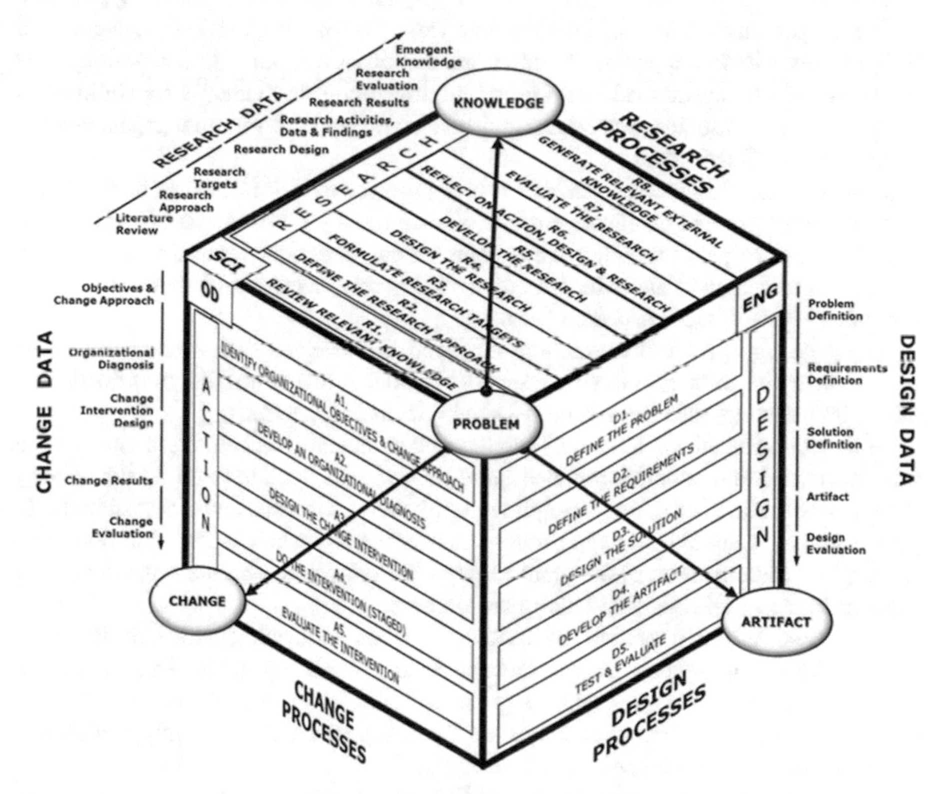

Fig. 4. ADR foundations: the science, OD, and engineering perspectives

This cube illustrates the confluence of process and data, along three interdependent dimensions.

It puts into perspective the path from an initial problem to its solution, in terms of promoting change, producing an artifact, and getting new knowledge.

Looking at the adjacent vertical faces of the "cube" – integrating the Action and the Design processes which configure the Organization Development and the Engineering cornerstones – we can envision the closure of this construct.

Topped by the Science cornerstone – representing the Research dimension – they integrate the essential foundations to progress to a more detailed analysis, design, and representation of their articulation.

4 Discussion, Reflection, and Main Conclusions

Organizational Research – as an inquiry practice for solving organizational problems, getting new knowledge, promoting deep change, delivering useful artifacts, and moving ahead – is a powerful instrument on the hands of researchers and professionals to promote research and development.

Particularly, a pragmatic application of Organization Development approaches combined with traditional Engineering practices – integrating Action Research and Design Science Research paradigms into organizations – can promote double-loop [42] generative [43] organizational learning, while developing individual's capabilities, as well as, positive attitudes, behaviors, and real contributes to promote organizational excellence and effectiveness.

However, for early-career academics and professionals – interested in their sustained development and on effective usage of such tools – it is important to have some kind of "route maps", providing them some initial directions, concerning the most relevant processes and data, which facilitate the accomplishment of their task within acceptable levels of rigor and relevance.

The main objective of this research has been to introduce such a pragmatic overview – under the form of a three cornerstone model – combining the traditional Scientific, Engineering, and Organization Development approaches.

In its root orientations, the model ought to be sufficiently clear and pragmatic to be well understood and adopted, but also quite general and flexible to allow for its easy adaptation to distinct situations, disciplines, application contexts, and circumstances. In general terms, it should be supportive – answering to "what to do" questions and targeting the identification of the main steps to be followed along the process and the associated data to be considered along the process.

The initial model aimed to be used as a basis to teach Qualitative Research Methods, by providing a simple and clear overview of AR and DSR. This has been a major design driver - responding to graduate (MSC and DBA) students' specific requests to have as pragmatic overview on the essential process to apply Action and Design Research inside organizational contexts.

It has been developed using the Design Science Research paradigm itself, based on a systematic literature review focused on the Process and Data scopes of AR and DSR.

It has produced a tri-dimensional top-level view (a cube) of the essential elements emerging from the main references in this methodological field.

It has been developed, and field-tested with success, by empirical research, along a strategic transformational change intervention within an IS/IT Unit of a major Bank [6]. This research has been the main focus of a successful Doctoral Dissertation in Information Science and Technology [5] and a partial view of the model, depicting the details of the Action Research components, has already been presented [7].

It has been, since 2014, effectively used, tested, and developed - as an useful educational instrument - to teach DBA candidates and IS/IT Management master students on Qualitative Research Methods. It has proven to be a straightforward and valuable instrument to facilitate a clear understanding of the essential elements involved in the conduction of Action Research and Design Science Research interventions within organizations.

As a didactical reference, the model does not exempted, either the necessary readings concerning the relevant literature in the specific field of application, or the use of complementary case studies to support students' learning.

However, it has revealed as a very useful instrument to facilitate the students' initial approach to applied research: providing a global overview, stimulating further individual study, and facilitating research proposal's elaboration.

As any model, it is a simplified and limited representation of the reality – providing a top level overview - as a departing point to approach its detailed data and process structures.

Its underlying process model for Action Research has already been developed, tested in practice, and presented [7].

However, there is still an opportunity for further research and development, particularly in terms of the design of the underlying data structure, and on the representation of the main events occurring along the research process which may trigger the associated action, design and research activities.

References

1. Robson, C.: Real World Research: A Resource for Users of Social Research Methods in Applied Settings, 3rd edn. Wiley, Hoboken (2011)
2. Simon, H.: The Sciences of Artificial, 3rd edn. MIT Press, Cambridge (1996)
3. Gregor, S., Hevner, A.R.: Positioning and presenting design science research for maximum impact. MIS Q. **37**(2), 337–355 (2013)
4. Vaishnavi, V., Kuechler, W.: Design Science Research Methods and Patterns: Innovating Information and Communication Technology. Auerbach Publications, Boca Raton (2015)
5. Henriques, T.A.: IT quality and organization development – using action research to promote employee engagement, leadership development, learning and organizational improvement. Ph.D. Dissertation in Information Science and Technology. ISCTE-IUL (2015)
6. Henriques, T.A., O'Neill, H.: IT quality and organizational development – using action research to promote employee engagement, leadership development, learning and organizational improvement. In: 2014 Annual Conference on British Academy of Management, BAM, Belfast, UK (2014)
7. Henriques, T.A., O'Neill, H.: A process model for organizational action research. In: 2018 Annual Conference on European Academy of Management, EURAM, Reykjavik, Iceland (2018)
8. Boonstra, J.: Some reflections and perspectives on organizing, changing, and learning. In: Boonstra, J. (ed.) Dynamics of Organizational Change and Learning, pp. 317–341. Wiley, UK (2004)

9. Burnes, B.: No such thing as … a 'one best way' to manage organizational change. Manag. Decis. **34**(10), 11–18 (1996)
10. Saunders, M.N.K., Lewis, P.E.T., Thornhill, A.: Research Methods for Business Students, 5th edn. Pearson Education, London (2009)
11. Yin, R.: Case Study Research Design and Methods, 4th edn. SAGE Publications, Inc., Thousand Oaks (2009)
12. Bryman, A.: Social Research Methods, 4th edn. Oxford University Press Inc., Oxford (2012)
13. Eriksson, M., Kovalainen, A.: Qualitative Methods in Business Research. Sage, Thousand Oaks (2008)
14. Creswell, J.W.: Research Design Qualitative & Quantitative Approaches. Sage Publications, London (1994)
15. Lewin, K.: Field Theory in Social Science. Harper & Row, New York (1951)
16. Dick, B.: Action research literature: Themes and trends. Action Res. **2**(4), 425–444 (2004)
17. Dick, B.: Action research literature 2004–2006: themes and trends. Action Res. **4**(4), 439–458 (2006)
18. Dick, B.: Action research literature 2006–2008: themes and trends. Action Res. **7**(4), 423–441 (2009)
19. Dick, B.: Action research literature 2008–2010: themes and trends. Action Res. **9**(2), 122–143 (2011)
20. Stringer, E.T.: Action Research: A Handbook for Practitioners. SAGE Publications, Inc., Thousand Oaks (1996)
21. Herr, K., Anderson, G.: The Action Research Dissertation – A Guide for Students and Faculty. SAGE Publications, Inc., Thousand Oaks (2005)
22. Greenwood, D., Levin, M.: Introduction to Action Research: Social Research for Social Change, 2nd edn. Sage, Thousand Oaks (2007)
23. Reason, P., Bradbury, H.: The SAGE Handbook of Action Research: Social Research for Social Change, 2nd edn. Sage, Thousand Oaks (2008)
24. Coghlan, D., Brannick, T.: Doing Action Research in Your Own Organization, 3rd edn. Sage Publications, Thousand Oaks (2010)
25. Coghlan, D., Brydon-Miller, M.: The SAGE Encyclopedia of Action Research. Sage Publications, Thousand Oaks (2014)
26. Holman, P., Devane, T., Cady, S.: The Change Handbook: The Definitive Resource on Today's Best Methods for Engaging Whole Systems, 2nd edn. Berrett-Koehler Publishers, Inc., San Francisco (2007)
27. Argyris, C., Schön, D.A.: Participatory action research and action science compared: a commentary. Am. Behav. Sci. **32**(5), 612–623 (1989)
28. Zuber-Skerritt, O., Perry, C.: Action research within organizations and university thesis writing. Learn. Organ. **9**(4), 171–179 (2002)
29. Gummesson, E.: Qualitative Methods in Management Research, 2nd edn. Sage, Thousand Oaks (2000)
30. Shani, A.B., Pasmore, W.: Towards a new model of the action research process. Acad. Manag. Proc. (1982)
31. Susman, G.I., Evered, R.D.: An assessment of the scientific merits of action research. Adm. Sci. Q. **23**(4), 582 (1978)
32. Kemmis, S.: The Action Research Reader, 3rd edn. Deakin University Press, Victoria (1998)
33. Mezirow, J.: Transformative Dimensions of Adult Learning. Jossey-Bass, San Francisco (1991)

34. McKay, J., Marshall, P.: Driven by two masters, serving both - the interplay of problem solving and research in information systems action research projects. In: Kock, N., Information Systems Action Research: An Applied View of Emerging Concepts and Methods (2007)
35. Peffers, K., Tuunanen, T., Rothenberger, M.A., Chatterjee, S.: A Design Science Research Methodology for Information Systems Research. J. Manag. Inf. Syst. **24**(3), 45–77 (2007)
36. Van Aken, J.E.: Management research based on the paradigm of the design sciences: the quest for field-tested and grounded technological rules. J. Manage. Stud. **41**(2), 219–246 (2004)
37. Hevner, A., Chatterjee, S.: Design Research in Information Systems: Theory and Practice. Integrated Series in Information Systems. Springer (2010)
38. Reeves, T.C.: Design research from a technology perspective. In: van den Akker, J., Gravemeijer, K., McKenney, S., Nieveen, N. (eds.) Educational design research, pp. 52–66. Routledge, London (2006)
39. Hevner, A.R.: A three cycle view of design science research. Scand. J. Inf. Syst. **19**(2), 87–92 (2007)
40. Offermann, P., Levina, O., Schönherr, M., Bub, U.: Outline of a design science research process. In: 4th International Conference on Design Science Research in Information Systems and Technology, vol. 11 (2009)
41. Sein, M.K., Henfridsson, O., Purao, S., Rossi, M., Lindgren, R.: Action Design Research. MIS Q. **35**(1), 37–56 (2011)
42. Argyris, C.: Double-loop learning, teaching, and research. Acad. Manag. Lean. Educ. **1**(2), 206–218 (2002)
43. Senge, P.: The Leader's New Work - Building Learning Organizations. In: Gallos, J.V. (ed.) Organization Development: A Jossey-Bass Reader, vol. 38, pp. 765–792. Wiley, San Francisco (2006)

Approaching Ethnographic Research About Human Interaction as Making Music Together

William K. Rawlins(✉)

Ohio University, Athens, OH, USA
rawlins@ohio.edu

Abstract. This essay develops the idea of perceiving and practicing ethnographic research through the lens of music-making. I propose and initially explore four tenets of approaching ethnographic research about human interaction as aesthetic activities of making music together. Accomplishing this stance involves: (1) active involvement and inclusion of all parties in the emerging process of composition, improvisation, and performance; (2) committed listening, attunement, and mutual responsiveness; (3) achieving reflexive immersion while embracing the ethical demands of assemblage; and (4) co-achieving rhythm and the connections of musical temporality and form. In doing so, I emphasize theoretical and lived understandings of an aesthetics of interpersonal communication. I argue that it is vital to recognize the interconnection of aesthetic activities with everyday life and the opportunity to learn and create possible understandings together in ethnographic inquiry.

Keywords: Aesthetics of interpersonal communication
Ethnographic research · Making music · Listening · Attunement
Assemblage · Rhythm

1 Introduction

In this essay I develop the idea of perceiving and practicing ethnographic research through the lens of music-making. I believe that our investigative conversations can aspire to the dialogical qualities of music-making that welcome us across cultural backgrounds to belong communicatively with others while valuing our singular qualities as individuals. Musical dialogue understands human identities as accomplished in concretely situated and embodied relationships simultaneously involving meaningful similarities and differences between people. Every person's presence is valued. I argue that aspiring to interact in the spirit of performing music together richly personifies the dialogical potentials of communication.

Accordingly, I propose and initially explore four tenets of approaching ethnographic research about human interaction as aesthetic activities of making music together. Accomplishing this stance involves: (1) active involvement and inclusion of all parties in the emerging process of composition, improvisation, and performance; (2) committed listening, attunement, and mutual responsiveness; (3) accomplishing reflexive immersion while embracing the ethical demands of assemblage; and (4) co-achieving rhythm and the connections of musical temporality and form. In doing so, I

A. P. Costa et al. (Eds.): WCQR 2018, AISC 861, pp. 202–209, 2019.
https://doi.org/10.1007/978-3-030-01406-3_17

emphasize theoretical and lived understandings of an aesthetics of interpersonal communication. As a result, I first will consider a commitment to aesthetic practices and ideals in reflexively developing understandings of and with other persons. Embodied ethnographic inquiry involves creative communicative endeavors accomplished with others. As such, there are constructive implications of cultivating imagination, creativity, dialogical engagement, listening, mutual attunement, rhythmic responsiveness, improvisation, answerability, and accountability to each other as communicators for solving problems and enhancing everyday communication. It is vital to recognize the interconnection of aesthetic activities with everyday life and the opportunity to learn and create possible understandings together in ethnographic inquiry.

2 Toward an Aesthetics of Interpersonal Communication

I develop here an account of *aesthetic* communicative practices in achieving and sustaining edifying ethnographic conversations. In proposing this outlook, I draw considerable inspiration and conceptual insight from John Dewey's *Art as Experience* [9]. Dewey observes, "art is the most effective mode of communication that exists". To view interpersonal communication in pursuing ethnographic understandings as an artful endeavor means "recovering the continuity of esthetic experience with normal processes of living". Artful endeavors are not rarefied achievements restricted to elite contexts. The aesthetic performance of everyday life involves "everyday making", an ongoing process of "making up (mental imagining) and making-real (material realization)" undertaken with others [14]. Although specific experiences of creative activity gravitate toward wholeness and a shared sense of fulfillment, such satisfaction also nourishes desire for more creative activity, self and world expansion [7, 9]. Aesthetic experiences are not finalized once and for all – continuity and change in their materials, meanings, participants, and contexts compose the pulse of possibility. In developing my full account, I will place Dewey's ideas in dialogue with those of Kenneth Burke, Mikhail Bakhtin, Martin Buber, Gregory Bateson, and Elaine Scarry, whose positions complement and extend Dewey's in exciting and edifying ways.

For Dewey "the work of art" presupposes reflexive communicative relationships. It involves activities of imagination, invention, and composition. At the same time, such work draws upon previous experiences, traditions, materials, and routines to create "fresh meanings" [9]. Creative expression is embodied activity utilizing readily available resources. It results in material, shareable objects for experience by self and others – pictures, tools, sound waves, songs, well-turned phrases, comforting gestures, turns at talk, pageants of welcome, dwellings, all manner of "artifacts" [14] and "equipment for living" [8]. Yet as creators do the work of artistic creation, they also reflexively undergo the incremental and culminating results of their efforts. Artistic creation is simultaneously recreation of the materials at hand, self, others, and worlds.

Meanwhile, the work of art is never fully constituted merely as a material realization of someone's imaginative efforts. Aesthetic creations require consummation in and through the experience of others. We need others to create aesthetically fulfilling expression. Bakhtin adamantly underscores this necessity for others to consummate aesthetic activity in *Art and Answerability* [1]. Our co-participation as co-creators and co-experiencers of cultural understandings as artful endeavors of interpersonal communication highlights the ongoing poiesis and praxis of rounding out our lives with others [11]. Just as we are reflexively doing and undergoing our own creative efforts, we serve in the potentially creative capacity of witnessing, appreciating and thereby consummating the artful deeds of others. Our reflexive experiences of our own (re) creations provide the lived potential for actively fashioned empathy for others' attempts to make something of themselves and their circumstances. As such, Deweyan aesthetics include a conception of appreciatively critical perception in experiencing the events of one's life. How might we cultivate such an outlook to notice the beauty and potential for renewal in our everyday moments learning with and seeking to understand others? In further development of these ideas, Bakhtin's meditations on the "excess of seeing" [1] and the "once-occurrent moment" [2], and Buber's [5] articulations of full "presence" in our encounters with other persons will help me elaborate a response.

In harmony with the work of art, form and rhythm arise through the co-activities of expression and perceptual experience by participants. In a general sense, whenever we enact or perceive everyday events as a gestalt or with "a sense of qualitative unity," form is present [9]. Like Bateson [3], Dewey insists that perception itself is never static. With varying degrees of conscious intent, our lived experiences are rendered coherent through our integration of them into configurations over time. However, artistic endeavors are distinct in their deliberate patterning of conditions to accomplish meaningfully unified wholes. Accordingly, Dewey defines artistic form as "the operation of forces that carry the experience of an event, object, scene, and situation to its own integral fulfillment" [9]. Artful form involves "an experience that is carried to consummation" [9]. How such form recurs becomes a matter of the "organization of energy" for Dewey, activities of arousing energy and bringing it to rest [9]. The door is open for Burke's explicitly communicative and audience-oriented definition of form as "the creation of an appetite in the mind of an auditor, and the adequate satisfying of that appetite" [7]. Bakhtin [1] thickens this plot intriguingly with his detailed consideration of form as experienced "within" by human beings in aesthetic activity, and form as experienced "without" and bestowed as a gift from others, his so-called "inner and outer form". So Bakhtin shares Dewey and Burke's concern with meaningful consummation as an essential feature of aesthetic form that arises through the perceptions and responses of others. We are vulnerable to each other's responses and in need of each other's consummation to co-create a well-formed learning moment or meaningful interview conversation. We honor each other through our active participation and acknowledgement.

3 Ethnographic Research as Making Music

I propose that the shared experience of making music together richly embodies aesthetic qualities of communication and simultaneously serves as a vivid metaphor for understanding and participating in interpersonal communication in ethnographic work. For example, I am continually struck by the accountability to each other and the mutual attentiveness toward the cumulative progression of a song that transpires when playing music. As I develop my consideration of co-creating communicative aesthetics and co-learning as a musical activity of ethnographic inquiry, I reflect upon my own lived experiences as a musician to develop examples of what I am discussing. While addressing rhythm, space, and time, Dewey remarks "music in its evident temporal emphasis illustrates perhaps better than any other art the sense in which form is the moving integration of an experience" [9]. Small [15] echoes this outlook in describing musicking. He argues that the process of making music "is part of that iconic, gestural process of giving and receiving information about relationships which unites the living world, and it is in fact a ritual by means of which the participants not only learn about, but directly experience, their concepts of how they relate, and how they ought to relate, to other human beings and the rest of the world." Conceiving of ethnographic conversations as opportunities to make music together offers compelling capacities for embodying dialogue in co-creating insights about cultural practices. I briefly describe here the aspirations of approaching ethnographic research as making music.

First, *pursuing ethnographic dialogue as musical requires active involvement and inclusion of all parties in the emerging process of composition, improvisation, and performance*. It requires exchanging and responding to everyone's contributions in sustaining our music-making as a communicative relationship with others. You have to show up in the space of performance. Emphasizing our co-produced conversation as a musical performance highlights Bakhtin's "once-occurrent moment" [2] and Buber's "personal making present" [6]. It highlights the unfolding significance of our meeting in *this* moment, gathering where we are to achieve its richest potentials. We should facilitate our respective capacities for emerging contributions in our conversations in the field and avoid assuming the primacy of our own turns at talk as "the researchers."

Meanwhile, there are historical and cultural grounds for our respective contributions. We need to allow our differences to motivate our desire to understand each other more fully. They constitute a significant reason for why our conversation is occurring in the first place. This being so, co-performing our knowledge-producing conversation in a dialogical spirit of making music involves Buber's dialogical requirement of simultaneously standing one's ground as a culturally constituted music maker and allowing others to happen to oneself [16]. Each of us has distinctive convictions that make us who we are, but that should not foreclose the opportunity to hear what the other has to offer.

Adopting this stance towards ethnographic inquiry suggests practices for actively communicating respect for our co-participants in research, encouraging their involvement as equals in the co-creation of understandings, and expressing gratitude for the insights they embody and impart through their contributions to every interaction we engage in together. Modeled on music-making, this approach makes ethnographic inquiry inherently more participatory. It reenvisions field interviews as mutually informative conversations and imaginative activities of building possibilities for co-learning and sharing understanding, however local or global in scope.

Second and crucially important, *pursuing ethnographic conversations as musical emphasizes the creative, ethical, and mutually affirming significance of engaged listening*. As previously mentioned, intermingled with those of others, we also hear our own contributions. Throughout their realization, aesthetic creations like music require consummation through the experience of others. Our embodied attempts at musical dialogue affirm the personhood of others. We need others to create aesthetically fulfilling expression – and nowhere more so than in the collective performance of music. Understanding art as communication, Dewey insists, "The hearer is an indispensable partner. The work of art is complete only as it works in the experience of others than the one who created it" [9]. Conceiving ethnographic conversations as dialogical music-making insists upon the vital role of listening carefully and responding in the accomplishment of our co-learning. Every person's presence is valued. Making music together demands *attunement* to and with others. It is vital to acknowledge everyone's contributions, experience our accountability to each other, as well as the collective affirmation and shared attentiveness toward the cumulative progression of a song that transpires when making music.

Fully embracing a commitment to listening heightens our effectiveness as ethnographers by enhancing our receptivity to other persons' words and nonverbal behaviors in speaking with and attuning ourselves to them. Doing so requires a commitment to be completely available while in the presence of another person. To achieve this, it is vital to minimize distractions and to be mindful of the time and place of our conversation in each other's immediate lives. We must work to optimize the circumstances in which our dialogue occurs. Actively passive, yet involved and responsive listening nourishes the other's willingness to speak with us [12]. In doing so, we also must appreciate the significance of silent moments, which give us all the chance to pause and reflect on what has and has not been said. Conversational participants should not experience tacit pressure to speak when they would rather hold their peace [12]. Just as in music-making, devoted listening simultaneously embodies an invitation to contribute and ongoing affirmation of each other's communicative choices.

Third, *while doing the work of musical creation, we also reflexively undergo the incremental and culminating results of our efforts in co-creating an assemblage.* In making something new together, we remake our own thoughts, feelings, and relationships with others, an aesthetic process John Dewey calls "doing and undergoing." This reflexive self and other-making constitutes an "assemblage" [17] that embodies "an ethical contract" built upon "a self-other obligation" [10]. This co-achieved ethic contrasts with moral standards imposed from outside of our ethnographic co-learning activities and conversations. Of course, as responsible inquirers, we enter the research context observing the advice of our colleagues on institutional review boards. However, when informed by an overarching metaphor of making music together, the truly constitutive practices of our co-learning are achieved *with* each other through the edifying practices of our actual interaction. Our reflexive experiences of our own creations enable actively fashioned empathy for others' attempts to make something of themselves and their circumstances. Including everyone in this way, we embrace and share the risks of creating and performing together. The space of such performance teeters on collectively performed possibilities unfolding in time. We are answerable and accountable to each other yet simultaneously invited to enter into the music-making of our ethnographic dialogue when and how we may.

Engaging ethnographic research as musical assemblage compels us to own and openly value the reflexive nature of our research activities. Just as making music together unfolds over time through playing certain notes and then experiencing their effects on us as parts of an emerging whole, as we co-create understandings with others, we are changed ourselves. We reflexively experience the incremental and palpable results of our mutual efforts. I believe that this orientation toward fashioning and experiencing understandings - that all parties have been involved in making - reduces our detachment and humanizes the process and products of ethnographic research. In co-creating knowledge, we simultaneously recreate selves, others, and the worlds in which we dwell. And we do so according to shared standards that we have established in our practices of learning together.

Fourth, one of music's most palpable dialogical capacities is that of shaping time and collective experiences of temporality. Each and *every ethnographic encounter embodies a particular rhythm of activity*. The rhythm we perform together co-fashions time. Such rhythm transpires in conjunction with aesthetic form. Considering it "a condition of form," Dewey defines rhythm as "ordered variation of changes" [9]. Seemingly as micro-moments of Burkean [7] form, "Each beat, in differentiating a part within the whole, adds to the force of what went before while creating a suspense that is a demand for something to come" [9]. Dewey further remarks, "each recurrence is novel as well as a reminder", continuing, "Every closure is an awakening, and every awakening settles something" [9]. As a drummer and a guitarist myself, witnessing this sensitivity to the interactive functions and realization of the meaningful pauses and soul-disclosing emphases of well-realized beats takes my breath away.

Bakhtin [1] also devotes much attention to rhythm as a co-produced embodied experience of moral significance in shaping human interaction—"a beautiful necessity" that we do not create for ourselves but engage in for others' benefit. In a passage of rare beauty, Bakhtin observes, "Wherever the purpose of a movement or an action is incarnated into the other or is coordinated with the action of the other—as in the case of joint labor—my own action enters into rhythm as well. But I do not *create* rhythm for myself: I *join* in it for the sake of the other. Not my own nature but the human nature in me can be beautiful, and not my own soul but the human soul can be harmonious" [1].

Understanding how such connection and disconnection recurs with our participants and for them in composing their own lived experiences and worlds becomes a matter of the "organization of energy" for Dewey – doing and undergoing, surrender and reflection, repetition and variation in rhythm, activities of arousing energy and bringing it to rest, resulting in "a changed suspense" [9]. Yet, what temporal regimes contextualize the organization of peoples' energies? The interactive practices of ethnographic research emerge from and engage with complex, lived rhythmic configurations. It is important to reflect on the multiple domains of activity that intersect in shaping the rhythmic composition of our ethnographic research. Meanwhile, we need to acknowledge what may be the primary endeavors patterning the life rhythms of those persons we are seeking to understand. How closely do the contingencies and temporal patternings of our own activities coordinate with or permit us to grasp the internally and externally patterned rhythms of the persons' activities we are seeking to comprehend? Moreover, rhythm involves the creation of anticipated continuities out of discontinuities. In doing so, consummation is key. How do we consummate our conversations with others in ways that invite more discourse? More sharing? More possibilities [13]? Are we implicitly compelling our participants to march to the rhythm of our own lives – both in their participation in our studies and in our renderings of what we have learned?

Thematizing form, rhythm, and shared music-making attunes us to both the temporal unfolding and the perceived wholeness of our communicative activities. In pursuing ethnographic research, I want us to attend carefully to how we embody aesthetically the lived experiences of multiple modes of temporality, such as calendar and clock time, sequenced and simultaneously configured events, the dramatic timing of occurrences, and the musicality of human interaction. Doing so honors the ways of life at stake in our research and the co-lived nature of the diverse temporal realities composing ethnographic inquiry.

4 Conclusions

I believe that working and speaking together in this fashion in ethnographic inquiry continually enacts the potential for activating consciousness. Approaching ethnographic research about human interaction as making music together potentially entails responsiveness, appreciative being, active listening and involvement, as well as reflexive communicative relationships in the time we spend together face-to-face in the field. We must show up in the moments we are in each other's presence to create fresh meanings.

In developing further this aesthetic account of ethnographic inquiry as music-making, I envision addressing other vital issues that I can only briefly mention here. First, I want to probe in this context what it means to communicate with *integrity*. Throughout *Art as Experience* [9], Dewey emphasizes the capacities of art to integrate form, subject matter, external materials, and selves' and others' experiences of human contexts. He is concerned with "the fulfillment of an experience in terms of the integrity of the experience itself"—performing an act that "is exactly *what* it is because of *how* it is done" [9]. By these lights, what does it mean to communicate and learn together with integrity? As communicators pursuing ethnographic understandings, how do we bring activities to fulfillment while preserving openness to other persons and to fresh possibilities? How do we accomplish aesthetic integrity without insularity? Second and related, how do we cultivate our respective *capacities to create* within the forms and rhythms we enter into and accomplish with others? How can we learn how to punctuate relationships and contexts differently to identify possibilities for doing things differently? How can we practice and encourage improvisation? How can we cultivate resourcefulness in the composition of our lives with others? Third, I want to explore *aesthetic practices as proposing alternative worlds for human being and becoming*. There are crucial connections among notions of taste and belonging to a community of aesthetic beings, that is, human beings. Burke describes "artistic truth" as "the exercise of human propriety" and as "the externalization of taste" [7]. Through our aesthetic activities we propose standards concerning what we believe is beautiful, in good taste, and worth sharing with others. Working with Hannah Arendt's writings, Beiner [4] states, "Taste 'decides not only how the world is to look, but also who belongs together in it.' It defines a principle of belonging, is an expression of the company one keeps, and as such, like politics itself, is a matter of self disclosure". To the extent that we share conceptions of the beautiful, we are able to create and belong together in a common world. An aesthetic object or event that we agree is worthy may invoke this world momentarily, or it may constitute a basis for extended association and altered ways of life. We offer up our practices, our products, and our physical attributes as embodied beings for confirmation and consummation by others according to evolving aesthetic standards—which simultaneously function as premises for sharing or denying access to a world. Consequently, proposing new aesthetics is proposing new worlds for human dwelling. It is therefore important to assert the interconnection of aesthetic activities with everyday life and opportunities.

References

1. Bakhtin, M.M.: Art and answerability. In: Holquist, M., Liapunov, V. (eds.) Art and Answerability: Early Philosophical Essays by M. M. Bakhtin. University of Texas Press, Austin (1990)
2. Bakhtin, M.M.: Toward a Philosophy of the Act. University of Texas Press, Austin (1993)
3. Bateson, G.: Steps to an Ecology of Mind. University of Chicago Press, Chicago (2000)
4. Beiner, R.: Political Judgment. University of Chicago Press, Chicago (1983)
5. Buber, M.: I and Thou. T & T Clark, Edinburgh (1937)
6. Buber, M.: The Knowledge of Man: A Philosophy of the Interhuman. Harper, New York (1956)
7. Burke, K.: Counter-Statement. University of California Press, Berkley (1968)
8. Burke. K.: The Philosophy of Literary Form. University of California Press, Berkley (1973)
9. Dewey, J.: Art as Experience. Perigee, New York (2005)
10. Elliott, D.J.: Music Matters: A New Philosophy of Music Education. Oxford University Press, New York (1995)
11. Peterson, E.E., Langellier, K.M.: The performance turn in narrative studies. Narr. Inq. **16**, 173–180 (2006)
12. Rawlins, W.K.: The Compass of Friendship: Narratives, Identities, and Dialogues. Sage, Thousand Oaks (2009)
13. Rawlins, W.K.: Brimming moments: rhythm, will, readiness, and grace. Deps. Crit. Qual. Res. **3**, 76–88 (2014)
14. Scarry, E.: The Body in Pain: The Making and Unmaking of the World. Oxford University Press, New York (1987)
15. Small, C.: Musicking: The Meanings of Performing and Listening. Wesleyan University Press, Middletown (1998)
16. Stewart, J., Zediker, K.: Dialogue as tensional, ethical practice. S. Commun. J. **65**, 224–242 (2000)
17. Turetzky, P.: Rhythm: Assemblage and Event. Strat. **15**, 121–138 (2002)

Qualitative Architecture. A Multidisciplinary Approach to the Provision of Social Housing. Case Study Cuenca - Ecuador

Marco Avila Calle(✉), Mauricio Orellana Quezada, Jorge Toledo Toledo, María de los Ángeles Tello Atiencia, and Federico Córdova González

Catholic University of Cuenca, Academic Unit of Engineering, Industry and Construction, Av. De las Americas and General Torres corner s/n, Cuenca, Ecuador
{mavila,morellanaq,jftoledot,mtello, ncordovag}@ucacue.edu.ec

Abstract. The cost of social housing is very high in Ecuador and in some Latin American countries, so the social class with low economic resources cannot access these housing programs, which is why self-construction has become the only viable alternative. to obtain housing in this social level. In the present investigation of qualitative and multidisciplinary approach, the social and habitability conditions presented by the population of low economic resources categorized in social stratum C- in the city of Cuenca are analyzed, in order to establish the spatial and constructive requirements, which allows generating design strategies to achieve a sustainable and affordable housing proposal for the social study group, guaranteeing covering the qualitative and quantitative deficit of housing in the city, thus improving the perception of the quality of life of the users.

Keywords: Qualitative research · Constant comparative · Social housing Architecture

1 Introduction

Qualitative research in the social sciences allows researchers to know firsthand the reality or nature of the phenomenon investigated, linking the human being as the central axis of the problem. There are many areas of professional work that make use of qualitative methodologies to study a social conflict; Communication, Linguistics, Sociology, Anthropology, Psychology, Marketing, Economics, Political Science, among others, have shown that qualitative research plays a major role in the solution of such conflicts.

According to the International Standard Classification of Education (ISCED) of UNESCO, Architecture is part of the Knowledge Area of Engineering, Industry and Construction [1] by its nature, however, in the process of research, design and construction in Architecture, qualitative and mixed methods are used to solve the problems and needs detected in society through the architectural work.

A. P. Costa et al. (Eds.): WCQR 2018, AISC 861, pp. 210–222, 2019.
https://doi.org/10.1007/978-3-030-01406-3_18

As the architectural work is a direct response to the needs of man, it must be conceived of the space designed as a modeling element of human behavior, so it is necessary to start any architectural project with user analysis, to establish the needs to be met, in This process qualitative research plays a key role.

1.1 The Problem

In the design process within any method of elected architectural design begins with the detection of user needs, to be able to propose the architectural program that guides the design. It is at this stage where the architect must consider multiple analysis techniques in order to abstract the primary requirements of potential users. In the professional context, the architect must face the reality of a society with multiple problems, that to identify them is not enough with quantitative research because human perception, customs, traditions, etc., can be different between individuals, homes and much more in a community, city or country.

For that reason, in many occasions the role of the architect is transformed into that of a sociologist, to understand social behaviors, in a psychologist, to be able to understand the individual attitudes, in an economist, to analyze the market and financing possibilities of the project, in a doctor, to detect the health problems inherent in the lifestyle of their clients, is that, in these research processes in the architectural field qualitative methods are of great interest. However, the professional of architecture must work multidisciplinary with other professionals to help the architectural work is the result of an integral diagnosis and not as a response to subjective conditions, then, we must consider that through the designs the daily work of the clients, if appropriate spaces are proposed, the activity will be adequate, but if deficient spaces are proposed, the activity will be deficient, and this is achieved only by knowing the nature and reality of the potential users.

The problem deepens when the work transcends the individual client and focuses on the collective, becoming social housing in one of the most complicated architectural projects to plan for being immersed in the social, economic, political, cultural, etc., in the present study analyzes the problematic of social housing in the city of Cuenca Ecuador, since the qualitative and quantitative deficit of social interest housing in Latin America is worrisome, the lack of access to basic infrastructure services afflicts many families of the region: 21% lack electricity and sanitary facilities. 12% of homes are inadequate construction materials, while 6% have dirt floors or overcrowded conditions, which are harmful to health [2]. From the point of view of the environment and health, the human being needs to live and develop in an environment that provides security, comfort and protection, considering that the environment is directly related to the health conditions of people which remain in the home daily at least 50% of their time in these spaces, housing is an element of vital importance in public health in terms of study of conditions for risk assessment and quality in health of those who inhabit these homes [3].

There are several aspects of vital importance that are considered to develop architectural proposals especially when addressing the issue of social housing, as the housing model must adjust to meet the basic needs of the human being, this space should be an instrument that facilitates biological performance, psychological and social, according to a study conducted on healthy housing indicates that housing, microhabitat of the species, is a determinant of health. Its component elements can contribute to the health/illness of its residents. The risk factors of housing for health can be of a physical nature (electromagnetic fields, noise, mechanical vibrations, micro-climate), chemistry (materials of risk - asbesto), biological (pollen, bioaerosols, rodents, arthropods) and psychosocial (stress-shortages, inappropriate family relationships) [3].

To the above, the socio-spatial segregation (Ghetto) is added, with serious injustices in terms of access to urban goods and services, which makes it impossible for certain population sectors to acquire decent housing [4]. The city of Cuenca in 2001 had a population of 612,565 inhabitants, for the year 2010 it was 702,893 inhabitants, that is, a 14.7% increase, from these data it is determined that the housing deficit is from 35,000 to 45,000 [5], being the most affected class C - considered low income in the city.

The problem posed is approached from two investigative approaches; qualitative and quantitative, with the first approach is to categorize the socio - spatial and constructive patterns of social housing in the city of Cuenca, while the quantitative approach will guide the research to confirm the patterns and define the recommendations for the design of housing projects in the city of Cuenca. When the congress on qualitative research is discussed in this article, the investigative work that gathers the perceptions, customs, traditions and lifestyles of the social stratum C-, in the city of Cuenca, Ecuador, is evidenced.

Social housing is conceived as an individual and complex entity today, therefore it is necessary to find a methodological support in which an integrative and multidisciplinary study is feasible, which nevertheless does not remain in the generalities, but may deepen in knowledge. Rodas Beltrán Ana Patricia [6] in her study: Habitability in social housing in Ecuador from the view of complexity: development of a system of analysis, proposes that a research on social housing cannot be approached from a simplistic view, should be through complex thinking, using as a methodological scheme the theory of complex systems. In addition, in his study defines the elements that make up the analysis system of social housing (see Fig. 1), in this research these elements are used as guides for the preparation of the questionnaire in the in-depth interview and are the main categories in the coding phase.

1.2 Objectives

General

Categorize socio - spatial and constructive patterns of social interest housing for class C-, through constant comparative analysis to propose recommendations in the design of housing projects in the city of Cuenca.

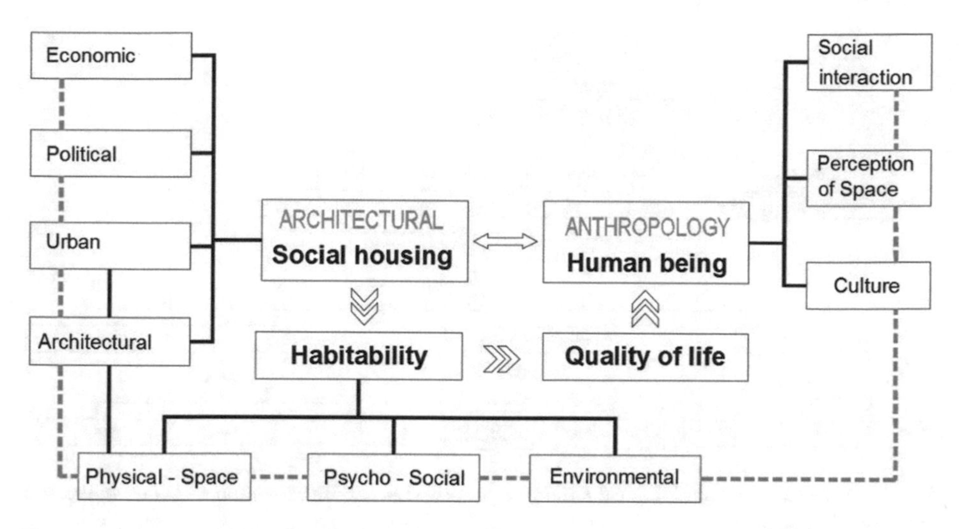

Fig. 1. Proposal of a complex analysis system for the study of social housing from a multidisciplinary approach.

Specific

1. Identify the settlements of social interest housing in the city of Cuenca.
2. Explain the socio-spatial and constructive patterns of low-income housing for class C - in the city of Cuenca.
3. Generate recommendations for the design of housing proposals for social class C - in the city of Cuenca.

2 Methodological Strategies

The research is of qualitative explanatory approach and uses the "Grounded Theory" initially presented by Barney Glaser and Anselm Strauss in "The Discovery of Grounded Theory" in the year 1967 [7], supported by the method of comparative analysis constant. Figure 2 shows the methodological scheme used in the investigation.

2.1 Constant Comparative Analysis Method

The chosen method allows the researcher to link directly with the reality of the users and guarantees that the information collected is reliable, relevant, relevant and effective because the method of constant comparative analysis seeks to build theory rather than prove it or discover it through the formulation of Sensitive, theoretical and practical and structural questions by comparing incidents in terms of their properties and dimensions, and in terms of their similarities and differences. This construction implies a new look in accordance with the ontological and epistemological assumptions of the researcher, that is, it is to explain in a new way the reality that is intended to be understood [8]. Several authors define the phases of qualitative research according to the method used,

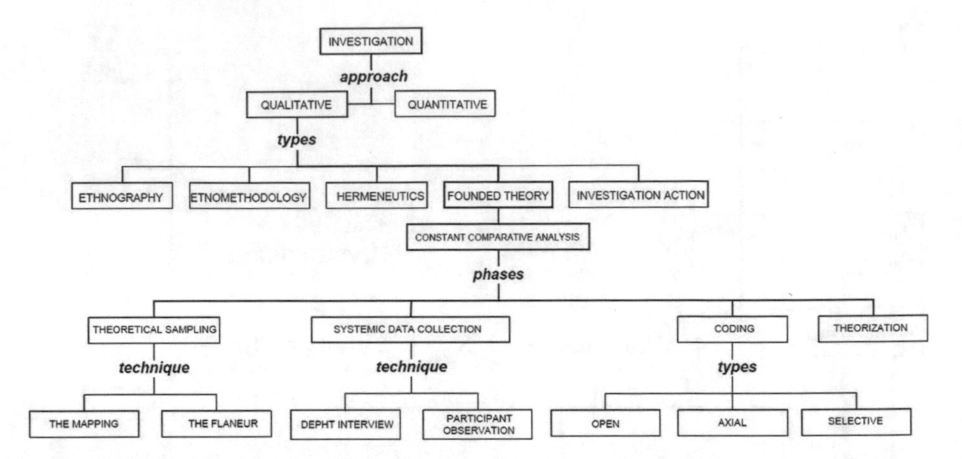

Fig. 2. Scheme of methodological strategies proposed in the investigation of social housing in the city of Cuenca.

however, all coincide in four non-linear phases, for this research case the following order is presented: (a) Theoretical sampling, (b) Systemic data collection, (c) Coding and (d) Theorization.

Theoretical Sampling: Unlike statistical sampling in qualitative research, theoretical sampling is cumulative. Data collection is guided by the concepts derived from the theory being built and based on the concept of "making comparisons" whose purpose is to go to places, people or events that maximize opportunities to discover variations among concepts and make the categories denser in terms of their properties and dimensions [7]. Three types of theoretical sampling are distinguished; Open, fluctuating and discriminative relational, each of them is directly related to the specific coding that you want to perform. In the present case study, the following techniques have been used to define the appropriate sampling:

The "mapping": One of the basic elements in the beginning of work of a qualitative nature has to do with the problem of placing oneself mentally in the terrain or scenario in which the research will be developed. To achieve this purpose, one of the starting processes is what Anglo-Saxon literature calls "mapping" and which is translated as mapping or mapping. When you want to guide yourself in an unknown place, you get a map or, failing that, prepare it when it does not exist or is not available [9]. In the context of the research, the concept of a map has been considered literally, since through basic sketches the sectors to be investigated were delimited. Cuenca is a city where social cohesion in the area of housing is very high, there are no specific sectors where the population is socially divided, rather, there is a passive coexistence between different social strata. In Fig. 3, through mapping, supported by the flaneur 11 sectors of the city were defined where there has been greater growth in terms of human settlement to make observations and depth interviews.

The "flaneur": Through this technique known as flanear you can observe the places as scenarios and experience their elements, colors, sounds and movements, by making spontaneous tours of the city, photographing nodes, milestones, or significant places [10].

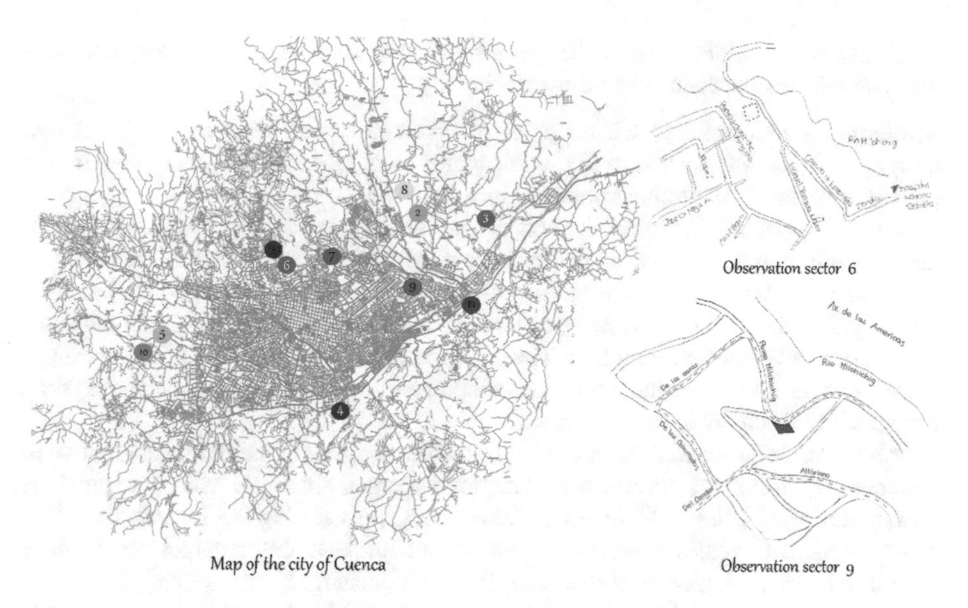

Fig. 3. Mapping applied to identify the sectors in which cases of social housing are found.

Figures 4 and 5 show the application of the flaneur in the selected sectors, within which the presence of social housing in the city of Cuenca can be evidenced.

Fig. 4. Flaneur applied in observation 01

Fig. 5. Flaneur applied in observation 10

The discriminative sampling: With the mapping and flaneur techniques, 11 sectors and 50 case studies where the data collection was carried out were concretely defined. It was not necessary to start with an open or relational sampling, because guided by the categories of the complex analysis system it was possible to deepen the study of these categories in the social housing, for which the type of defined sample was the discriminative one that is associated with the Selective coding and its principle is to maximize opportunities to verify the argumentation or constructed argument, as well as the relationships between existing categories [9].

Systemic Data Collection: It is the phase in which the researcher, through his research design, collects information using appropriate techniques and instruments. In the present case study, the following techniques are used:

Depth Interview: This interview consists of a structured conversation with each of the key informants, through an interview guide, which contains open questions or topics to be discussed derived from the indicators that wish to be explored [10]. For the investigation, the depth interview was elaborated and applied to the observational cases defined in the theoretical sampling. The 20 open questions of the interview questionnaire were in direct relation with the components of the complex analysis system proposed by Rodas Beltrán Ana Patricia.

Participant observation: Participant observation emerges as a different alternative to conventional forms of observation. The fundamental difference with the previous observation model lies in a characteristic concern, for carrying out its task from "inside" the human realities it intends to address, in contrast to the "external" view, those of non-interactive forms of observation [9]. This technique was applied during the investigation and the field journal and the photographic archive were used as instruments, which allowed to collect information to contrast it with the one collected in the depth interview, in Fig. 6 an extract of the architectural survey of the newspaper is presented of field and analysis of the photographic archive made in the observation of sector 9 in the city of Cuenca.

Fig. 6. Photographic analysis of materials in observation 09

Coding: coding is the process by which the information obtained is grouped into categories that concentrate the ideas, concepts or similar topics discovered by the researcher. Codes are labels that allow units of meaning to be assigned to descriptive or inferential information compiled during an investigation [11]. The present investigation was guided by the categories already defined by the complex analysis system such as Habitability, Social Housing, Human Being and Quality of Life, the subcategories correspond in the same way to those indicated in Fig. 1 linked to each category. With the help of the ATLAS software. The collected information was coded by grouping them selectively as shown at the end in Figs. 7, 8, 9 and 10 to be interpreted later.

Theorization: theorization is the process by means of which alternative explanations are constructed and assumed, relying, for this, on the reading and sustained interpretation of the data generated by the research; always looking for a better, more convenient and simple explanation of said data [9]. To carry out the theorizing process in the case of the study, the chaining of data obtained from selective coding is applied and the socio - spatial and constructive patterns of social housing in the city of Cuenca - Ecuador are interpreted.

3 Presentation and Analysis of Results

Social housing in the city of Cuenca has characteristics as similar aspects inherent to the problem of social housing in Ecuador and in some Latin American countries. Undoubtedly, the social group immersed in this problem is of limited resources, and is within the category of stratification of socioeconomic level called Level C- by the Ecuadorian Institute of Statistics and Censuses (INEC). After carrying out the systemic collection of data through the depth interview and participatory observation, the information obtained was codified and processed by the ATLAS.Ti Software in which, starting from the open coding previously established by the complex analysis system, the following results were obtained within the components: Habitability, Social Housing, Human Being and Quality of Life.

3.1 Habitability

Figure 7 shows that social housing in Cuenca is a product of self-construction with recycled materials such as; the wood, block, plastic, metal gates and zinc in the cover, in some cases have been able to count on scarce economic resources product of the saving, they are interested in acquiring the bond for the housing that the governmental body provides, but unfortunately they are not favored, your home is made in stages and in some cases is linked to a workshop or business space to which the owner is dedicated.

For the social group of study, the most important spaces of your home are the bedrooms, the kitchen, and the bathroom, although a part of the owners say they feel comfortable with the bedroom spaces they have, as much they would like to have a room. More, the family composition is between three to five members, but the vast majority feels that their housing does not cover the needs of the family group for various causes of constructive and spatial nature, they would like to improve the interior environments rather than worrying about external aspects.

A high percentage of these homes due to the economic condition of the users, have problems of pests such as flies and rats, linked the problem also by the breeding of animals that is observed in a good part of this social group, another reason which this type of housing is not integrated into the neighbourhood, for the inconvenience caused to the neighbours to say of the owners themselves.

Regarding basic services, social housing in the city of Cuenca has more problems due to the economic conditions of the owners who do not hire the service, rather than the provision of these services by the municipal entity. Although the observed sectors have infrastructure works, only 64% of the homes observed have water, electricity, and sewage, have electricity and water 27%, only light 9%, while garbage collection services have the 100%. Of the sample observed, in the bathrooms they have 91% toilet, the rest use latrine and 45% have a shower; 100% of homes use a gas stove.

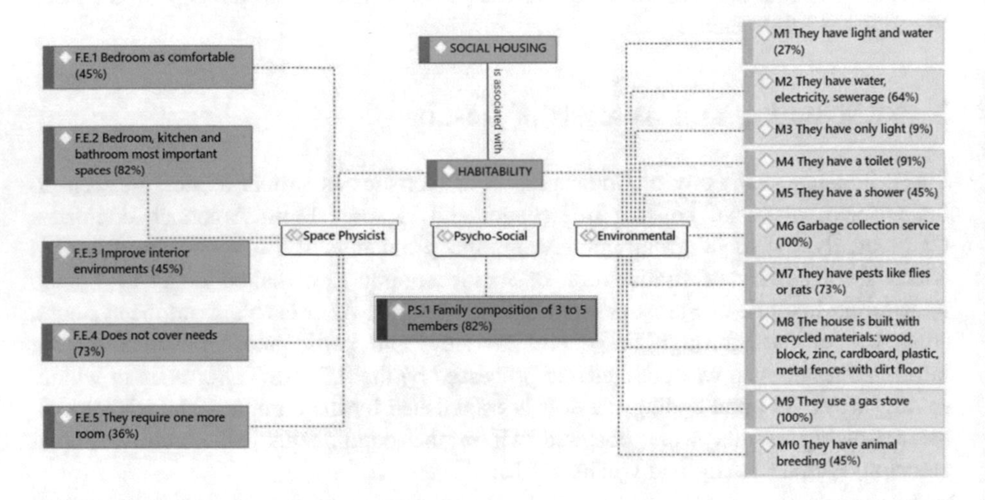

Fig. 7. Results of the perception of habitability, from the environmental (health), psycho - social, physical - spatial point of view in social housing in the city of Cuenca.

3.2 Social Housing

Figure 8 shows that the houses are mostly accessed by a dirt road, they do not have an enclosure, many do not have a front door and the kitchen is accessed first, the social area is made up of a only environment that constitutes the dining room and kitchen, have separation with the bedrooms that are arranged laterally to the social area and that are usually two although the family group is large, mostly have a single bathroom within the housing, in a small percentage they build it outside in the yard where the laundry and cellar works in some cases.

A small percentage of this group would have access to a loan if the payment conditions of the debt in terms of time and amount were convenient, since they are people who have an informal job and contribute financially to both father and mother in some cases.

The houses are built without municipal permits, but most of them are in land owned by inheritance, there are no invasion problems in the city.

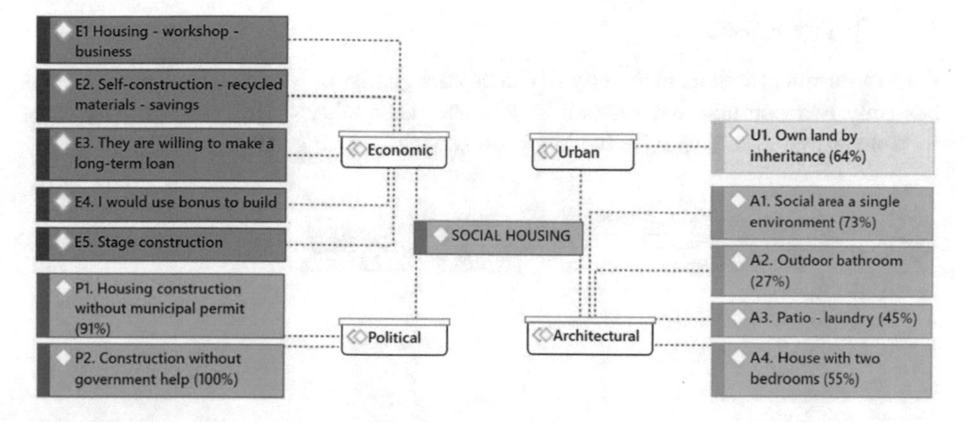

Fig. 8. Results of the perception of the economic, political, urban and architectural component of social housing in the city of Cuenca.

3.3 Human Being

100% of those that make up this group of socioeconomic level with which the social housing is linked to profess the catholic religion, the daily routine to say of those involved is getting up at dawn, father and mother go out to work, children study at both primary and secondary levels in a few cases at a higher level; One constant is that children have lunch with their mother alone, the whole family group meets at night.

The constructive characteristics that denote the particularity of these houses do not exist an articulation with the neighbourhood, it is necessary to mention that these buildings can be found in any sector of the city where they have neighbourhood social groups of medium economic extracts and even medium high, this information is summarized in Fig. 9.

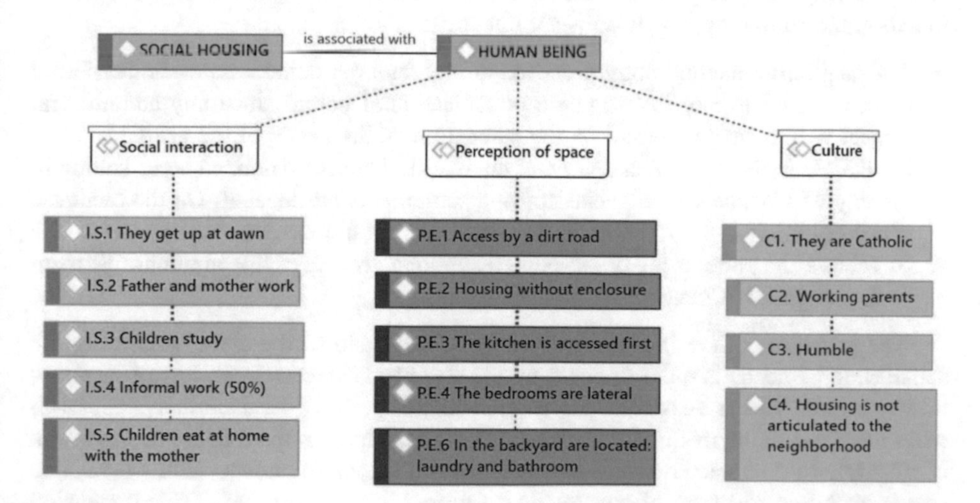

Fig. 9. Results of the perception of the cultural component, social interaction, and perception of the space of users of social housing in the city of Cuenca.

3.4 Quality of Life

For low-income families in the city of Cuenca the quality of life from the housing focus has only two conditioning factors: 73% relate the quality of life with having comfortable spaces and 27% with having basic services, see Fig. 10.

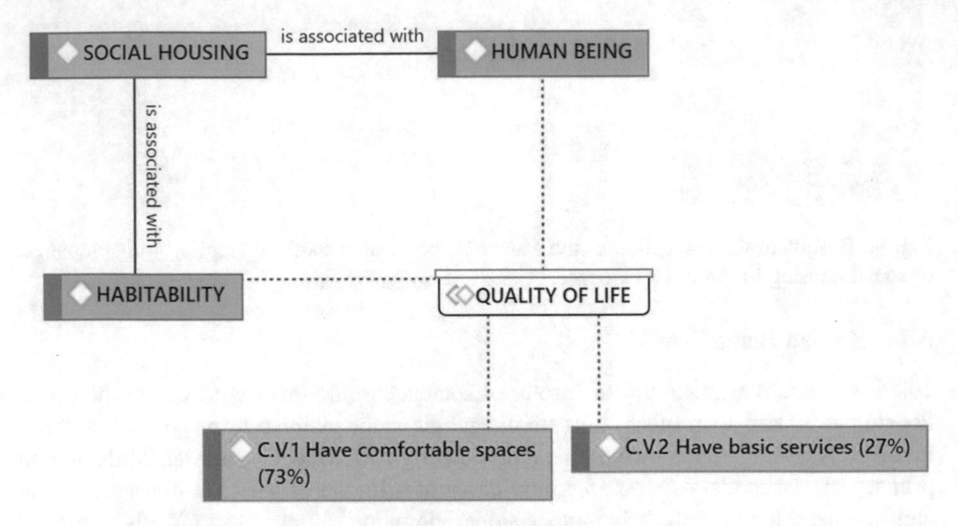

Fig. 10. Results of the perception of quality of life in social housing in the city of Cuenca.

4 Conclusions and Recommendations

After carrying out the qualitative research in the social housing of the city of Cuenca - Ecuador, the following conclusions are obtained:

- The qualitative methodology proposed in this research demonstrates the feasibility of being used in any investigation of architectural nature since any architectural project will always be based on the perception of the needs of the users.
- Unlike other cities in the country and the world, there is no defined sectorization in the city of Cuenca where social housing settlements are located. On the contrary, social cohesion is very visible in the urban and rural areas.
- In Table 1 the patterns are observed in the modus vivendi of the families of Stratum C- in the city of Cuenca.

The problem of housing has a diversity of dimensions, therefore, measuring the satisfaction of needs is quite difficult, on the one hand there is a problem of quantity, how many and who have access to housing, which is perhaps the quantitative approach more analyzed in the medium, but equally or perhaps more important is the dimension of quality: what characteristics it has and what functions the house fulfills (available space, materials and type of construction, services, equipment, etc.). This dimension involves several aspects subject to social and cultural definitions and perceptions but

Table 1. Socio - spatial and constructive patterns

Socio - spatial patterns	Constructive patterns
– Family composition is 3 to 5 members. – The most important spaces are the kitchen, bathroom and bedroom. – They are interested in comfort in interior spaces rather than exterior ones. – The perception of quality is based on having basic services and comfortable spaces. – The access to the house is through the kitchen that joins the social area and the bedrooms, the bathroom is mostly outside the house as well as the laundry. – The family meets at night mother and father work and the children study.	– Floors: wood and earth – Walls: wood, block, plastics and recycled metals. – Cover: Zinc. – Finishes: Does not exist. – Structural System: Wooden boards, adobe walls. – There are many cases of houses that have been inherited from bahareque that are in precarious housing. – The land where the construction is located is owned by the users by inheritance or donation, there is no invasion.

that must also start from minimum standards, within this analysis qualitative research is undoubtedly the ideal to obtain this information even more because in Ecuador, it has not been reached a consensus among the different actors involved in the problem of social housing, in terms of establishing these standards, this qualitative research allows a first approach to the problem of quality of social housing in the city of Cuenca with the main actors immersed in the problem that users are.

As recommendations, the following strategies are proposed that will improve the quality of life within the housing projects of social housing proposed by the government:

- Create training campaigns in the area of construction at C- levels to support the self-construction of safe houses.
- Normalize self-construction as a government strategy for the provision of low-cost social housing, as long as there is control by the relevant entities to guarantee the safety of the inhabitants.
- To authorize universities and non-governmental institutions, through projects linked to society, to provide training and construction of social housing with building systems based on recyclable materials and insurance for families, these building systems must be checked scientifically its effectiveness and safety.
- Create a system for recycling and storing materials, so that through optimal processes they can certify their quality to be used within self-construction at low cost.
- Declare as a public policy the provision of basic water, electricity and sewerage service, in cases declared as social interest housing, thus ensuring the reduction of possible pockets of unhealthiest and improving the perception of quality of life of users.
- Create government campaigns to mitigate pests such as flies and mice to prevent the spread of diseases.

References

1. Códigos Unesco: Nomenclatura para los campos de las ciencias y las tecnologías, 1–57 (n.d.). https://upct.es/contenido/doctorado/Documentos/2012/CODIGOS_UNESCO.pdf
2. Estudio del BID: América Latina y el Caribe encaran creciente déficit de vivienda | IADB (n.d.). https://www.iadb.org/es/noticias/comunicados-de-prensa/2012-05-14/deficit-de-vivienda-en-america-latina-y-el-caribe%2C9978.html#getNews(9969,”). Accessed 1 Feb 2018
3. Pérez, C.B.: Vivienda saludable: un espacio de salud pública. Revista Cubana de Higiene y Epidemiologia **50**(2), 131–135 (2012)
4. Inmobiliario, M.: Boletín CF + S 29/30, Junio 2005
5. Muñoz, O., Patricia, P., Ochoa, R., Maritza, J.: Estudio de Factibilidad Financiera para la Construcción y Comercialización de casas, Ubicadas en el sector de Challuabamba en la ciudad de Cuenca Estudio de Factibilidad Financiera para la construcción y, 111 (2011). http://www.google.com.ec/url?sa = t&rct = j&q = &esrc = s&source = web&cd = 5&cad = rja&uact = 8&ved = 0CEMQFjAE&url = http://dspace.ups.edu.ec/bitstream/123456789/1294/14/UPS-CT002241.pdf&ei = QDRdU4OmN8ThsAStzIGwCA&usg = AFQjCNGTHZxGEA7Du4Arh96knIL4qruFaQ&bvm = bv.65397613,d.cWc
6. Rodas, A.P.: La habitabilidad en la vivienda social en Ecuador a partir de la visión de la complejidad: elaboración de un sistema de análisis. In: X Seminario Investigación Urbana y Regional. POLÍTICAS DE VIVIENDA Y DERECHOS HABITACIONALES. Reflexiones Sobre La Justicia Espacial En La Ciudad Latinoamericana, pp. 1–10 (n.d.)
7. Strauss, A., Corbin, J.: Bases de la investigación cualitativa: técnicas y procedimientos para desarrollar la teoría fundamentada (2002). https://doi.org/10.4135/9781452230153
8. Carrillo Pineda, M., Leyva-Moral, J.M., Medina Moya, J.L.: El análisis de los datos cualitativos: un proceso complejo. Index de Enfermería **20**(1–2), 96–100 (2011). https://doi.org/10.4321/S1132-12962011000100020
9. Sandoval Casilimas, C.: Investigación cualitativa. Módulo (1996). ISBN: 958-9329-18-7
10. García, J.A.: Métodos y técnicas cualitativas en la investigación de la ciudad, 79–85 (n.d.)
11. Hernández Sampieri, R., et al.: Capítulo 14. Recolección y análisis de los datos cualitativos, Metodología de La Investigación (2006)

Good Night Stories for Rebel Girls: Disrupting a Computing Engineering Class in the Higher Education Context

Patricia López Estrada(✉)

School of Languages and Social Sciences, Instituto Tecnológico de Costa Rica, San Carlos Campus, San Carlos, Costa Rica
plopez@itcr.ac.cr

Abstract. This study presents the experience of a feminist pedagogical practice of inclusive readings as well as a space for critical questioning stemming from the reflective literary analysis from the reflective literary analysis of *Good Night Stories for Rebel Girls Volume 1* in an English class for the Computer Engineering major at *Instituto Tecnológico de Costa Rica, San Carlos* Campus. Data were collected through weekly reflective journals during a university semester and analyzed using domain analysis [1, 2]. The aim of the study was to describe the students' constructivist perceptions about the learning experience about the readings. This study highlights the inspiration behind the stories, a sense of determination in the characters, and the acquisition of knowledge through the stories; simultaneously, it emphasizes the denial of gender equality, the acknowledgement of gender inequality, and the celebration of an innovative pedagogical practice in higher education.

Keywords: Critical pedagogy · Feminist pedagogy · Empowerment pedagogy Higher education · Reading · Writing

1 Introduction

The 21st century has triggered the reinvention of critical and feminist pedagogy as mechanisms to emancipate students in society submerged in sexist prejudice and social stereotypes. Inclusion, critical thinking, dialogic interaction, reflection, and diversity support these pedagogies as feasible alternatives from students' realities in the classroom [3–7]. Critical pedagogy focuses on aspects that promote criticality, students' commitment, a sense of belonging, identity, the construction of knowledge through dialogic conception, and the development of learning autonomy [3].

Empowerment pedagogy promotes students taking ownership of their own learning process. This pedagogy challenges students to enthusiastically become aware and create their own knowledge [3, 8, 9]. Freire [3] emphasizes that empowerment involves change. It is in this change that thinking differently takes place, and accommodating to new views as well as opening up to new insights are mandatory.

This study demonstrates the application of a significant learning experience of critical thinking, through the use of empowerment pedagogy and feminism in the

A. P. Costa et al. (Eds.): WCQR 2018, AISC 861, pp. 223–234, 2019.
https://doi.org/10.1007/978-3-030-01406-3_19

context of higher education. The teaching practice consisted of the reading of 40 stories of real girls from the book *Good Night Stories for Rebel Girls Volume 1* in an English class at *Instituto Tecnológico de Costa Rica* (TEC) in its San Carlos' Campus.

1.1 Context

Instituto Tecnológico de Costa Rica (as well as its different headquarters and academic centers in the country) presents a pattern of attraction, admission, permanence, and graduation that is higher for male students [10, 11]. This trend is associated to several aspects, such as a high concurrence of men in administrative and academic positions, as well as in the student body, and an organizational culture which isolates women or sees them as inferior with a lesser cognitive and academic capacities for the majors offered at TEC. Those majors are in the fields of science, engineering, and technology ("masculinized" majors). In a study about Gender Inequality at *Instituto Tecnológico de Costa Rica*, in the 2011–2014 period, Rodríguez [11] states:

> The average of women who accomplished admission at ITCR in the 2011 to 2014 period was of 41.2%, which represents a difference to men of 17.6%. While this implies the inclusion of women in the admission process in the institution, it does not imply gender parity. If we compare the initial data to the amount of women who requested to take the admission test, we find that, in the end, only 26% of them were admitted into the institution. In the case of men, that same percentage is of 27.7%. However, when those numbers are analyzed regarding the effective registration figures in the institution, the difference between men and women becomes more visible. This is how from the total of men who requested to take the admission test, approximately 12% managed to be admitted in an academic program within the institution, while for women, the amount dropped to 8.2% [11, p. 22].

In the regional San Carlos' headquarters, this problem is more evident regarding the amount of men (blue) and women (orange) who were new students in 2011 to 2014, as presented in the following figure [11].

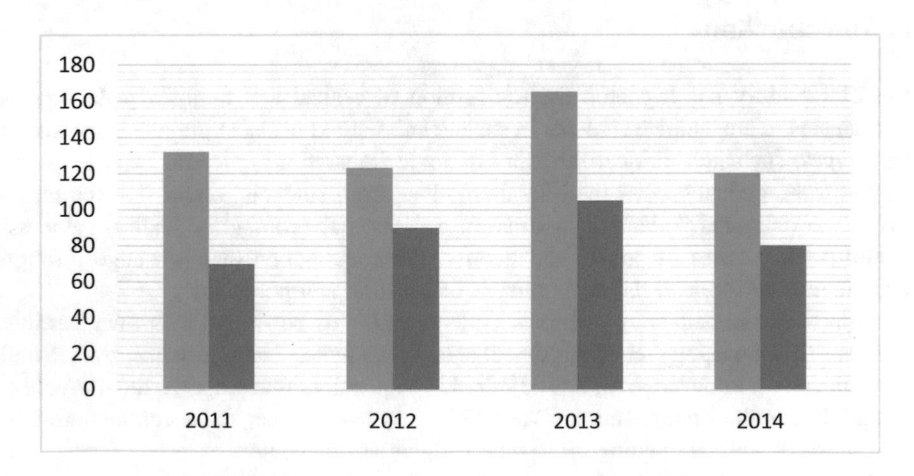

Fig. 1. Rodríguez, 2015, p. 27

More specifically, Computer Engineering offered the following data regarding the admission of the major. In 2017, out of thirty students, twenty-six were men and four were women. Similarly, new students in 2018 were thirty-two, out of which twenty-seven were men and five were women. Data from the School of the Computer Engineering major of the San Carlos' Campus indicate that the graduations in the years 2016 and 2017 had a similar trend. In 2016, out of twenty-six students who graduated, twenty-two were men and four were women. The same pattern repeated itself in 2017, where out of thirty-nine students who graduated, thirty-three were men and six were women. The teaching faculty of the School of Computer Engineering, as of the first semester of 2018, has five female professors (one female secretary in the administrative part) and thirteen male professors.

This study was carried out in an advanced English class made up of eleven students, ten males and one female, ranging from 18–22 years old from the Computer Engineering major during the second semester of 2017 (fifteen weeks of classes). The class took place once a week for a total of 3 hours. The major requires students to take four English courses in their program, which integrate two macro linguistic abilities: listening and speaking. The students have to do a diagnostic test at the beginning of the program. If they pass the test, they go directly to the major's English courses (four in total); students who fail the test have to take a Basic English course to level their language proficiency. The courses' programs went through a revision process with the proposal of new focuses in English for authentic communication and with contents from the Common European Framework of Reference for Languages [12].

The reading of the girls' stories was presented as a proposal to the Computer Engineering students at the beginning of the second semester of 2017. The activity consisted of the reading aloud of two stories to the students per class during the first five weeks. Then, the amount of readings increased to three readings from week five to fifteen, for forty stories total read in class during the semester.

The book used in class is titled *Good Night Stories for Rebel Girls Volume 1*. The writers of the book are Elena Favilli and Francesca Cavallo; the book was published in 2016 and the editing process was accomplished through online donations. Sixty women from all over the world did every illustration in the book. Additionally, another woman, translator Ariadna Molinari, did the books' translation into English. The book, written in one page long stories, tells the true stories of 100 women with extraordinary accomplishments despite their adverse conditions regarding socioeconomic and geopolitical factors. However, it highlights their vulnerability due to being female.

The stories of the rebel girls gather women from all around the world who come from different social spheres such as sports, politics, culture, economy, science, and technology, among others. Some examples of the characters in the stories are Amelia Earhart, Frida Kahlo, Isabel Allende, Mala Yousafzai, Margaret Thatcher, Maria Montessori, Maya Angelou, Michelle Obama, and Simon Biles, among others. The main objective of publishing the book was the dissemination of role models for girls around the world, in order to counteract the expansion of female stereotypes encouraged or promoted by traditional fairy tales, as well as presenting stories were women are heroes, regardless of the adversity they have to face.

The professor who taught the class was the researcher of the study. The professor has been teaching at TEC since 2012, and most of her English teaching is language-based. However, the emphasis of the pedagogical practice of this class was content-based. The professor sought to challenge the students' knowledge by providing them with readings to question and broaden their knowledge about feminism and society. The knowledge that is not the norm in their classes, their major, or their personal lives. The readings offered a distinctive perspective that included women as visible, empowered individuals to further the students' own ways of thinking and perceiving.

The intention behind reading the stories in the class was promoting a significant learning experience with the critical internalization of processes of inequality in a classroom regarding masculinized majors at TEC. The practice focused on the creation of a space to sensibilize students in a male-dominated class. It also sought to raise awareness of the gender and power struggles society currently faces. In addition, it aimed at exposing students to real stories in order to ignite their curiosity and make them conscious of gender justice in the way knowledge and education are presented in a technological major in higher education.

2 Theoretical Framework

Empowerment pedagogy refers to a profound recognition of the competencies in order to "build a more humane social order, to engage in meaningful learning, and to connect with others to collectively recreate social justice" [7, p. 10].

Empowerment pedagogy and feminist pedagogy are inherently bonded and they both aim at empowering students to be critical and reflective of their own perceptions and social assumptions; they work as emancipatory pedagogies [4]. It is in this emancipatory space, that teachers and learners in the classroom can be transformed since knowledge is de-constructed and co-constructed [13]. Students are encouraged to be reflective and question the status quo and any imposed knowledge so that they can construct their own understanding of social issues. Louise-Lawrence [6] states that when students are given the freedom to analyze, their views change since they internalize their beliefs of the world.

Feminist pedagogy strives to help students and teachers learn to think in unique ways, especially "ways that enhance the integrity and wholeness of the person and the person's connections with others" [7, p. 9]. It ultimately aims at a transformation of the academy and seeks to promote a classroom environment, which leads to that transformation [7, p. 10]. For Briskin and Coulter [14], feminist pedagogy "acknowledges that the classroom is a site of gender, race, and class inequalities, and simultaneously a site of political struggle and change. It recognizes that teaching and learning have the potential to be about liberation" (p. 251).

Feminist principles aim at considering students' experiences and making gender visible in the classroom [4, 15]. It is in the classroom practices that knowledge is questioned and through dialogue and critical discussion, new understandings are created and perceptions are transformed.

According to Shrewbusy [7], feminist pedagogy must use the following empowerment classroom practices:

1. Enhance the students' opportunities and abilities to develop their thinking about the goals and objectives they wish and need to accomplish individually and collectively
2. Develop the students' independence (from formal instructors) as learners
3. Enhance the stake that everyone has in the success of a course and thereby make clear the responsibility of all members of the class for the learning of all
4. Develop skills of planning, negotiating, evaluating, and decision making
5. Reinforce or enhance the self-esteem of class members by the implicit recognition that they are sufficiently competent to play a role in course development and are able to be change agents
6. Expand the students' understanding of the subject matter of the course and of the joy and difficulty of intense intellectual activity as they actively consider learning goals and sequences (pp. 10–11).

Other feminist values include:

> the personal is political, egalitarianism, reflexivity, social action, debunking of the banking system of learning, analysis of power differentials, challenging traditional assumptions regarding sources and/or definitions of knowledge, incorporating lived experience into the classroom, and giving voice to those who have been marginalized [5, p. 437].

Feminist educators should rely on pedagogical practices that promote critical thinking as a mechanism to raise consciousness to injustice and question the system, thus working towards social change [5]. They also must be aware that feminism implies tackling power issues, in the sense that some relations of power are not just and it is within this struggle that feminism comes to play [15]. Webb, Allen, and Walker [16] suggest the professor-student relationship must promote a sense of classroom community that is respectful of diversity, enhances acquired experiences, and dares to challenge traditional opinions.

In a study conducted by Silva Flores [17], about how feminist academics perceived their teaching based on their lived experiences, it is concluded that academics embraced feminism as a mandatory teaching practice and they practice feminism taking into account their ingrained theoretical and emotional beliefs. One of the participants in the study asserts that feminist pedagogy means building safe places to address inequalities and reflect about inclusion not only in teaching and learning, but also in all interpersonal interactions in the classroom. Another participant in the same study [17] stresses the importance of multiplicity of perspectives as well as how essential it is for different voices to be presented in order to decentralize the dominant voices that currently exist in the classroom and in society. Some other participants address the importance of providing students with a space to form their own opinions and think in different ways, thus making them responsible for their own learning. Silva Flores [17] furthers that her study shows that feminist pedagogical practices can positively affect students' lives. The author concludes that there are some challenges that feminist university teachers go through; for example, students who question the need to talk about women's inequality since they argue it does not exist nowadays. This type of students' resistance to feminism has also been documented in other studies [18].

3 Methodology

The data used in this study were collected through two primary sources: journals written by the students after each story read during the semester and a student's reflective writing analyzing this pedagogical practice at the end of the semester. Both the after story writing and the final reflection were in free form with a simple structure. Once the reading was finished, students wrote their immediate reactions individually for five minutes. Students were instructed to write what caught their attention the most about the readings. All writing done by the students was collected by the instructor of the class, who also conducted the analysis of the data.

During the semester, students had verbal discussions in five occasions about the reading. In other three occasions, students used technology (smartphones and computers) to find more information (photos, biographies, videos, articles, etc.) regarding the protagonists of the stories. This contributed instrumentally by complementing the stories with valid information for the students' writings. Every writing was done in class (as an additional assignment every class) and had no value in the evaluation of the course.

The data collected from the daily writings as well as the reflective writings were based on the constructivist perception of the students in the class. The written reactions were free form and, despite the fact that topics unrelated to the stories tend to come up, students' insights emphasized on topics of gender equality and inequality. The main goal of the writings was to have students present their particular and unique perceptions regarding the readings.

Constructivism plays a main role because it celebrates knowledge constructed through the personal and social experiences of each individual. In this way, realities can differ among individuals without having to choose one as superior for comparative purposes. Realities and perceptions are simply different since they are linked directly to each person's experiences and vision of the world [1]. This descriptive study seeks to highlight students' perceptions from their unique processes of meaning construction and the social realities of their personal, social, and academic environment.

The qualitative study is based on inductive analysis (*domain analysis*), specifically content analysis [1, 2]. Inductive analysis fits within constructivism since the primarily focus of the current study is descriptive. This analysis grouped contents to obtain significant categories in the study stemming from data diversity. There was a systematic examination for patterns of meaning which were taken from specific elements to create greater categories within the data [1]. The steps proposed by Hatch [1] include:

- read the data and identify frames of analysis ("meaningful units")
- create domains based on semantic relationships discovered within frames of analysis
- identify salient domains
- assign them a code (and put others aside)
- reread data by refining salient domains and keeping record of where relationships are found in the data

- decide if the domains are supported by the data by examining examples that do not fit with or run counter to the relationships in the domains
- complete an analysis within domains, and search for themes across domains.

The researcher first did vast and constant critical readings of both the daily writings and the reflective writings with the objective of inducing patterns from the data. Initial data coding took place with the "meaningful units" [1, p. 163], which later created bigger semantic units.

From the nine possible semantic relationships, as suggested by Spradley [2], the current study focused on the semantic relationship of attribution (X is a characteristic of Y, X is an attribute of Y). In this relationship, X was an attribute of student's perceptions. These relationships originated domains that linked semantic relationships with specific elements to create greater categories. Domains that showed patterns and connections were selected. Then, the domains that contributed to describing the perceptions of the students in regards to the readings of the stories became the salient domains. These domains resulted in the conclusions that sum up the students' constructivist appreciations regarding the learning experience about the readings of *Rebel Girls* in an English class for the Computer Engineering major.

4 Results

Results encompass both the daily writing assignments and the final reflective writing. The perceptions from the data include the inspiration behind the stories, a sense of determination in the characters, and the acquisition of knowledge through the stories. These results were linked to the denial of gender equality, the acknowledgement of gender inequality, and the celebration of an innovative and differentiating pedagogical practice.

On the data analyzed in both writing tasks, the rhetoric highlights words reiteratively to refer to the girls in the stories as inspiring, incredible, extraordinary, brave, warriors, athletic, genius, independent, strong, daring, determined, and empowered. The results of the written assignments were as follows.

4.1 A Source of Inspiration: "A New Kind of Princess, Opposite to What Most People Think."

Most students referred to the female characters in the stories as heroines. Such a notion is based upon the fact that – through the readings – women overcome more than internal conflicts. They fight against impositions whether personal, familiar, religious, or social. The characters are also perceived as heroines due to their resilience when facing adversity in the context that they lived (some of them have passed away). Their strength is evidenced in the stories through emotional, physical, and cognitive issues. What stands out is that they are considered heroines because they left a mark; they broke the glass ceiling (flying an airplane, climbing a mountain wearing a dress, etc.). Those women took the first step so other women today can follow their footsteps (refusing to enter a marriage for convenience, refusing to use clothing socially acceptable for women to wear, studying or working for something that was forbidden or denied to them such as science and technology).

Students indicated that these heroines are the "new princesses;" these princesses think, act, and dress in opposition to what most people consider princess material. The stories generated inspiration in the students because the girls got out of their comfort zone. They believed in a different world full of possibilities and opportunities, where they would not be judged by their condition (being female); they believed in a world where making a difference is worth it, where being different can be brave.

However, there was a denial of gender equality. Out of the eleven students, five agreed that while the stories were true-life stories and very inspiring, gender inequality and discrimination do not exist today. Partly, they consider that – while women in the stories are strong, brave, and fight for what they want, so are men. To them, this comes from the fact that both men and women can do anything they set out to do, that it is not a gender issue, but one of effort and personality. While students indicate that women have to struggle constantly due to male domination, they emphasize that the opposite scenario (a world where men have to struggle constantly due to female domination) would not make a perfect world either.

What would make sense to them is a system where there is gender equality. Students suggest that gender does not determine abilities or cognitive skills. They believe that men and women have the same opportunities. A student referred to a famous phrase, "behind every great man, there's a great woman," which implies that the power comes from men. Positioning a great man appears first while a woman has a more secondary and dependent function. Another student commented the following: "Women should not be restrained or limited because their potential could be like that of Mozart, Einstein or Plato." His phrase exemplifies the comparison to superiors with all male role models.

4.2 A Sense of Determination: "You Have to Do What You Have to Do."

Tying into heroine appreciation, students agree that these girls' stories emphasize a sense of determination represented through the constant struggle to which girls are exposed at an early age. This same determination stems from both an internal and an external struggle against stereotypes, oppression, and discrimination within a specific society in a specific time. Students assert that possibilities and opportunities do not just appear; they have to be pursued, created, and celebrated bravely. This bravery is what drove girls in the book to follow their dreams, to make a difference, and not to allow one person or system decide what they could or could not do. The girls' accomplishments in the stories stemmed from boldness and conscious decision-making to break the paradigm, as well as firmness in convictions, which do not necessarily represent the status quo.

There was a recognition of gender inequality by the six remaining students, who are aware of gender differences and how those differences generate inequality. They consider that inequality existed in the past and it also exists in the present, that the system is not equal for everyone, especially for women. Students concur that the messages in the stories are real and powerful, and that they generate social conscience regarding women's bravery in spite of adversity.

They reflect that women and men are not considered equal and that too much time has gone by without giving women the respect they deserve, as well as the influence and value of women in history. To them, the status quo has remained unchanged and women are still considered vulnerable and inferior, in comparison to men. For students, there has to be a change in the way in which women who have made history are showcased. There is an urgent need to break the paradigms and the stereotypes, to be more progressive, more open to change, and fairer to the females. Some students have referred to the fact that most of the awards in science and technology and of the knowledge taught at the university are exclusively given or attributed to men, that they hear little about women and their accomplishments, and that there are real accomplishments achieved by women that have been kept as "hidden stories."

4.3 Acquisition of Knowledge: "This Is the First Time I Have Heard About the Contribution of a Woman."

Students unanimously agree that this was the first time they heard about these girls and their stories. They reiterate that they learned about the first female president, female scientist, female professional motocross athlete, female Muslim in the Olympics, female lawyer and doctor, female programmer, and female aviator, in this class.

Students claimed that they had no prior knowledge about the girls' stories, and reflected on the amount of discoveries and global accomplishments which women have done for years and have gone unnoticed for generations. Likewise, students agree that girls (now women) need recognition, in order to be role models for other girls. Some of the students even questioned why women's contributions have been kept "secret;" they state that history itself has hidden women through time. From a critical perspective, students write that in their classes, they are exposed to contributions to science and technology achieved by men, such as Mozart, Einstein, John von Neumann, and Alan Turing, among others.

All students pointed out that this English class was the first one in which they were exposed to real stories of heroines in diverse disciplines like science, technology, sports, medicine, law, economy, and music, among others. Students maintain that the stories were symbolic, and through the writing process, they were able to reflect and be critical towards the stories, their context, the struggles, and the social battles that became personal to the rebel girls. At the same time, students celebrated that the execution of this pedagogical practice was entertaining, innovative, and differentiating. They suggested that these stories should be taught in other educational institutions, not just of higher education, but also of primary school and high school, where girls and boys can work on the topic from an early age.

5 Discussion

The results of this qualitative research are very interesting, even with a reduced sample and a constrained application time. The dynamics of the English class, while occurring once a week for three hours, managed to be interactive and a promoter of knowledge creation through reflection and critical thinking. The short stories worked very well in the context of the class; the pragmatic practice caught the students' attention in the class.

Students' use of positive adjectives is an important result in the study since they might contribute to the eradication of inequality and inclusion. Students constantly referred to the girls of the stories as inspiring, amazing, extraordinary, warriors, athletes, independent, strong, bold, and emancipated. Furthermore, they considered the girls as cool "empowered girls," with a stroke of genius in their thinking and their actions. Occasionally, students referred to the girls as "wild" and "dangerous," all words used with a positive connotation.

Students were aware that the stories were real and that everything in them was new knowledge for them. This was one of the most significant results; that no students knew anything about the forty women and their accomplishments. The reading of the stories was an experience that presented knowledge about women's accomplishments (in the athletic, medicinal, scientific, technologic, cultural, and political fields).

As part of feminist pedagogy, one of the classroom-based practices resulted in students challenging traditional assumptions regarding knowledge, and creating their own understanding about the role of women in the stories. The classroom environment was conductive to students sharing their opinions and feeling emancipated from their ingrained beliefs about social, power, and gender issues. They constructed their own understanding of social issues. Students explored their beliefs and were critical about the fact that in their university classes (in other subjects) only male names are brought to the attention of students, as if knowledge relies solely on male discoveries. Reflecting upon that made students confront and at times change their beliefs about the world.

Students reflected on the notion of "princess". They recognize the princesses in the stories as heroines in their own right, with their own "castles," empowered by the clothes they wear, the style of their hair, the goals they achieve, the dreams they have, without their gender being important. Tied to this, the sense of fearlessness of the girls captured students' attention. While the girls in the stories were aware of the norms and the social barriers they were subjected to, they were determined to overcome them. To accomplish their goals and to break a paradigm was their most daring act, which they undertook bravely and proudly.

The dichotomy present in the students, regarding gender equality denial and the recognition of gender inequality, was essential for this study. Students presented divided perceptions regarding equality. Approximately half the students agreed that – while the stories are real and inspiring – lack of gender equality is not a real problem in our society. To an extent, students perceived the stories as fictional; they were seen as examples of "realities" that do not exist. Although, it has to be highlighted that the class had ten male students and one female student, and the Computer Engineering major has thirteen male professors and just five female professors; furthermore, the percentages of both admission and graduation are clearly dominated by men. On the other hand, the rest of the students consciously expressed how gender inequality is perpetuated in all social, academic, and professional levels. They argue that the stories are genuine because they reflect the struggles of many women through history.

A relevant aspect is the recognition that the students give to this practice by the professor, which they qualify as a transformative experience within their university context. Transformation took place and gender was made visible in the English classroom. It was in this classroom practice that knowledge was critically questioned and new understandings were generated by the students; their gender perceptions were transformed. Students concurred that this pedagogical practice must be replicated in other educational contexts to enhance students' reflective and critical skills. Another aspect to consider was how students felt empowered through meaningful learning and engagement in genuine discussion on building a more humane social order. They reflected on the importance of connecting with others to promote social justice collectively in the classroom, and in society.

6 Conclusions

This study has to be complemented by including more participants and doing in depth interviews of focus groups to internalize students' perception regarding what they conceive as processes of gender equality and inequality in the classroom and in society.

However, despite the small sample of participants, the limited timeframe for this pedagogical practice, and being this the first time put into practice, the stories of rebel girls fostered a sense of hope. Each week, students would walk into the class expecting to learn about a new story of a heroine. The weekly reading promoted knowledge construction and the development of learning autonomy.

The practice of this critical pedagogy does not pretend to be a mandatory guide for other teachers to follow. It does not seek to incentivize the implementation of feminist pedagogy or empowerment pedagogy in other classes or contexts of higher education. However, the pedagogical practice is a wake-up call for educators, not only in higher education or in institutions that offer engineering, scientific, and technical majors with a tendency to attract male students, to consider making a change in their pedagogical practices. Such a change would elevate the value of students—male and female—empowerment to come up with their own perceptions.

Classrooms must become spaces to rethink societal power relationships and to challenge knowledge portrayed as hegemonically male-oriented. Teachers can make a change to support the belief that critical and reflective pedagogy can be transformative agents in a world that claims for social justice. Any teaching act that promotes a better, more just world must be celebrated.

This process of change has to be a constant process of raising awareness for all educators in every educational institution and for all levels (primary school, high school, and higher education). It must also involve both female and male students, since it constitutes a reflection mechanism and the development of opinions about pertinent topics in a collective community. The final objective lies in generating a space of criticism and individual conscience for the construction of a society that is more egalitarian, inclusive, and fair.

References

1. Hatch, J.A.: Doing Qualitative Research in Education Settings. State University of New York Press, Albany (2002)
2. Spradley, J.: The Ethnographic Interview. Holt, Rinehart & Winston, New York (1979)
3. Freire, P.: Pedagogy of the Oppressed. Herder and Herder, New York (1970)
4. Forrest, L., Rosenberg, F.: A review of the feminist pedagogy literature: the neglected child of feminist psychology. Appl. Prev. Psychol. **6**, 179–192 (1997)
5. Hahna, N.D.: Towards an emancipatory practice: incorporating feminist pedagogy in the creative arts therapies. Arts Psychother. **40**, 436–440 (2013)
6. Louise-Lawrence, J.: Feminist pedagogy in action: reflections from the front line of feminist activism - the feminist classroom. Enhancing Learn. Soc. Sci. **6**(1), 29–41 (2014)
7. Shrewsbury, C.M.: What is feminist pedagogy? Women's Stud. Q. **3–4**, 8–16 (1993)
8. Freire, P.: Pedagogy of Hope. Continuum Publishing Company, New York (1994)
9. Cummins, J.: A theoretical framework for bilingual special education. Except. Child. **56**(2), 111–119 (1989)
10. Queralt Camacho, L.: Trayectorias laborales y personales de mujeres profesionales de carreras tradicionalmente masculinas: el caso de las egresadas del Instituto Tecnológico de Costa Rica. Tesis de Maestría para optar por el grado de Máster en Género y Políticas de Igualdad. Universidad de Valencia (2007)
11. Rodríguez Fernández, A.: Estudio sobre Brechas de Género en el Instituto Tecnológico de Costa Rica. Periodo 2011–2014. Editorial Tecnológica de Costa Rica, Cartago (2015)
12. Common European Framework of Reference for Languages: Learning, Teaching, Assessment. Council of Europe, Strasbourg (2001)
13. Hahna, N.D.: Conversations from the classroom: reflections on feminist music therapy pedagogy in teaching music therapy. Dissertations Abstractions International: Section A, Humanities and Social Sciences, vol. 72, no. 7, pp. 4–13 (2011)
14. Briskin, L., Coulter, R.P.: Feminist pedagogy: challenging the normative. Can. J. Educ. **17** (3), 247–263 (1992)
15. Fisher, B.M.: No Angel in the Classroom. Rowman & Littlefield Publisher Inc., Plymouth (2001)
16. Webb, L.M., Allen, M.W., Walker, K.L.: Feminist pedagogy: identifying basic principles. Acad. Exch. Q. **6**(1), 67–72 (2002)
17. Silva Flores, J.: Feminist pedagogy in the UK University classroom: limitations, challenges and possibilities. In: Light, T.P., Nicholas, J., Bondy, R. (eds.), pp. 33–55. Wilfrid Laurier University Press, Canada (2015)
18. Langan, D., Morton, M.: Through the eyes of farmers' daughters: academics working on marginal land. Women's Stud. Int. Forum **32**(6), 395–405 (2009)

Using CAQDAS in Visual Data Analysis: A Systematic Literature Review

Ana Isabel Rodrigues[1,3](✉), António Pedro Costa[2], and António Moreira[3]

[1] Department of Management, Polytechnic Institute of Beja, Beja, Portugal
ana.rodrigues@ipbeja.pt

[2] Ludomedia and Research Centre "Didactics and Technology in Teacher Education" (CIDTFF), University of Aveiro, Aveiro, Portugal
pcosta@ludomedia.pt

[3] Research Centre "Didactics and Technology in Teacher Education" (CIDTFF), Department of Education and Psychology, University of Aveiro, Aveiro, Portugal
moreira@ua.pt

Abstract. The use of Computer-Assisted Qualitative Data Analysis Software (CAQDAS) is very recent when compared with the history of qualitative data analysis, which began in anthropological literature during the 20 century. More recently and framed by the use of the visual element in qualitative methods, researchers have a set of data at their disposal with visual support, allowing the introduction of new interpretive elements that enrich the analysis and understanding of their object of study. This paper aims to systematically review the literature, examining the current state of the art of visual methods and visual data analysis, focusing on the use of CAQDAS. To this end, relevant journal articles will be analysed in the future, with the identification of some important issues as well as gaps in existing knowledge. This analysis will provide valuable input for the development of research suggestions and directions for future work in this area.

Keywords: Systematic Literature Review · Visual research · Visual methods Visual data analysis · CAQDAS

1 Introduction

It is in the face of new challenges to qualitative research that new opportunities emerge. For Minayo [1], we can never forget that theories and methods are related to the reality of the world. And because the world has changed a lot in the last twenty years, qualitative research is going its own way and adapting itself to the "new" reality. A new context, framed by a visual culture paradigm, is revealed to researchers who, as we know, and especially in qualitative research, are much more sensitive to what surrounds them. Based on this, there is growing evidence that visual methods and visual data mark a shift in qualitative research. A new reality characterized by visual culture requires new approaches, new methods and new techniques. With all these, the element

A. P. Costa et al. (Eds.): WCQR 2018, AISC 861, pp. 235–247, 2019.
https://doi.org/10.1007/978-3-030-01406-3_20

of data analysis emerges as a fundamental issue and the use of Computer-Assisted Qualitative Data Analysis Software (CAQDAS) packages began to offer similar possibilities for visual qualitative data analysis. Since this is a recent field of research, this paper emerges from the following needs: (i) to summarize existing evidence concerning visual methods and the use of CAQDAS tools and their increased uptake for audio-visual analysis; (ii) to identify where there are gaps in current research in order to help determine where further investigation might be needed; and (iii) to help position new research activities. Based on these reasons, the introduction of the use of systematic evaluation processes, in particular Systematic Literature Review (SLR) will help to obtain an objective summary of research evidence concerning this topic by producing better quality reviews and evaluations. This paper is an ongoing study – a preliminary approach – that attempts to share and present the goals and procedures taken until this stage.

2 An Approach to Visual Research

Humankind is becoming increasingly image-based. According to Mirzoeff [2], visual culture is everywhere: surrounding us all with still-and moving images (e.g. computers, television, mobile phones, social media); and concurrently nowhere: all mediatized representations are mixed, where one *medium* becomes the content of another. The image itself might inform, elucidate, enlighten and, at the same time, the visual evolves in an "unsubstantial distraction from the real" [2]. That is the goal of visual culture field, what Mitchell calls "the visual construction of the social field" [cited on 2]. For that reason, the use of the visual element in qualitative research is materialized in the so-called "visual movement" [3], with the roots of its application in the field of visual anthropology. It was only in the 1990s that the use of visual methods was widely diffused in the social sciences [4]. Researchers today have a set of data at their disposal with visual support - paintings, photographs, films, drawings and, diagrams - allowing the introduction of new interpretive elements that enrich the analysis and understanding of its object of study. The image informs, elucidates, documents, and adds value and meaning to the phenomenon itself. Qualitative research has a new and insightful universe in terms of analysis and interpretation of social reality [5, 6].

The use of visual methods thus emerges as a field of study, which Rose [7] defines as "Visual Research Methods" (VRM), which is in an initial, but developing phase of qualitative research [4]. Traditionally, VRM comes from Visual Anthropology [8, 9] and later from Visual Sociology [10]. They have recently gained their place in a variety of social sciences such as geography, health and, psychology. The use of visual data can include photography or video among other forms and can be incorporated into qualitative research projects in various ways [11]: analysis of visual data (photos, videos, posters, etc.), adoption of photographs and videos to trigger discussions and conduct interviews on certain subjects or the production of photographs or videos as elements of study by the researcher or participant himself. Nowadays, VRM is already an area recognized by the scientific community, while still in an initial stage [7].

In this sense, research can use images available for analysis; the images themselves can be created/produced by the researcher or by the individual being studied; or else the visual element may appear as a data for analysis or as an element that induces and conducts the investigation process. Image informs, elucidates, documents and adds value and meaning to the phenomenon itself [5, 12]. In general terms, it is possible to identify two main dimensions in the adoption of visual elements [5, 12]:

1. The first refers to the creation of images (visual data) such as videos, photographs or drawings by the researcher him/herself to document or analyse aspects of social life and social interaction. The researcher makes his/her notes, records of what he/she observes, and analyses using visual elements;
2. The second concerns the collection and study of images produced and/or "consumed/observed" by the research subjects. In this case, the research project is more "visual" and there is a greater social and personal connection with these images on a part of the subject being studied.

Next, a systematization of some important thoughts concerning the use of visual methods and visual data is presented:

- The use of VRM allows the visualization of the intangible dimensions of human activity [13];
- The use of visual data can encourage participants to engage in research more creatively [14];
- The VRM allows attention to be focused on what is truly important to the participant, including aspects that might not have occurred to the researcher [15];
- Conversations are full of verbal references to images and the use of words alone cannot express all the elements [11]. There are authors [16] who advocate that there should be no separation between the use of text and image; they should be seen in a complementary way;
- Images seem to release words, for example, in the context of Photo-Elicitation. Images can help participants express complex understandings about their experiences and validate textual data [10];
- Images combined with interviews can add enriching data [15].

In sum, as stated by Clark and Morriss [17] "the visual turn has generated a plethora of approaches, from data gathering to the visual representation of big data, to the exploration of visual manifestations of social phenomena" (p. 31). As mentioned by the authors, VRM provides a more 'authentic' perspective compared to more conventional research methods. In this line of thought, [18] recommend that visual images can exemplify "experience, humanity, and meaning… and thus… edify the significance in the humanness and affectivity of research participants" (p. 433). Therefore, new opportunities and insights emerge for qualitative studies.

3 Visual Methods and Data Analysis: The Use of CAQDAS

As Minayo [1] points out, "we can never forget what theories and methods have to do with the reality of the world" (p. 31). The fact is that the reality of the world has changed dramatically regarding the visual and visuality. Trying to explore the nature of understanding social phenomena through visual approaches, means and materials have gained new insights. For Atkinson and Delamont [19], there are various social phenomena that can be captured visually and analysed in terms of their manifestations. For example, Clark and Morriss [17] in their study about the use of visual methodologies in international social research over the past 10 years, conclude that numerous methods and techniques have been applied, including photography (photovoice and photo-elicitation); visual mapping and visual timelines; drawing and painting; collage; film and video; vignettes, etc. According to the authors, there is a dual typology of *pre-existing visual materials*; and the use of *researcher-instigated visuals* that are provoked/produced by the researcher himself or participant-generated (films, photos, drawings, etc.).

Some examples of methods and techniques that use visual data to obtain information are presented here. The goal is simply to provide a sketch of the field and identify some important methodological illustrations. With reference to Photo-Elicitation, this started to be used as a method by the anthropologist Collier [8] who studied the phenomenon of migration caused by economic and technological changes. He emphasized the fact that the use of photos evokes participants' memories, allowing for deeper and richer interviews. Sociologists Harper [10, 20] and Banks [21] have contributed greatly to the recognition of photo-elicitation as a visual method, which is based on the simple idea of applying and conducting an interview using photographs as a stimulus. In fact, photographs (more than words) evoke deeper elements of human consciousness.

It is also worth mentioning Reflexive-Photography, which is considered a participatory and self-reflexive visual method or strategy to obtain participant's opinions, ways of thinking and feelings. This method is based on a set of photographs collected, analysed and commented on by the participant in conjunction with the researcher [22], and has a great potential for crossing with other approaches. It has been used in fields such as education [23, 24] health [25] and in recent areas such as tourism [26]. A final remark on another example: Photo-Essay. This is considered a participatory technique or visual strategy to obtain participant's opinions, ways of thinking and feelings, based on a set of photographs collected, analysed and commented in by the participant in conjunction with the researcher. Photo-Essay is strongly associated with Visual Anthropology due to the use of classic photographic essays [27].

With all these new qualitative materials the analysis element emerges as a fundamental issue. As is well known, specialized software for the analysis of qualitative materials and data has been in development for the last thirty years. According to Silver and Patashnick [28] from the mid-1980s, Computer-Assisted Qualitative Data Analysis Software (CAQDAS) packages began to offer similar possibilities for qualitative data as had been developed for quantitative data from the mid-1960s. CAQDAS has gone through several phases of evolution, and is currently (2010–2015) in phase VI, called "A fractured future", which is characterized by the development of web 2.0/3.0 platforms that work in network, marking the beginning of CAQDAS 2.0 [29]. Still according to Silver [29], several CAQDAS packages have appeared on the market with regularity and programs that do not have users end up disappearing. Each software has features that are distinctive and can be used by researchers for different purposes. To Weitzman [30] at the time of the decision-making process of choosing a CAQDAS package there are key questions that need to be answered:

(a) What type of user are we? Our level of knowledge and use is fundamental in choosing the most appropriate CAQDAS;
(b) Do we choose a CAQDAS package to respond to a project based on a short-term vision or to respond to more medium- to long-term research objectives?
(c) What type of database (sources) will we use and what kind of project will we work on?
(d) What type of data analysis (textual or visual) do we plan to use in the project?

The following table highlights eight CAQDAS packages that allow the analysis of visual data: ATLAS.ti, Dedoose, HyperRESEARCH, MAXQDA, NVivo, QDA Miner, Transana and webQDA (see Table 1).

Table 1. Brief description of eight CAQDAS packages that perform visual data analysis

Software	History, Data Types and Formats
ATLAS.ti	Initially developed at the Technical University of Berlin as a multidisciplinary project (1989–1992). The prototype was later developed by Thomas Muhr and the company ATLAS.ti was created in 2004. The files are kept in an external system. ATLAS.ti works with a multitude of different types of multimedia files. Video, audio and images files can be directly associated with a project and treated in the same way as textual files. Multimedia formats: .jpg, .jpeg, .gif, .bmp, .wav, .avi, .mpg, .mpeg, .mp3, .mp4, .wmv). It allows direct import of images, audio or video files from some platforms and social networks. Multimedia files are treated like any other document, and it is possible to encode them, extract results and annotate them
Dedoose	Web-based software developed by academics from the University of California, USA, with the support of the William T. Grant Foundation, the successor to EthnoNotes. Multimedia files are treated like any other document, and it is possible to encode them, extract results and annotate them. Media formats: jpgs, gifts .mpeg1, .mpeg2, .avi, .mp3, .wav audio
HyperRESEARCH	Initially developed in 1990 in Boston, USA. Since 1991 it has been developed by ResearchWare Inc. In 2005 HyperTRANSCRIBE was launched, a tool for transcribing video and audio files. Video and audio files and images can be directly associated with a project and treated in the same way as textual files. Multimedia formats: .jpeg, .wav, .avi, .mpeg, .mp3
MAXQDA	Launched for the first time in 1989, it was originally developed by Udo Kuckartz to manage large amounts of political textual data and was extended to academia. It allows direct import of images, audio or video files from some platforms and social networks. Multimedia files are treated like any other document, and it is possible to encode them, extract results and annotate them

(continued)

Table 1. (*continued*)

Software	History, Data Types and Formats
NVivo	Tom Richards and Lyn Richards originally created the NUD * IST at the University of La Trobe, Australia in the early 1980s. Later QSR International was created and started producing the NUD * IST, whose name has since been changed to NVivo. Video and audio files and images can be directly associated with a project and treated in the same way as textual files. Multimedia formats: .jpeg, .wa, .wma, .avi, .mpeg, .mpg, .mp3, .mp4, .3gp, .mts, .m2ts
QDA Miner	Developed in 2004 by Normand Peladeau, it is marketed by Provalis Research. It allows direct import of images, audio or video files from some platforms and social networks. Media files are treated like any other document, and you can encode them, extract results, annotate or specify them. Multimedia formats: .bmp, .jpeg, .gif, .png
Transana	Software originally created by Chris Fassnacht, is now being developed and maintained by David K. Woods at the University of Wisconsin-Madison, USA (Center for Education Research). Its main focus is on audio and video, rather than on text, contrary to other solutions. Multimedia formats: .mpeg1, .mpeg2, .avi, .mp3, .wav audio. Managing your database makes it easy to organize and store large amounts of scanned video
webQDA	Launch of the first version in 2010. Web-based software to support the analysis of qualitative data in a collaborative and distributed environment. Multimedia files are treated like any other document, and it is possible to encode them, extract results and annotate them. Multimedia formats: .jpg, .png, .mpp3, .wav, mp4, .ogg, .webm. You can use YouTube videos or Dropbox public sharing videos

Source: Elaborated by the authors, adapted from Lewins and Silver [31]. The official website of each software was also considered.

Despite all the development in the use of CAQDAS, a discussion of the practical or technical procedures for analysing qualitative data using software is lacking in the social science literature [28], mainly related to audio-visual data analysis. These authors conclude that "in the context of the number of qualitative studies conducted, they are infrequent, and it is notable that it is only in the last ten years that they have appeared, while CAQDAS packages have been developing for nearly thirty years". In relation to visual data, Silver and Patashnick [28] claimed that due to their multidimensionality, visual elements can only be described and not transcribed. In their study the authors conclude that[1]:

> In comparison to image/video storage/editing software, CAQDAS packages have a number of advantages, most notably with respect to analytic tools such as code-based retrieval options. However, there are several ways in which CAQDAS developers could usefully draw upon other

[1] Available at http://www.qualitative-research.net/index.php/fqs/article/view/1629.

tools in refining their products for audio-visual analysis. In particular are visual annotation tools, sorting, sequencing, and (re)presentational tools. Annotation tools are potentially key means by which to explore and interpret audio-visual data, yet, as they stand in CAQDAS packages, are blunt tools for fine purposes. In particular, there is a need for the development of visual annotation tools whereby freeform shapes can be used to mark up images, and which can be hidden or revealed using layer tools.

In sum, a more explicit dialogue between researchers about their analytic procedures for handling audio-visual data is needed, as the above authors have concluded. Additionally, a more explicit dialogue between researchers and software developers is essential. There seems to be a vast field of research to explore concerning the use of CAQDAS in visual research, methods and data. For this reason, a state-of-the-art approach is assumed to be a suitable starting point.

4 Methods

As with other disciplines that employ systematic reviews, there are a number of different reasons why this type of literature review should be undertaken [32]. In this sense, the present study emerges from the following needs: (i) to summarize existing evidence concerning visual methods and the use of CAQDAS tools and their increased uptake for audio-visual analysis; (ii) to identify where there are gaps in current research, in order to help determine where further investigation might be needed; and (iii) to help position new research activities. Based on these reasons, the introduction of the use of systematic evaluation processes, in particular, of Systematic Literature Review (SLR) will help to obtain an objective summary of research evidence concerning this topic by producing better quality reviews and evaluations. One of the most cited definitions [33] describes SLR as a:

> specific methodology that locates existing studies, selects and evaluates contributions, analyses and synthesizes data, and reports the evidence in such a way that allows reasonably clear conclusions to be reached about what is and is not known. A systematic review should not be regarded as a literature review in the traditional sense, but as a self-contained research project in itself that explores a clearly specified question, usually derived from a policy or practice problem, using existing studies (p. 671).

This study intends a future analysis, synthesis and presentation of a comprehensive SLR of visual methods, visual data analysis and the use of CAQDAS on visual data. Based on this literature review that is to a great extent descriptive and inductive in nature, the authors expect to evaluate the state of the art regarding what type of visual methods and visual data analysis have been used in qualitative research and, the multiple functionalities of several CAQDAS tools at the level of visual data analysis. There are several solutions on the market that analyse visual data, allowing work with various visual data sources such as still images (photographs, drawings, paintings) and dynamic images (the case of videos), provided they are in digital format.

In the next stage of this study, the structure of this paper is expected to be as follows: Sect. 1 will define and theoretically contextualizes the main topics: visual research, methods and the use of CAQDAS; Sect. 2 will describe the research methodology applied for this paper, which will follow a five-step SLR approach;

Sect. 3 will present the findings of the SLR; Sect. 4 will discuss the future results obtained by the SLR, presenting meaningful research suggestions and directions for future work; and Sect. 5 will provide limitations and conclusions of the paper.

The SLR will adopt a five-step approach as defined by Denyer and Tranfield [33] and Khan *et al.* [34]:

Step 1: Question formulation/framing questions for a review;
Step 2: Locating studies/Identifying relevant work;
Step 3: Study selection and evaluations/assessing the quality of studies;
Step 4: Analysis and synthesis/summarizing the evidence;
Step 5: Reporting and using the results/interpreting the findings.

4.1 Step 1: Question Formulation/Framing Questions for a Review

The following research questions were determined for this SLR:

RQ1: What type of visual research and methods have been used in the qualitative research field?
RQ2: What type of visual data analysis has been used in qualitative research?
RQ3: How have CAQDAS tools been used for visual methods and data analysis and what are their main functionalities?

4.2 Step 2: Locating Studies/Identifying Relevant Work

The initial search using bibliographic database of eight qualitative research journals was undertaken. For the moment, the following journals were considered based on their reputation: International Journal of Qualitative Methods, Qualitative Research, Qualitative Inquiry, Qualitative Social Work, Qualitative Health Research, FQS Forum: Qualitative Social Research, Qualitative Research Journal, and Studies in Qualitative Methodology. The search strategy and results are presented in Table 2.

Papers were selected for retrieval if they were judged to include data about visual methods and the use of CAQDAS in this field. The search considered the words "Visual Research", "Visual Methods" and CAQDAS, applied to visual field, in the title, keywords and abstracts of the selected papers. This allowed amplifying the research field under study.

4.3 Step 3: Study Selection and Evaluations/Assessing the Quality of Studies

The inclusion of these criteria helps address the research questions. To focus the review, a 30-year time horizon was first established (1990–2018), since the 1990s marks the use of visual methods in social sciences and also more active use of CAQDAS in qualitative research. Only articles published in peer-reviewed journals in English were considered, since the quality control of search results can be enhanced due to the rigorous process to which such articles are subject prior to publication. In the next stage of the present study, articles will be then excluded or included based on the

Table 2. 1st Search Strategy and results

Source (Type of Journal)	Keyword/n° articles	Date of search
International Journal of Qualitative Methods "Visual Research" OR "Visual Methods" AND "CAQDAS"	Visual Research Visual Methods (38) CAQDAS (5)	28.05.2018
Qualitative Research (Sage) "Visual Research" OR "Visual Methods" AND "CAQDAS"	Visual Research OR Visual Methods (43) CAQDAS (6)	30.05.2018
Qualitative Inquiry (Sage) "Visual Research" OR "Visual Methods" AND "CAQDAS"	Visual Research OR Visual Methods (26) CAQDAS (5)	30.05.2018
Qualitative Social Work (Sage) "Visual Research" OR "Visual Methods" AND "CAQDAS"	Visual Research OR Visual Methods (18) CAQDAS (2)	01.06.2018
Qualitative Health Research (Sage) "Visual Research" OR "Visual Methods" AND "CAQDAS"	Visual Research OR Visual Methods (36) CAQDAS (2)	01.06.2018
FQS Forum: Qualitative Social Research "Visual Research" OR "Visual Methods"; "CAQDAS" AND "Visual methods"	Visual Research OR Visual Methods (77) CAQDAS and Visual Methods (7)	02.06.2018
Qualitative Research Journal (EMERALD) "Visual Research" OR "Visual Methods" AND "CAQDAS"	Visual Research OR Visual Methods (63) CAQDAS (2)	02.06.2018
Studies in Qualitative Methodology (Emerald) "Visual Research" OR "Visual Methods" AND "CAQDAS"	Visual Research OR Visual Methods (18) CAQDAS (3)	03.06.2018

Source: From the authors

relevance of the abstract reviewed by the team. Disagreements will be resolved by discussion among the researchers.

The following keywords were considered: firstly "Visual research" AND "visual methods" in order to respond to the first research question ("What types of visual research and methods have been used in the qualitative research field?"). The goal is to extract and analysed the main research methods and techniques that have been used in qualitative research. Next, the search strategy using the keywords "CAQDAS" AND "Visual methods" was tackled, with the aim of being more selective in terms of results to answer research question 2 and 3 ("What type of visual data analysis has been used in qualitative research? How have CAQDAS tools been used for visual methods and data analysis and what are their main functionalities?"). Surprisingly, no results at all were obtained in all journals, except in FQS Forum Journal (7 articles), which suggests that this might be considered a "non-results-based study". A preliminary and exploratory analysis of the titles and abstracts might be understood that there seems to be a relative lack of detailed documentation about analytic and technical procedures or critiques and comments about software utility. Corroborating Silver and Patashnick's [28] conclusions (the only paper found until now that really focuses in detail on audio-visual analysis using software), this might represent an important gap in the literature and a failure to be transparent and reflexive. This is a still ongoing study.

4.4 Step 4: Analysis and Synthesis/Summarizing the Evidence

Information from papers (titles and abstracts) will be extracted and coded using CAQDA, in this case webQDA software. Literature reviewers can use computer-aided text analysis (CATA) tools for content analysis to organize and manage data and to code bibliographic categories to make the review process more systematic, faster and more reliable [35]. For instance, it seems possible to undertake a content analysis of titles and abstracts using categories such as: visual research (type of approaches?), visual methods (type of method; characteristics and features; visual data analysis technique; type of visual data; source of production - generated by the researcher or participant?; combined with other data?), and type of functionalities and analysis of visual data in CAQDAS. These are topics to be examined more deeply in the future by the research team.

Arising from RQ1 "What type of visual research and methods have been used in the qualitative research field?", the team decided to have a more systematized approach that considered using content-analysis as a type of analysis of the relevant work extracted from the SRL. For that, webQDA [36], a Computer-Assisted Qualitative Data Analysis Software (CAQDAS) is being used. The process is based on the following steps:

1. Textual data preparation (this corresponds to the pre-analysis phase [37] where the titles, abstracts and keywords from all sources of data (8 journals in total, as Table 2 demonstrates) are organized in several types of documents in order to facilitate the coding and analysis stage. During this stage it is possible to undertake a preliminary reading of the data which helped, not only to assess the quality of the studies (step 3 of the SLR), but also to prepare the analysis phase (step 4 of the SLR);
2. A classification of the textual data, framed by the research question, in order to answer the following questions: "What are the main paradigms and approaches considered in terms of visual-based research?"; "What are the main types of visual methods that have been used and for what purposes?"; "Are they combined with other methods?"; "What type and source of visual data have been used? Photos, videos? Researcher-generated or participant-generated?".
3. Coding and analysis, in order to help answer steps 4 and 5 of the SLR (summarize the evidence and interpret the findings). An initial coding was used [38] since it "is intended as a starting point to provide the researcher with analytic leads for further exploration." (p. 81). But, as recommended by Clarke (2005), a period of "digesting and reflecting" on the data is important (cited in [38]).

After organizing the results into several documents as internal sources (with webQDA), titles, abstracts and keywords were read, and a free coding process will be undertaken, in order to answer the questions previously mentioned. Only the format types "titles", "abstracts" and "keywords" will be considered at this stage, and not all the full paper.

4.5 Step 5: Reporting and Using the Results/Interpreting the Findings

It is not yet possible to determine and present results from this study at this stage. However, at this moment it is expected to identify the most commonly reported approaches currently used in visual research field, in order to fully understand the role of visual element in various studies (as the main data? or as complementary to textual data?). In addition, it is expected to determine the most commonly visual methods used in qualitative studies (Photo-elicitation? Photovoice? Videography?). Finally, with the findings from a SLR, a clearer picture about the use of CAQDAS in visual data analysis will be forecasted.

5 Conclusions

Based on empirical findings from previous studies, the goal of this study is to analyse, synthesize and present a comprehensive SLR on the topic of visual research methods and the use of CAQDAS and their main functionalities for visual data. The authors believe that the SLR methodology is undoubtedly a suitable tool for moving away from descriptive reviews of the literature, with contributions from main findings of the literature and the identification of a possible lack of results or non-results, and the establishment of a basis for future research. A search strategy was clearly defined for identifying relevant literature at this stage, based on qualitative research published in journals. The intention in the future is to spread the review in various fields of research, in order to obtain more integrated results. Hopefully, the results obtained in future work will assist researchers and software developers to develop new tools and methodologies in the audio-visual field for use with CAQDAS.

References

1. Minayo, M.C.S.: Fundamentos, percalços e expansão das abordagens qualitativas. In: Costa, A.P., Neri de Souza, F., Neri de Souza, D. (eds.) Investigação Qualitativa: Inovação, Dilemas e Desafios, vol. 3, pp. 17–48. Ludomedia, Oliveira de Azeméis (2016)
2. Mirzoeff, N.: An Introduction to Visual Culture, 2nd edn. Routledge, New York (2009)
3. Heisley, D.D.: Visual research: current bias and future direction. Adv. Consum. Res. **28**, 45–47 (2001)
4. Athelstan, A., Deller, R.: Visual methodologies (Editorial). Grad. J. Soc. Sci. **10**(2), 9–19 (2013)
5. Rodrigues, A.I., Costa, A.P.: A imagem em investigação qualitativa: análise de dados visuais. In: Amado, J., Crusoé, N.M. de C. (eds.) Referenciais Teóricos e Metodológicos de Investigação em Educação e Ciências Sociais, 1st edn., pp. 195–218. Salvador da Bahia: Edições UESB (2017)

6. Rodrigues, A.I., Souza, F.N., Costa, A.P. (eds.): Análise de dados visuais: desafios e oportunidades à investigação qualitativa (Editorial). Rev. Pesqui. Qual. **5**(8), Brasil (2017). http://rpq.revista.sepq.org.br/index.php/rpq/issue/view/8/showToc
7. Rose, G.: Visual Methodologies: An Introduction to Researching with Visual Materials. Sage, London (2016)
8. Collier Jr., J.: Visual Anthropology: Photography as a Research Method. Holt, Rinehart, & Winston, New York (1967)
9. Collier Jr., J., Collier, M.: Visual Anthropology: Photography as a Research Method. University of New Mexico Press, Albuquerque (1986)
10. Harper, D.: Talking about pictures: a case for photo elicitation. Vis. Stud. **17**, 13–26 (2002). https://doi.org/10.1080/14725860220137345
11. Pink, S.: Visual methods. In: Seale, C., Gobo, G., Gubrium, J.F., Silverman, D. (eds.) Qualitative Research Practice, 1st edn., pp. 361–377. Sage, London (2004)
12. Rodrigues, A.I.: Codificação de dados visuais: o uso de imagens como exemplo, Webinar 30.10.2017 (2017). https://youtu.be/_AGKjaYpO-o, https://doi.org/10.13140/rg.2.2.19604.81287
13. Whincup, T.: Imagining the intangible. In: Knowles, C., Sweetman, P. (eds.) Picturing the Social Landscape: Visual Methods and the Sociological Imagination, pp. 79–92. Routledge, London (2004)
14. Seale, C.: Researching Society and Culture. Sage, London (2004)
15. Samuels, J.: When words are not enough: eliciting children's experiences of buddhist monastic life through photographs. In: Stanczak, G.C. (ed.) Visual Research Methods: Image, Society and Representation. Sage, London (2007)
16. Frith, H., Riley, S., Archer, L., Gleeson, K.: Imag(in)ing visual methodologies. Editorial. Qual. Res. Psychol. **2**(3), 187–198 (2005)
17. Clark, A., Morriss, L.: The use of visual methodologies in social work research over the last decade: a narrative review and some questions for the future. Qual. Soc. Work. Res. Pract. **16**(1), 29–43 (2017). https://doi.org/10.1177/1473325015601205
18. Russell, A., Diaz, N.: Photography in social work research: using visual image to humanize findings. Qual. Soc. Work. **12**(4), 433–453 (2013). https://doi.org/10.1177/1473325011431859
19. Atkinson, P., Delamont, S.: Analytic perspectives. In: Denzin, N.K., Lincoln, Y.S. (eds.) Collecting and Interpreting Qualitative Materials, 3rd edn, pp. 285–311. Sage, Thousand Oaks (2008)
20. Harper, D.: Visualizing structures: reading surfaces of social life. Qual. Sociol. **20**, 57–74 (1997)
21. Banks, M.: Visual Methods in Social Research. Sage, London (2001)
22. Harrington, C., Lindy, I.: The use of reflexive photography in the study of the freshman year experience. Paper presented at the annual conference of the Indiana Association for the institutional research, Nashville, March 1998
23. Schulze, S.: The usefulness of reflexive photography for qualitative research: a case study in higher education. S. Afr. J. High. Educ. **21**(5), 536–553 (2007). https://doi.org/10.4314/sajhe.v21i5.50292
24. Steyn, G.: Using reflexive photography to study a principal's experiences of the impact of professional development on a school: a case study. Koers **74**(3), 437–465 (2009)
25. Amerson, R., Livingston, W.G.: Reflexive photography: an alternative method for documenting the learning process of cultural competence. J. Transcult. Nurs. **25**(2), 202–210 (2014). https://doi.org/10.1177/1043659613515719

26. Cahyanto, I., Pennington-Gray, L., Thapa, B.: Tourist–resident interfaces: using reflexive photography to develop responsible rural tourism in Indonesia. J. Sustain. Tour. **21**(5), 732–749 (2012). https://doi.org/10.1080/09669582.2012.709860
27. Lyon, D.: Conversations with the Dead. Henry Holt and Co., New York (1971)
28. Silver, C., Patashnick, J. Finding fidelity: advancing audio-visual analysis using software. FQS Forum: Qual. Soc. Res. **12**(1) (2011). Art. 37
29. Silver, C.: CAQDAS at crossroads: choices, controversies and challenges, In: Costa, A.P., Reis, L.P., de Souza, F.N., Moreira, A. (eds.) Advances in Intelligent Systems and Computing, Computer Supported Qualitative Research, pp. 1–13 (2018)
30. Weitzman: Software and qualitative research. In: Denzin, N.K., Lincoln, Y.S. (eds.) Handbook of qualitative research, pp. 803–820. Sage, London (2000)
31. Lewins, A., Silver, C.: Using Software in Qualitative Research: A Step-by-Step Guide. Sage, London (2007)
32. Budgen, D., Brereton, P.: Performing systematic literature reviews in software engineering. In: Proceedings of the 28th International Conference on Software Engineering - ICSE 2006. ACM Press, New York (2006)
33. Denyer, D., Tranfield, D.: Producing a systematic review. In: Buchanan, D.A., Bryman, A. (eds.) The Sage Handbook of Organizational Research Methods, pp. 671–689. Sage, Thousand Oaks (2009)
34. Khan, K., Kunz, R., Kleijnen, J., Antes, G.: Five steps to conducting a systematic review. J. R. Soc. Med. **96**(3), 118–121 (2003)
35. Costa, A.P., de Sousa, F.N., Moreira, A., de Souza, D.N.: Research through design: qualitative analysis to evaluate the usability. In: Costa, P.A., Reis, P.L., Neri de Sousa, F., Moreira, A., Lamas, D. (eds.) Computer Supported Qualitative Research, pp. 1–12. Springer, Cham (2017)
36. Souza, F.N. de Costa, A.P., Moreira, A.: webQDA - Qualitative Data Analysis (version 3.0). Micro IO e Universidade de Aveiro, Aveiro (2016)
37. Bardin, L.: *Análise de Conteúdo* [Content Analysis]. Edições, Lisboa, p. 70 (1979)
38. Saldaña, J.: The Coding Manual for Qualitative Researchers. Sage, London (2009)

Dialogical Thematic Analysis of Conversation in Tutors Forums Online

Claristina Borges da Silva(✉)

University of Brasília, Brasília 70 910 900, Brazil
claristinas@gmail.com

Abstract. This article presents the technique of Dialogic Thematic Analysis of the Conversation as a method of analysis of four topics of a forum specially created for research purposes in a virtual learning environment and aims to analyze the concepts of self-positioning, the identification and the identity of the tutors in graduate course, through dialogic interactions. The dialogue between these three concepts will allow us to identify the tutors' perceptions about the competences required by the new learning culture resulting from the development of digital technologies that modify the traditional roles of the teacher. The Dialogic Thematic Analysis is presented as a method of verbal data analysis in qualitative researches and is structured according to the following procedures: the transcription of the interviews; the definition of the analytical unit; intensive reading of transcribed material; the organization of statements in themes and subtopics (analysis of recurrences, relationships and similarities of meanings in statements); and the elaboration and analysis of semiotic maps. According to the presuppositions of the Bakhtinian dialogism, in the dialogical thematic analysis the enunciations are taken as analytical units. In the case of the forum, object of this research, we analyze the enunciations understanding them as the slogans, topic proposed by the researcher to start a discussion in the forum as well as the answers and/or propositions of the others. In this way, according to the principle of responsiveness we identify and analyze the positions of the self through the alternation of the speakers.

Keywords: Distance education · Tutoring · Dialogical self

1 Introduction

The mode of e-learning has become increasingly present in contemporary society marked by the development of digital technologies. In the perspective of expanding the performance of universities, the use of virtual learning environments in teacher training has been promoted by several institutions with a view to news forms of organization of the educational space. Thus, software such as Moodle, acronym for "Modular Object-Oriented Dynamic Learning Environment", have been used to produce and manage these educational activities, based on the Internet and/or local social networks.

Among the resources and activities by the Moodle platform is the Forum, an important discussion activity that provides interaction and exchange of ideas and information among participants, tutors and students, in a course/discipline. In it,

A. P. Costa et al. (Eds.): WCQR 2018, AISC 861, pp. 248–254, 2019.
https://doi.org/10.1007/978-3-030-01406-3_21

interactions are made through asynchronous communication and messages are shared through various forms such as texts, image files, videos, etc. Forums are therefore considered the means that offer the best communicative capacity to support the development of collaborative actions.

The mediator activity of the tutor in the forums provides new and meaningful reflections on the way of learning and the role of the teacher, whose performance is based on competencies as moderator or facilitator. According to Brazilian legislation [1], the tutor must act in the area of knowledge of their respective training, as support to the activities of teachers and pedagogical mediation with students.

2 Research Context and Objectives

In this research, the positions of the tutors in the context of graduate courses or complementary training courses offered by the Institute of Psychology of a university were approached. The research was carried out in a virtual learning environment through a Forum specially created in the Moodle platform.

We seek to analyze the positions of the four tutors to identify the representations and meanings of self and others in the production of knowledge. Through the personal and professional examination of their profiles we seek to know their trajectories, qualifications and experiences, as well as to know the expectations that these tutors had of the development of the didactic units with their students with the intention of verifying how they perceived themselves to act as tutors, their identification with the course and how its position outlined its identity. This research was based on theories that guide Cultural Psychology [2–5], Self-Theories and Positioning Theories [6, 7], The Dialogical Perspective [8–10]. It is a qualitative research whose method used for data analysis was the method of Dialogical Thematic Analysis that has been used by the research group Thought and Culture of the Ágora Psyché Laboratory of the Department of Psychology of the University of Brasília [11–16].

The decision by the learning forums was motivated by the fact that these tools store all discursive production, indicating personal and collective positions in knowledge production [6, 7], their changes and transitions in a cultural perspective [3–5] and dialogical [8, 9, 17, 18]. In this research, the perspectives of heterogeneity, equifinality and multifinality presented in Zittoun's approach to problem of the use of ergodicity as a methodology in human development research [10]. These perspectives were added to the method of Dialogical Thematic Analysis. With this perspective we wanted to go beyond the question-answer model, stimulating the writing through the interactions of the researcher in the course of the propositions of the topics as well as the possible interactions stimulated by the creation of new discussion topics proposed by the participants themselves, considering the unfolding of the narrative of the participant, according to the model proposed by [19].

Four tutors invited by the researcher and who agreed to participate in the study participated in the study. These tutors were selected, in a public call to act in the course, according to the following qualifying prerequisites: Degree in Pedagogy, Psychology, Speech Therapy and/or Human Sciences, Health or Social areas; minimum experience of 1 year in the teaching of Basic and/or Higher Education.

The generative issues that initiated the narratives and interactions in the four topics of the forums revolved around the themes presented in Table 1:

Table 1. Forum

Forum	Research on training and qualification of tutors
Topic 1	Production of personal and professional memorials
Topic 2	Relation between the e-learning mode and the face-to-face modality
Topic 3	Perceptions of identity and identification with the course
Topic 4	Expectations about the course prior to its beginning, in the course and at the end

Methodology

The Dialogical Thematic Analyses presents itself as a method of verbal data analysis in qualitative researches [16] and is structured according to the following procedures: the transcription of the interviews; the definition of the analytical unit; intensive reading of transcribed material; the organization of enunciations in themes and subthemes (analysis of recurrences, relationships and similarities of meanings in enunciations); and the elaboration and analysis of semiotic maps.

The research group Thought and Culture of the Ágora Psyché Laboratory of the Department of Psychology of the University of Brasília [16] has been using the Thematic Analysis Method in combination with the approaches of Bakhtinian dialogism in order to highlight the dynamics of interactions and constructions of meanings in the analyses of narratives of participants of the studies in the dialogical perspectives. In the research presented in this article, adaptations were made in this method because the communication in the online forum occurred asynchronously, in written form and did not allow the researcher to perceive signs such as gestures, physiognomic expressions or pauses in reflections.

Analysis of Information

The procedures described above were adapted in the present research for the analysis of the data and involved the printing of the proposed topics of forum made in full in the format of Nested Responses. This format allows the clear visualization of the themes or slogans and the answers in a backward structure, that is, a feature of Moodle software that shifts the answers to the right of topic to which they respond and that facilitates the identification of the message that was answered.

After the transcription of the topics of the forum and during the intensive reading, we followed [19] orientation to construct a coding reference in narrative interviews, adapting the written texts collected in the forum. These authors as well [16] recommend that the qualitative text be reduced gradually, that is, in two or three rounds of series of paraphrases that will be paraphrased in synthetic sentences that will later be paraphrased in some keywords aiming, in this way, generalization and condensation of meaning.

Table 2 lists the themes and subthemes that emerged from the forum whose questions involved the following aspects related to the trajectories of the tutors: How has your trajectory been as a tutor? When did you start acting in e-Learning? Did you

Table 2. Themes and subthemes (recurrences and similarities)

Themes	Subthemes
Training Tutor	Courses; Learning throughout practice
skills	Mediation of conflicts; Planning and time organization Skills; Assiduity; Proactivity; Creativity; Ability to read and write
Functions	Coordinator; Advisor
E-learning's advantages	Context Discussions of texts in the forums; Collaborative learning; Dialogue with people from other units of the Federation; Relativization of space and time
E-learning's difficulties	Low socialization; lack of autonomy; non-compliance with deadlines by some students

do any training in the area? If yes, what kind and when did you do? Tell me your story. Thus, the themes and subthemes that emerged from the slogan posted by researcher were identified. Different colors were used to highlight recurrences in narratives and arrows with double tips to indicate relationships and similarities [16].

According to the presuppositions of the Bakhtinian dialogism, in the dialogical thematic analysis the enunciations are taken as analytical units [16]. In the case of the forum, object of this research, we analyze the enunciations understanding them as the slogans, that is, the topic proposed by the researcher to start a discussion in the forum, as well as the answers and/or propositions of the other participants. In this way, according to the principle of responsiveness [8, 17] we identify and analyse the positions of the self through the alternation of the speakers.

Dialogue is understood as the classic form of discursive communication due to its characteristics of precision and simplicity, since each replica has specific conclusions when it expresses a position of the speaker and that, on provoking response, provides a responsive position [17]. In this perspective, we consider the conception of self, described by Hermans [6], based on the propositions of James e Bakhtin [20, 21] and that refers to the dynamic multiplicity of positions that the individual occupies in the diverse social contexts in which interacts. This concept is based on the establishment of a relationship between the self and the personal and collective culture.

Based on Linell's examples and propositions [18], Hermans [6] explains that the notion of social power or dominance is an intrinsic feature of dialogic processes and is intimately associated with the position each person occupies in a particular institution. Linell [18] distinguishes four different dimensions involved in the dominance of interaction: interactional dominance, dominance topics, amount of conversation, and strategic moves.

Analysis of Interactions

Following the procedures of the Thematic Dialogical Analysis, a map of meanings, Fig. 1, was elaborated from the analysis of recurrences and similarities in the interactions in the forum and presented in Table 2. Through the analysis of the map, it was possible to identify in the relations of interactions in a form of abstract, the outline of some results, such as self-positioning of identification, identification with the course and identity as tutor or teacher, that is, what it is to be a tutor in this course.

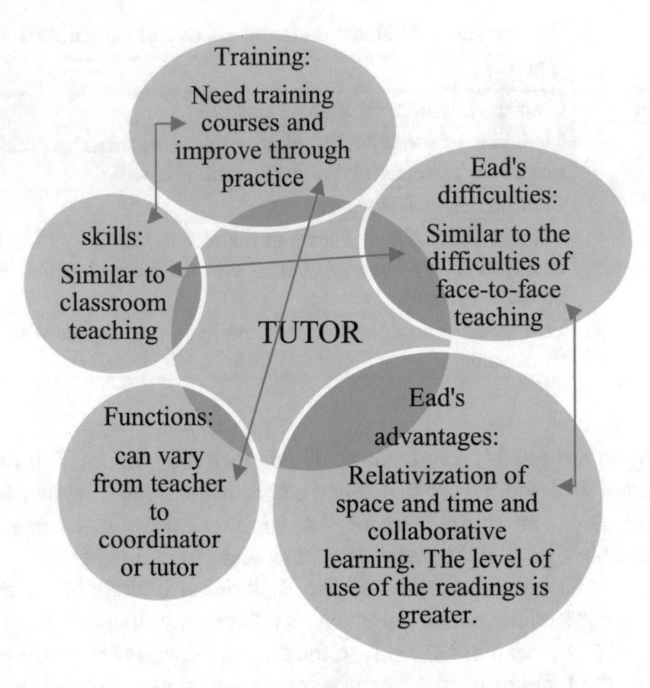

Fig. 1. Map of meanings.

The analysis and discussion of the map involved two distinct moments: the first moment refers to the external influences in the conceptions about what is to be a tutor and the second moment refers to the self-perceptions of the tutor. The construction of narratives in terms of polyphony of spatial oppositions allows a particular idea to be treated in the context of the inner and outer dialogues, revealing a multiplicity of perspectives: the intersection, the consonance or the interference of discourses [21]. In this polyphonic construction, a specific theme does not have a fixed character, self-contained, immutable or of continuous meaning.

In this way, it was possible to observe in their narratives the tutor's positions reflect how they are positioned by the norms that govern their function such as formation and skills. In addition, the tutors also position each other, thus influencing the perceptions of each.

In the second moment, we identify the internal dialogues reflected in the narratives when it was possible to perceive how each one positions himself or herself and distinguishes them from others, that is, some perceive themselves only as tutors who support teachers while others perceive themselves as teacher and/or authors. These conclusions are related to the fact that the modalities of face-to-face teaching and e-learning have common characteristics and divergent characteristics that also influence the identification with the course, as well as the identity of the tutors being that they are delineated and are being constructed in a hybrid way in the course of their performances.

In a subtle way, which still requires deepening, the researcher and author of this article included in the methodology of the Dialogical Thematic Analysis the approach of Zittoun et al. [10] that point out the problem of the use of ergodicity as statistical methodology in the research of human development in traditional academic psychology and explain the failures of this method from the epigenetic perspective that, coupled with the landscape metaphor, challenges the traditions of statistical inference psychology and population genetics. Heterogeneity, equifinality, and multifinality are the factors that explain these failures: heterogeneity considers the fact that human beings are all different and that each life course is unique, equifinality based on the fact that different paths of development can lead different individuals to achieve the same points in the course of life and the notion of multifinality that describes the notion that one and the same behavior can serve different goals at the same time.

3 Conclusions

Dialogical Thematic Analysis of Conversation can be considered as a useful method in the field of qualitative research. Its flexibility feature allows the identification and analysis of recurring themes in a conversation from which it is possible to delineate the perceptions that are being built in the interactions.

In this way, the study of narratives in research can make evident the dynamics of interactions and constructions of meaning [16].

In this research we analyze the narratives of the participants, as well as their interactions mediated by the online forums from the perspective of the Dialogical Thematic Analysis and considering the three factors of the epigenetic perspective: heterogeneity, equifinality, and multifinality. In this way we were able to identify, throughout the analysis, how the themes related to the identify of the tutor and the identification with the course arose according to the dialogical perspectives in the interactions as well as in the self-perception of the participants.

References

1. BRASIL: Resolução no 1, de 11 de Março de 2016. Diário Oficial da União, Seção 1, 23–24 (2016)
2. Vigotsky, L.S.: Obras escogidas, vol. III. Visor Distribuciones, Madri (1995)
3. Bruner, J.: Actos de significado. Alianza Editorial, Madri (1991)
4. Rosa, A.: Entre la explicación del comportamiento y el esfuerzo por el siginficado: una mirada al desarrollo de las relaciones entre el comportamiento individual y la cultura. Rev. Hist. Psicol. **21**(4), 77–114 (2000)
5. Pino, A.: As marcas do humano: às origens da constituição cultural da criança na perspectiva de Lev S. Vigotski. Cortez, São Paulo (2005)
6. Hermans, H.J.M.: The dialogical self: toward a theory of personal and cultural positioning. Cult. Psychol. **7**(3), 243–281 (2001)
7. Harré, R., van Langenhove, L.: Positioning Theory: Moral Contexts of Intentional Action. Blackwell Publishers Ltd., Oxford (1999)

8. Bakhtin, M.: Para uma filosofia do ato responsável. Pedro & João Editores, São Carlos (2010)
9. Linell, P.: What is dialogism? Aspects and elements of a dialogical approach to language, communication and cognition. Linköping University, Sweden (2003). Lecture first presented at Växjö University, October 2000. https://manoftheword.files.wordpress.com/2013/10/linell-per-what-is-dialogism.pdf
10. Zittoun, T., Valsiner, J., Vedeler, K., Goncalves, M., Ferring, D.: Melodies of life. Developmental Science of the Human Life Course. Cambridge University Press, Cambridge (2013)
11. Barbato, S.B., Caixeta, J.E.: Novas tecnologias e mediação do conhecimento em atividades colaborativas no ensino superior. Linhas Críticas **20**(42), 363–382 (2014). http://periodicos.unb.br/index.php/linhascriticas/article/view/11624
12. Beraldo, R.M.F.: Dinâmicas de intersubjetividade em atividades colaborativas em contexto mediado por fórum online no ensino médio. Tese de doutorado, Universidade de Brasília, Università degli Studi di Parma, Emilia- Romagna, Itália (2017). http://repositorio.unb.br/bitstream/10482/23756/1/2017_RossanaMaryFujarraBeraldo.pdf
13. Caixeta, J.E., Borges, F.: Da Entrevista Narrativa à Entrevista Narrativa Mediada: definições, caracterizações e usos nas pesquisas em desenvolvimento humano. Fronteiras **6**(4), 67–88 (2017). http://doi.org/10.21664/2238-8869.2017v6i4
14. Mieto, G.S.M., Barbato, S.B., Rosa, A.: Professores em transição: produção de significados em atuação inicial na inclusão escolar. Psicologia, Teoria e Pesquisa 32(n. especial), 1−10 (2017). http://dx.doi.org/10.1590/0102-3772e32ne29
15. Peres, S.G.: O processo de significação da professora contadora de histórias e a interação com crianças no contexto da biblioteca escolar. Dissertação de mestrado, Universidade de Brasília, Brasília, Brasil (2017). http://repositorio.unb.br/bitstream/10482/23071/1/2017_SilvanaGoulartPeres.pdf
16. Silva, C., Borges, F.: Análise temática dialógica como método de análise de dados verbais em pesquisas qualitativas. Linhas Críticas, **23**(51), 245−267 (2017). http://periodicos.unb.br/inde.php/linhascriticas/article/view/28451/20063
17. Bakhtin, M.: Estética da criação verbal, 4a edn. Martins Fontes, São Paulo (2003)
18. Linell, P.: The power of dialogue dynamics. In: Marková, I., Foppa, K. (eds.) The Dynamics of Dialogue, pp. 147–177. Harvester Wheatsheaf, New York (1990)
19. Jovchelovitch, S., Bauer, M.W.: Entrevista narrativa. In: Bauer, M.W., Gaskell, G. (eds.) Pesquisa Qualitativa com Texto, Imagem e Som: um manual prático, pp. 90–113. Vozes, Petrópolis (2002)
20. James, W.: The Principles of Psychology. Henry Holt, New York (1890)
21. Bakhtin, M.: Problems of Dostoevsky's poetics. Ardis, Ann Arbor (1973)

Woman's Satisfaction with Her Water Birth Experience

Mariana Gonçalves[1], Emília Coutinho[2(✉)], Vitória Pareira[3], Paula Nelas[2], Cláudia Chaves[2], and João Duarte[2]

[1] Centro Hospitalar do Baixo Vouga, E.P.E., Aveiro, Portugal
[2] Escola Superior de Saúde de Viseu do Instituto Politécnico de Viseu, Viseu, Portugal
ecoutinhoessv@gmail.com
[3] Escola Superior de Enfermagem do Porto, Porto, Portugal

Abstract. Maternal and obstetric health specialist nurses focus their attention on the human responses to life transitions, particularly in physiological situations. The appearance of associations, in Portugal, for physiological birth defence and for water births defence was the reason for this research. This is a qualitative, phenomenological study from the Max Van Manen perspective with the aim to understand the experience of mothers who had one or more water births in Portugal. Thirteen semi-structured interviews were conducted with women who had at least one experience of childbirth in the water. The data were collected between July 2017 and February 2018. From the qualitative analysis of the data emerged seven categories. Considering the dimension of this study only two of these categories are possible to be presented in this article, namely: the benefits of water; and the woman's satisfaction manifestations with the experience of physiological labour in water.

Keywords: Childbirth · Water · Natural childbirth · Satisfaction Experience

1 Introduction

According to the International Confederation of Midwives [1] normal birth is defined as a dynamic process in which labour commences, continues, and is completed with the infant being born spontaneously at term, in the vertex position at term, without any surgical, medical, or pharmaceutical intervention; where foetal and maternal physiologies and the psychosocial context are respected. Maternal and Obstetric Health Specialist Nurses (EESMOS) are professionals with specific background and training designed to support and promote physiological delivery and also to detect any deviations from "normal" and intervene whenever needed [2]. Thus, the conduct of the EESMOS should be based on privileging natural childbirth care, providing safety in care and adopting an expectant attitude [2].

Water birth, due to its intrinsic characteristics, is part of a philosophy of normal/natural/physiological delivery. In this sense, in many countries of the world, it is closely linked to the practice of maternal and obstetric health specialist nurses, who

A. P. Costa et al. (Eds.): WCQR 2018, AISC 861, pp. 255–265, 2019.
https://doi.org/10.1007/978-3-030-01406-3_22

have tried to meet the expectations of women and use the benefits of water to provide them with comfort and well-being.

In Portugal, only one public institution, Hospital de São Bernardo in Setúbal, offered, in its services, the opportunity for the woman/couple to experience a water birth. This hospital in November 2008 became a pioneer in the use of water during labour and delivery, being the first public institution in the Iberian Peninsula to provide this experience to the woman/couple [3]. In June 2014, this practice was questioned and suspended in this institution [3]. Exactly in the same month when the water birthing practice ended at the São Bernardo Hospital, a civic movement emerges, driven by the passion for childbirth in water known as Mães D'Água [3]. This crusade arose with a group of mothers who had experienced the water birth in the Hospital de São Bernardo and that, after the suspension of this practice, gathered more than 4,900 signatures for the maintenance of the water births in that hospital, as well as the extension to other hospitals of the National Service (SNS), without success. Although this cause has been shelved, the movement continues to work daily to: inspire and empower women; promote natural/water childbirth; and to generate a change in the Portuguese NHS by bringing back the water births to public hospitals and making it a reality accessible to all pregnant women/couples choosing this form of birth and delivery [3].

In this sense, and to promote the improvement of maternal-foetal health care in Portugal, it is considered fundamental to understand the experience of mothers who had one or more births in water in Portugal.

2 Methods

The present study, carried out by six researchers, with the intention of meeting the above-mentioned objective was guided by the research question "What is the experience of mothers who had one or more water births in Portugal?". Based on constructivist assumptions, this study was developed under the qualitative-interpretative paradigm [4], attempting to capture the participants' perspective to understand the phenomena. Several theorists have focused on the analysis of the qualitative concept as referring to quality rather than quantity, of the object being studied is centred on what or who and that could not be uncovered and even measured using quantitative approaches [5].

Among the several qualitative options available, the phenomenological method was selected because it is better able to respond to the research question and the purpose of the study. Since phenomenology is the hope to understand the world [4], it will be the phenomenology that will best help to understand the experience lived by women during their water births. The steps of Van Manen's hermeneutic phenomenology [6–8] guided this study, in a flexible and dynamic way as recommended by the author, interacting with each other to clarify the essential nature of the phenomenon. The study started by internalizing the first step Turning to the nature of the experience, with attention to: selecting the phenomenon of interest, selecting the perspective of study of the phenomenon, selecting the context where the phenomenon occurs, selecting the participants of the study. In the second step Inquiring the lived experience, attention was given to: entering the study context, obtaining the experiential descriptions about the care, leaving the context of the study. In the third step, a phenomenological hermeneutic reflection was developed. It was then time to move on to the fourth step by developing a phenomenological hermeneutic writing. In the fifth step, a response to a relationship oriented towards the phenomenon was attempted. Finally, in the sixth step Counterbalancing the context of the research was considered. Although Van Manen [6] does not present any further steps in method terms, in his writings he mentions the Ethical Principles, which were rigorously respected and in the Communication of the research, which was endeavoured to achieve with this and other publications, but also with the feedback to the participants.

The study participants are women who experienced water birth in Portugal. The selection of the women was intentional, through the registration in an online questionnaire, which thus demonstrated their willingness to participate in the study. It was only after this expression of interest from each woman that efforts were made to contact the participants through the addresses they provided. In this contact an interview was scheduled, considering the availability and the preferred location for each woman. The guiding question for all the interviews was: Tell me about your experience of water birth in Portugal? Participants spoke freely on the topic. All interviews were conducted by the same researcher. The verbatim of the interviews was done and analysed by all the researchers as the interviews were carried out. The number of participants was determined by the data saturation [9], i.e., when the researcher no longer finds explanations, interpretations or descriptions of the phenomenon under study. In the present study 13 women who had one or more water births in Portugal participated.

In the development of the study, the evaluation/critique criteria of qualitative research studies proposed by Morse [10] and Richards and Morse [9] were also considered, such as credibility, confirmability, meaning, standardization, saturation, transferability.

As presented above, in the steps of Max Van Manen's hermeneutic phenomenology, in terms of qualitative analysis of the data, the phenomenological reflection of a full description of the experience lived in the search for phenomenological themes or structures of meaning was used. In this work of analysis, the author's orientations were followed by isolating in the text the thematic aspects hidden in three approaches: by the holistic or sententious approximation, by emulating the phrase that translated the meaning of the whole; by the selective or prominent approach to see which phrases or expressions were truly fundamental or translators of the phenomenon or the experience lived; and finally by the detailed approach or line by line questioning the contribution of each sentence to the understanding of the phenomenon or experience described [6].

To achieve a better data organization and visualization, the NVivo11 Pro program was used, supporting the gathering, systematizing and categorizing the testimonies collected in the interviews. Sources were imported, the nodes created and coded, and data analysis was performed. The final tree comprised seven major categories, all related to the woman's experience of water birth, with two categories presented here.

Max Van Manen's [6] reflections on ethical aspects were followed, as well as the evidences of Amado [5] regarding "research ethics" and "researcher ethics". Also Fortin and Gagnon [11] were a reference for the significance given to the importance of respecting the participants in the study not only to provide all the necessary clarifications but also to explicitly present the objectives of the study, the care to be taken in the involvement of each one in the process, the ethical sense in the treatment of the data itself, as well as in ensuring informed consent and request written consent, from the participants, of their voluntary participation, informing them of the possibility of giving up at any time during the study. Respect for the most rigorous ethical principles was always present in both the respect of the participants and the respect for the intellectual property of others who were the target of this research, as recommended by Amado [5]. The communication of the study to the National Data Protection Commission and the obtaining of the respective authorization (3179/2017) was felt as an ethical necessity by the research group because this study is part of the health field, an area deemed as sensitive. Informed consent was signed by all study participants and the researcher- The right to self-determination, intimacy, anonymity and confidentiality, and the right to fair and equitable treatment as mentioned by others [12] were also respected. The consent of each participant was also obtained for recording the interview, ensuring not only the faithful collection of the data, but also facilitating the establishment of an empathic relationship during the interview.

3 Results

In the results presentation, in the following tables, the letter n means the number of participants and the abbreviation ru stands for registration units, that identifies the number of registration units that contributed to the construction of the categories and subcategories. The following categories are presented: water benefits; and manifestations of satisfaction with the experience of physiological labour in water by the woman.

On the category of Benefits of water, Table 1 highlights the pain relief subcategory, referred by 9 of the 13 mothers participating in the study, followed by an opportunity to see the child being born, the entrance into the water to provide the high moment of the experience, relaxation, greater mobility, lightness, feeling of safety, Increase the couple's closeness, facilitating childbirth, childbirth usually occurs, tranquillity, create conditions for the woman to succeed, and lower severity of the perineal laceration. Less mentioned by women were the subcategories: skin-to-skin contact, softness, comfort, shorter labour time, improved breathing, cleanliness, maintenance of perineal integrity, better body knowledge, freedom, pleasure, memories, trust and privacy.

Some expressive testimonies of this category/subcategories

> *"when we reached that peak of contractions, with the water we got a moment of relief"* (M1);
> *"And when the baby was born, it was amazing to be able to see, first, the baby being born"* (M5);
> *"Then I went into the water, and it was like this ... Barbara Harper names it 'aaaaaah moment'. That is the moment when we sit in the water and the hot water meets our skin from the pelvis to waist, and we go 'aaaaaah'... It really feels like 'Ok, I'm where I needed to be'."* (M3);
> *"water makes all the difference because it really soothes the body and relaxes"* (M7);
> *"posturing, I spent a lot of time in weird positions in the water, as like, with my legs to one side, with my feet in front, crossed. And I could not do it, while pregnant with contractions, out of the water. It gave me freedom to assume postures that I would not be able to, out of water"* (M9);
> *"A person enters [in the water] and feels ... feels much lighter soon. It's lightness"* (M10);

Table 1. Benefits of water.

Category	Subcategory	N	Ru
Benefits of water	Pain relief	9	30
	Provide the opportunity to see the child being born	9	13
	Entering the water is the highlight of the experience	8	13
	Relaxation	7	12
	Greater mobility	7	9
	Lightness	6	9
	Protection feeling	5	9
	Increase the couple's closeness	4	9
	Facilitate the birth	6	8
	Birth is slow	4	8
	Tranquillity	4	6
	To create conditions to enable the woman	4	5
	Lower severity of perineal laceration	5	5
	Promotion of skin to skin contact	4	4
	Softness	4	4
	Comfort	3	4
	Less time in labour	3	3
	Improve breathing	1	3
	Cleaning	2	2
	Maintenance of perineal integrity	2	2
	Better body knowledge	2	2
	Liberty	2	2
	Pleasure	1	2
	Safety	1	2
	Awaken good memories	1	1
	Trust	1	1
	Privacy	1	1

"It's as if I felt that there [in the pool] I was protected" (M7);
"The part that I highlight most, and G. [companion] is also of that opinion, it is the involvement that the water birth and the preparation itself, grows between the couple. It made us closer in a way ... If we were already close, due to being the moment that it was, the water brought us closer still. Because there were many exercises [in the water] that was the mother and the father" (M1);
"It was not fast, obviously because it was a four-and-a-half kg baby, right? But the memory that remains is, on one hand, how the water helped him [newborn] to slide" (M3);
"this issue of him being born in the water, to have waited for his moment. His head came out, then he stayed there for a while, then came another contraction, he came out, right to the top of me, we were there, with all the calmness of the world ... It's completely different!" (M10);
"I felt at peace during childbirth." (M2);
"I think the water gave me that, you see, it brought me ... it gave me and allowed my body to do what was necessary so that in this case the baby would also do her work as she had to do" (M6);
"the nurse examined me told me that I had a little tear and that I was going to go and be stitched" (M4);

"M. [new-born] was right on my chest from the first moment" (M4);
"The water made things much more, much smoother" (M6);
"the pool is round, I felt some comfort" (M7);
"Labour was relatively fast. Seven hours or so" (M1);
"My ability to breathe became greater" (M9);
"Everything I saw as the most practical advantages that I had seen with Cornelia's water birth that came to fruition, which is: cleanliness. That is, there is no visible blood, of the sort ... blood! [Laughs] For those who are not from the health area, blood is ... So there was no blood, the baby looked like it had literally been born laundered. [laughs] So it was so super clean" (M3);
"I did not tear, which was a wish of mine. I did not tear" (M8);
"[The birth in water] gave me more confidence still and made me... made me know myself a little better. That after all I am able to respect my body more, to perceive what was happening. To think that 'Ok, this hurts, it's not because I'm going ... it's not because I'm going to die. It hurts for a reason. ' So getting to know my body better, [to] enable to control it better" (M1);
"it gave me some freedom. The water ended up giving me some freedom there" (M12);
"in the photographs, my face is in pleasure, because for me it was literally to feel an area of your body never opened before opening at that moment. And to open it, I say it as a positive thing. Your hip looks like its dancing; it really does a "toc!"... Open! It is extraordinary" (M3);
"[I felt] super safe" (M8);
"It was funny because I know as soon as I entered the pool, so the most unconscious part of the experience I had with Watsu, that I had felt a very caring experience, a lot of inner self connection, a very special silence, that somehow was activated with the water" (M8);
"[Water made me feel] confident" (M2);
"I feel that the water gave me all the privacy I could wish for" (M8).

As for the manifestations of satisfaction with the experience of physiological labour in water by the woman, cf. Table 2, the most mentioned was to feel that it was a positive experience pointed out by 12 of the 13 mothers, followed by trust in nature, prefer natural/physiological childbirth, have a feeling of victory for having succeeded, consider EESMO the most capable to accompany, to feel that labour has profoundly transformed her, to feel satisfied with natural/physiological childbirth, to consider birth in water as natural, to express satisfaction in postpartum and to wish to repeat the experience of childbirth. Less mentioned by women were the subcategories: feeling special, believing that they had made the right choice, and needing to recommend the use of water in labour and delivery.

Some expressive testimonies of this category/subcategories are presented below:

"My water birth was an absolutely wonderful thing, it was wonderful, it was extraordinary it was everything I wanted and a little more" (M2);
"The next day I was sitting, suckling my daughter, with an oversized aura, super proud of me, my experience, and all that. And I think it's fundamental to have a positive birth experience." (M11);
"in normal labour, there is no reason to anticipate the work, to speed things up, or to intervene, when the body knows what it does" (M5);
"Whenever I thought of having a child, I always thought I would like to have it as natural as possible" (M4);
"What do I compare to... to any situation of a great victory in life, a moment of great victory. Because we really can achieve something that only depends on us" (M1);
"The birth in the water made me feel that I did not need any of that: just a pool, a quiet space, and I could do it. And I did it!" (M4);

Table 2. Women's Manifestations of satisfaction with the experience of physiological labour in water.

Category	Subcategory	N	Ru
Women's Manifestations of satisfaction with the experience of physiological labour in water	Feeling like it was a positive experience	12	36
	Rely on nature	4	13
	Prefer natural/physiological childbirth	9	11
	Have a sense of victory for having succeeded	7	11
	To consider EESMO as the most capable	6	11
	Feeling that childbirth has transformed her deeply	6	10
	Feeling satisfied with natural/physiological delivery	6	9
	Consider that childbirth in water is natural	7	8
	Verbalize postpartum satisfaction	5	8
	Want to repeat the birthing experience	5	7
	Feeling Special	3	4
	Believing to have made the right choice	3	3
	Need to recommend water use in labour and delivery	2	2

"As I had read, and seen, and informed about the natural childbirth movement, which is naturally headed by nurses, midwife nurses or midwives (where they are), it makes sense for these professionals to deal with normal births/natural/physiological" (M3);
"I do not think I have had anything in my life that would make me ... at least a moment so, so striking, or so fissuring almost ... it is almost as a life until childbirth and a life after childbirth" (M5);
"To be able to feel every phase of childbirth ... every phase of labour and after birth, and to feel, in this way, how the body works, how the body reacts is very interesting" (M5);
"I think it was important to have felt pain, and I think it's really important to feel pain in childbirth and I think the thing that makes labour easy with the epidural, cuts the process, because if I do not feel pain I might not be ready for what comes next because it is much harder to have a new-born than to give birth. [laughs] And overcoming it ... And overcoming myself with every contraction, and knowing myself to let go, to let it pass, to transform myself, is ... I think it's a fantastic women's capacity" (M9).
"A reminder that there [in the water] I had discovered that birth could be natural and could be safe because it was natural, and it would probably be safer the more natural it would be. [laughs] And it was so super nice" (M8);
"And so my postpartum was excellent" (M11);

"I wanted to have another child, but only if it was the same, because everything was so perfect" (M1);
"I did not have a second water birth just because it was not possible, because when M. [second daughter] was born they were no longer being performed ... but I would undergo a second water birth experience because I already have the knowledge of the first" (M12);
"I felt special. I felt that I was not just another woman who was having a child" (M11);
"I realized that it made all the difference, and I realized that it had been a great choice for both me and the baby" (M5).

4 Discussion

Both categories presented in the results, benefits of water, and women's manifestations of satisfaction with the experience of physiological labour in water, will be discussed separately according to the authors consulted.

Water Benefits
In this study, the category with the highest representation for mothers is the category on "Water Benefits", in which all of them refer to at least one benefit.

Labour pains can be perceived in different ways by the woman, although they are generally perceived as something painful and unpleasant [13]. With the experience of water delivery, in our study, pain relief emerges as the most referenced benefit for women, similar to other studies [14], suggesting the need for women to develop strategies for the pain relief in childbirth and the use of water is one of them. Some authors [15] used in their study a visual analogue pain scale to assess the pain perceived by women and concluded that women who delivered on water had a lower pain rate compared to women on land, both with and without epidural.

Midwives Alliance North America and Citizens for Midwifery in a 2016 position statement [16], in addition to reporting the least pain sensation corroborates some of the benefits of women's water immersion, such as greater comfort and mobility, promoting relaxation and reducing the duration of labour.

The same authors also refer the promotion of skin-to-skin contact and the increase of oxytocin and endorphin levels, hormones responsible for the pain reduction, making it bearable by a feeling of well-being, calmness, connection, transcendence and ecstasy [17].

Schafer [18] points out the increase in perineal integrity as a water benefit, reported by two women. Additional studies referred by Dekker [19] have shown that water-borne women are more likely to maintain intact perineum; five studies also show a higher rate of first and second degree lacerations in women who delivered in water [19].

The effect of free gravity on water gives the woman a feeling of lightness [17] perceived by six women as a beneficial sensation.

The Midwives Alliance North America and Citizens for Midwifery [16] point to an easier descent and rotation through the pelvis as a water benefit for the new-born, a benefit perceived by mothers as a birth facilitator.

According to Balaskas [17] a more private environment where the woman/couple can experience privacy, relaxation, and safety can also be created in the hospital layout using a bathtub for this effect, as referred by some women participants of this study

Women's Manifestations of Satisfaction with the Experience of Physiological Labour in Water

Regarding the category of women's satisfaction with the experience of physiological labour in water, it is important to mention that all the women in the study reported feeling satisfied with the experience of physiological labour in water.

The most commonly referenced subcategory for women, with a total of 12 women in 36 registration units, is "Feeling was a positive experience", a description of the experience that has been corroborated by WHO [20] with the need to provide women/couples positive experiences.

According to Balaskas [17] active birth is a mental attitude that "involves acceptance and belief in the natural function and involuntary nature of the birth process." The two subcategories then referred to most by women are "Rely on Nature" and "Prefer Natural/Physiological Childbirth".

It is also important to mention the subcategory "Have a feeling of victory for having succeeded" as the most significant, with 11 registry units and 7 participants mentioning it; also emerging is that mothers consider the Maternal and Obstetric Health Specialist able to accompany them in this process of childbirth. In terms of legislation, Law 9/2009 of the Portuguese Republic, "assigns to EESMOS/Midwives the competences for the full exercise of the surveillance of labour as well as assistance to normal birth" [2].

It should also be noted that six participants in the study revealed satisfaction associated with the physiological process of childbirth. Scientific evidence shows that women's satisfaction with childbirth is related to natural childbirth and especially to the care provided [21]. In this sense, nurses who specialize in maternal and obstetrical health must take into account the women expectations regarding their practices [22].

In the group of participants, of one study [23], five mothers reported being satisfied with postpartum, because of the delivery time decrease and less blood loss, which leave women with more energy.

According to Barbara Harper [24] the satisfaction of the woman/couple is visible in all stages, so, as referred by Lopes [25] one has to plan timely this singular event of human life with the woman/couple

Regarding the subcategory "Desire to repeat the delivery experience", the mothers show their satisfaction with this experience, corroborated by a study [14] that concluded that 72.3% of women who had water births declared that they would certainly choose this method to give birth again. Although maternal and obstetric care is free in Portugal, water delivery in Portugal is not currently available in public health institutions [3]. This inaccessibility to a free water birth is highlighted by some women in the study, as evidenced by the following registration unit "I did not have a second water birth just because it was not possible, because when M. [second daughter] was born they were no longer being performed … but I would undergo a second water birth experience because I already have the knowledge of the first" (M12).

5 Conclusions

When answering to the research question "What is the experience of mothers who had one or more water births in Portugal?" it was possible to reveal the benefits of water perceived as such by women with particular emphasis on the fact that they consider water as the key element of their childbirth experience, allowing pain relief, the opportunity to see the child born, the promotion of relaxation, and considering entering the water as the high point of the experience, which is in accordance with the conclusions of the studies presented in the literature; it was also possible to disclose the women's satisfaction woman in the experience of childbirth in the water, and that women are satisfied to trust and prefer physiological birth because they feel it was a positive experience, to convey a feeling of victory and to consider the nurse specialist in maternal and obstetric health as the most capable professional to accompany them.

Contributions from this study: it allowed to uncover the most significant knowledge about women's perception of the benefits of water births and the women's satisfaction with their experience of water birth in Portugal, giving visibility not only to this experience but also to qualitative research, namely the phenomenology that made it possible.

Limitations of the study: Due to its size, it is not possible to present the entire study.

Acknowledgment. IPV - Polytechnic Institute of Viseu CI & DETS - Center for Studies in Education, Technologies and Health, RESMI - Higher Education Network for Intercultural Mediation, Sigma Theta Tau International, Phi Xi Chapter.

Contributions. Conception and study design: MG, EC, VP; data analysis: MG, EC, VP; manuscript preparation: MG, EC, VP, PN, CC, JD; and discussion, revision, submission: MG, EC, VP, PN, CC, JD.

Funding. This study is financed by national funds by FCT - Foundation for Science and Technology, I.P., the project UID/Multi/04016/2016. Funding supported by FCT and CIDETS - Center for Studies in Technology and Health Education, Portugal.

References

1. ICM Homepage: Position Statement: Keeping Birth Normal. https://internationalmidwives.org/assets/uploads/documents/Position%20Statements%20-%20English/Reviewed%20PS%20in%202014/PS2008_007%20V2014%20Keeping%20Birth%20Normal%20ENG.pdf. Accessed 10 Jan 2017
2. Rodrigues, S.: Parto normal. In: Néné, M., Marques, R., Batista, M.A. (eds.) Enfermagem de Saúde Materna e Obstétrica. Lidel, Lisboa (2016)
3. Mães D' Agua Homepage: Uma breve história do Parto na Água em Portugal. http://maesdagua.org/uma-breve-historia-do-parto-na-agua-em-portugal/. Accessed 15 Mar 2018
4. Munhall, P.L.: Nursing Research. A Qualitative Perspective. Jones & Bartlett Learning, Sudbury (2012)
5. Amado, J.: Manual de Investigação Qualitativa em Educação, 2nd edn. Imprensa da Universidade de Coimbra, Coimbra (2014)

6. Van Manen, M.: Researching Lived Experience: Human Science for an Action Sensitive Pedagogy, 2nd edn. Althouse Press, London (1997)
7. Van Manen, M.: Phenomenology of practice. Phenomenol. Pract. **1**(1), 11–30 (2007)
8. Van Manen, M.: Phenomenology of Practice: Meaning Giving Methods in Phenomenological. Routledge, London (2014)
9. Richards, L., Morse, J.M.: Readme First for a User's Guide to Qualitative Methods, 3rd edn. Sage, Los Angeles (2013)
10. Morse, J.: Aspectos essenciais de metodologia de investigação qualitativa. Formasau, Coimbra (2007)
11. Fortin, M.F., Gagnon, J.: Fondements et étapes du processus de recherche, 3ème edn. Cheneliére Education, Québec (2015)
12. Amado, J., Vieira, C.C.: Salvaguarda das questões éticas na investigação e no relatório. In: Amado, J. (ed.) Manual de Investigação Qualitativa em Educação, 2ª edn. Imprensa da Universidade de Coimbra, Coimbra (2014)
13. Coutinho, E.C., Silva, A.L., Pereira, C.M.F., Rouxinol, D.F.C., Parreira, V.B.C., Nelas, P.A. B., et al.: Experiences of being a mother: meanings of childbirth, pain and birth. Front. J. Soc. Technol. Environ. Sci. **3**(3), 259–274 (2014)
14. Torkamani, S., Kangani, F., Janani, F.: The effects of delivery in water on duration of delivery and pain compared with normal delivery. Pak. J. Med. Sci. **26**(3), 551–555 (2010)
15. Mollamahmutoglu, L., Moraloglu, O., Ozyer, S., Su, F.A., Karayalcin, R., Hancerlioglu, N., et al.: The effects of immersion in water on labor, birth and newborn and comparison with epidural analgesia and conventional vaginal delivery. J. Turkish Ger. Gynecol. Assoc. **13**(1), 45–49 (2012)
16. Midwives Alliance North America, Citizens for Midwifery Homepage: MANA and CFM Position Statement on Water Immersion During Labor and Birth 2016. https://mana.org/healthcare-policy/mana-homebirth-position-paper. Accessed 9 Jan 2018
17. Balaskas, J.: Parto Ativo. 4 Estações Editora, Lda, São Pedro do Estoril (2017)
18. Schafer, R.: Umbilical cord avulsion in waterbirth. J. Midwifery Women's Health **59**(1), 91–94 (2014)
19. Dekker R Homepage: Evidence on the Safety of Waterbirth. Evidence Based Birth 2014. http://activebirthpools.com/wp-content/uploads/2016/11/Evidence-on-safety-of-water-birth.pdf. Accessed 9 Jan 2018
20. WHO Homepage: WHO recommendations Intrapartum care for a positive childbirth experience. http://apps.who.int/iris/bitstream/handle/10665/260178/9789241550215-eng.pdf. Accessed 9 Jan 2018
21. Ordem Enfermeiros Homepage: Livro de Bolso Enfermeiros Especialistas em Saúde Materna e Obstétrica/Parteiras 2015. http://www.ordemenfermeiros.pt/publicacoes/Documents/LivroBolso_EESMO.pdf. Accessed 9 Jan 2018
22. Coutinho, E.C., Rocha, A.M.A., Silva, A.L.: Expectations of a group of Portuguese pregnant women in the districts of Viseu and Aveiro regarding motherhood. Ciência & Saúde Coletiva **21**(8), 2339–2346 (2016)
23. Enning, C.: O Parto na Água. Editora Manole, São Paulo (2000)
24. Harper, B. Homepage: Gentle Birth Choices. Internet Archive: Healing Arts 2005. https://www.amazon.com/Gentle-Birth-Choices-Barbara-Harper-ebook/dp/B004KSQMEQ/ref=reader_auth_dp. Accessed 25 Apr 2018
25. Lopes, M.O.: Plano de Parto. In: Néné, M., Marques, R., Batista, M.A. (eds.) Enfermagem de Saúde Materna e Obstétrica. Lidel, Lisboa (2016)

Stories on the Internet: Challenges for Qualitative Research and the Example of Ethics

Judith C. Lapadat(✉)

University of Lethbridge, Lethbridge, AB T1K 3M4, Canada
judith.lapadat@uleth.ca

Abstract. The Internet is a huge repository of personal stories and ethnographic data. Why, as qualitative researchers in the social sciences, have we been so slow off the mark in studying the global social database that is the Internet? Our procedural and ethical guidelines for web research are in their infancy. The topics that we write about and online sites we study often are narrow in scope and stale-dated by the time they are published. Speaking as a qualitative researcher, early-adopting Internet user, and online researcher currently studying the blogosphere, I outline some of the challenges that contribute to the gap between the potential and current state of qualitative cyber research, reflect on the example of research ethics, and share an example from my research on blogs.

Keywords: Qualitative research · Internet research · Research ethics Blogs

1 Introduction

In addition to the many functions it serves, the Internet is a huge repository of personal stories and ethnographic data. Many of these data are publicly available and freely provided by the originators to the world at large, with few usage constraints. Ethnographers and other qualitative researchers in the social sciences, for the first time in history, have stories from all around the world instantly at their fingertips, along with live access to the individuals providing the accounts [16]. These stories are interactive in nature; develop over time leaving a persistent textual, auditory, and/or visual record; and often exist within organic cyber communities that themselves morph and grow. The individual storytellers and their online communities are linked to each other in a complex social network. Not only is all of this available for study, but qualitative researchers now have access to sophisticated miniature mobile computers (cell phones) for gathering electronic data and engaging with the online world, along with other electronic tools for analyzing, interpreting, and sharing the material that they are studying.

So why, as qualitative researchers, have we been so slow off the mark in studying the global social database that is the Internet? Our procedural and ethical guidelines for web research are in their infancy. The topics that we write about and online sites we

A. P. Costa et al. (Eds.): WCQR 2018, AISC 861, pp. 266–278, 2019.
https://doi.org/10.1007/978-3-030-01406-3_23

study often are narrow in scope, idiosyncratic, exotic, piecemeal, and stale-dated by the time they are published. In this paper, which is not empirical in nature but rather provides an explication of methodological issues, I focus on concerns specific to qualitative research in web contexts. Speaking as a longtime qualitative researcher, early-adopting Internet user, and online researcher currently studying the blogosphere, I outline some of the challenges that contribute to the gap between the potential and current state of qualitative cyber research, reflect on the example of research ethics, and share an example from my research on blogs.

2 Some Challenges of Doing Digital Research

There are a number of challenges that face qualitative researchers who are doing digital research in the social sciences. The scope of the Internet, the way our lives are intertwined with digital media, the rapidity of technological change as contrasted with the slow pace of change in academia, and the impact of mass access in altering the degree of respect accorded to credentials and formal expertise are all factors that impede scholarly social science research on online environments.

One obvious but hard to address problem when considering the Internet as a field for research, is that it is so big. The Internet is so enormous in size, its complexities are so multi-faceted and interconnected, and it is used by so many people globally for so many purposes, that defining the boundaries of the field for the purposes of a study can seem overwhelming.

Furthermore, people's online lives have become intertwined with their "In Real Life" (IRL) physical and material lives to the point that most people's cyber selves are an integral part of their lived experience. What we know, who we relate to, and what we do every day is mediated and informed by electronic media and takes place in cyberspace as much as in physical space, often simultaneously. People are immersed in and shaped by their engagement with electronic media in multiple ways throughout the day, most of it passing below notice. As I have argued elsewhere [21], each individual's daily pattern of engagement with the Internet is unique, creating both a digital fingerprint [25], and what we have called an "electronic biography" [21]. A digital fingerprint refers to the unique textual choices an online communicator makes in the performance of identity, for example, while playing a gaming character [25]. The related but more expansive notion of electronic biography refers to an online user's "situated usage patterns and choices" [21] in terms of all the ways in which they interact with social media and online environments on a day-to-day basis.

For research purposes, the idiosyncratic nature of individual cyber practices and embeddedness complicates study design. Also, as researchers, we cannot stand apart from the Internet processes and communities we research, as we become embedded and complicit as soon as we enter an online community. Even in moments when we are not online, the concerns and rhythms of our daily lives are shaped by the Internet.

To further complicate the challenge of conducting ethnographic or other social qualitative net research, the Internet – the ways in which it is being used, the number of people engaged in it and time spent online, the data it holds, its technological characteristics, and its affordances – are in a state of rapid change and expansion.

The magnitude of its pace of change grossly exceeds the pace at which the theories and practical methodologies of academic research are changing.

In the early days of the development of the World Wide Web, the conservatism of academia bred suspicion of the new medium. Online environments were seen as exotic and dangerous, especially for youth [13]. Less than two decades ago, scholarly publications warned other scholars against engaging in online teaching. Citing content off the Internet, or publishing in newly emerging online journals or in other online venues potentially had negative consequences for an individual's academic career progression. University Research Ethics Boards (REBs), or Institutional Research Boards (IRBs) and granting councils did not have guidelines in place pertinent to Internet research, so researchers struggled to obtain funding or to be granted ethics clearance for online studies. As a consequence, many academic scholars were slow to embrace and begin to study the social milieu of the Internet.

Even today, as researchers, we are wont to approach research question formulation and study design informed by the "old" paradigms and methodological constraints that shaped our academic disciplines. Although some qualitative researchers are doing outstanding trailblazing digital research, nevertheless the majority of researchers in any given department or discipline are doing studies firmly rooted in concerns, methodologies, and field sites drawn from the pre-Internet era.[1] One of these trailblazers, Gee [12], has argued that immersion in a semiotic domain such as video games is prerequisite to becoming literate in it, and that in order to study and write about a phenomenon, one must be literate in it. To the extent that qualitative researchers in the social sciences turn away from studying the expression of identity, social interactions, and formation of communities online [25], they are missing one of the most profound social changes in our contemporary time [16]. In that nearly everyone in the world today lives in a society mediated by the Internet, social research that ignores digital components is likely to be dated and to lack applicability to contemporary social experience.

Disciplinary boundaries are being challenged by the organic growth, interactivity, and cross-fertilization of thought that is possible in this worldwide participatory medium. On the Internet, university researchers are not necessarily seen as more credible; in true democratic fashion, anyone can present their point of view and those perspectives that are liked, shared, commented on, and linked to the most are the ones that shape popular opinions and beliefs. Scholars immersed in both academia and Internet research have described the conflicting role expectations and rules of engagement in these two very different worlds [2, 12]. Academic net researchers are caught up in rapid Internet-mediated social and technological change, and so they struggle to function within the inherent conservatism of disciplinary models and institutional systems. As an example, consider the slow pace of formal research. From drafting a proposal to obtaining funding to collecting data and analyzing and writing it

[1] As an example, at the 14th International Congress of Qualitative Inquiry in 2018, which is one of the largest annual qualitative research conferences in North America, only 19 sessions in the conference program subject index related to digital research, listed under the topics of "Technology" or "Digital Tools." Therefore, out of 415 sessions at the conference in total, less than 5% had a significant focus on digital research.

up, and then submitting it for peer-reviewed publication can take several years. Meanwhile, someone else already has written about that issue on their blog or a wiki site, and the Internet has moved on.

3 Guidelines and Procedures for Internet Research: The Example of Ethics

As social science research has evolved, researchers have developed methodological innovations to address theoretical and empirical dilemmas that have arisen. The manner in which qualitative research paradigms and practices have promoted different ways of conceptualizing ethics provides a good example of this change process. Cyber research continues to stretch our notions of ethical research practice even further.

3.1 Ethics in Qualitative Research

Tracy [27] has argued that there is value in developing criteria for judging the quality of qualitative research studies, and that these criteria also are useful in guiding research study design. Specifically, she has argued that there are some universal hallmarks that can be applied across research paradigms in assessing qualitative research quality. These "big-tent" criteria focus on the end goals of research, while simultaneously being inclusive of the wide range of paradigms and approaches that characterize qualitative inquiry. Tracy says that "high quality qualitative methodological research is marked by (a) worthy topic, (b) rich rigor, (c) sincerity, (d) credibility, (e) resonance, (f) significant contribution, (g) ethics, and (h) meaningful coherence" [27].

But, however useful research end goals might be, the devil is in the details. For the purposes of this paper, I will take ethics as an example.

The formation of REBs/IRBs and the research ethics codes that they adopted originally were intended to govern biomedical and psychological laboratory research [5]. Christians [5] says that these processes of ethics evaluation and ethics codes were bureaucratic at their core and represented a "value-free" notion of ethics, so expanding their scope to naturalistic and interpretive paradigms resulted in a poor fit for qualitative research [24], because qualitative research is contextual, reflexive, from a particular point of view, and embraces a value stance. As a result, qualitative researchers often were required to change their research design (e.g., provide a list of interview questions in advance), or they found their research stigmatized by research review panels [26], or they discovered that the ethics checklists offered little relevant guidance for the types of ethical dilemmas that their research presented [6]. (Also see [18, 19]).

In recent years, qualitative researchers have called for ways to address research ethics that better align with the theoretical frameworks of qualitative studies, and that provide ethical guidance and moral sensitivity throughout the research process. In qualitative inquiry, ethics should be "particularized, infused throughout inquiry, and requiring a continued moral dialogue" [4]. We need an approach in which "research ethics are layered, iterative, and formative, and ethical inquiry is accomplished through the reflexive deliberations of the researchers themselves" [19].

Such refinements in how we, as qualitative researchers, think about ethics and enact our research in ethical ways, are welcome methodological innovations. They have been developed, in part, as a response to the gaps in bureaucratic institutional ethics review processes. But, they have also come into being through researchers' thoughtful reflections on their own ethical choices and missteps in research. For example, Ellis's [7–10] advice on the importance of relational ethics derives from her own groundbreaking work in autoethnographic research. Through her experiences transitioning from doing ethnographic work to using her personal life experiences as the research data, she discovered that she faced ethical dilemmas and repercussions when she used community members' stories and then left the field; or when she wrote about intimate others, for example her elderly mother's ill body [7]; or when she wrote about sensitive personal experiences such as food disorders collaboratively with graduate students [11]. An ongoing relational ethics dilemma in autoethnography is the problem of how autoethnographic scholars should handle the stories of intimate others in their lives when they tell their own stories. Tullis [28] has developed some guidelines specific to ethics in autoethnography, and Lapadat [18, 19] has proposed that adopting collaborative approaches to doing autoethnography can avoid some of these ethical tensions.

The fact is, those researchers who push the edge sometimes will stumble into murky theoretical and methodological territory. That murky margin is also the most promising place for new breakthroughs [17]. Just as reflexivity became an important component of ethical research once qualitative researchers began to question the negative ethical implications of researching others from a position of greater power, taking their stories, and speaking for them, and just as autoethnography arose as a more extreme version of reflexivity [19], so too does qualitative research on the Internet offer new challenges and opportunities for ethical research practice.

3.2 Ethical Guidelines in Internet Research

Qualitative researchers' new ways of thinking about ethics and our new strategies for enacting ethical research practices in cannot be directly applied to Internet research in a straightforward way, however. The reason is that the Internet, by its nature, introduces different constraints, and also offers new possibilities that themselves will present new ethical challenges – ones that have not yet been addressed in offline research. It is not sufficient to simply apply existing techniques to the online environment [16].

It is hard to develop general ethical and procedural guidelines for Internet research because the Internet is a resource shared across many geographic, cultural, political, and legal domains. Assumptions that are considered to be shared knowledge in a mainstream North American context might be unfamiliar or considered culturally inappropriate in other contexts. For many of us, as qualitative researchers, we are just beginning to learn about our rights and responsibilities as global citizens, and we might find it hard to shake off a first-world mindset of entitlement, colonial myths of cultural superiority, and habits of cultural appropriation.

As another example, political or legal jurisdictions might enact different laws about how Internet data may be used. For example, in both Canada and the United States, copyright law protects intellectual property from being used without permission, whether in print, on the Internet, or in other formats. However, details of the copyright protection differ. For example, in Canada, an original work is protected by copyright, even if the copyright is not registered and the copyright symbol is not used, whereas in the United States, and in a number of other countries, a copyright must be registered to protect the work [22]. The European Union is in the process of updating EU copyright rules, and some of the requirements may significantly limit open creation and sharing of information on the Internet [29].

Similarly, different countries may have different requirements regarding privacy and data protection. Data uploaded to the Internet in a particular country can be stored on servers located in other countries. In Canada, there has been considerable debate about the implications of the USA Patriot Act on the privacy of Canadian data when data crosses the border, for example due to outsourcing of database management, cross-border businesses, and with the advent of cloud storage. Researchers can work with informants from all over the world via the Internet, but because of the variation in privacy laws in different countries and regions, it might not be possible to protect the data or ensure confidentiality. Often researchers are working with pseudonymous online informants and might not know the countries in which they are located.

So, there is all that qualitative data at our fingertips. But can we use online texts in our research without unwittingly violating a legal guideline, a cultural more, or an ethical expectation?

Our theoretical assumptions and practical aims also complicate matters. Depending on their purpose for interacting via the Internet, people adopt particular conceptual models that guide their functional practice of Internet use and their forms of interaction on it. For example, a media artist will conceptualize and use the affordances of the Internet in different ways than a commercial company will. The artist might want to combine images gleaned from the Internet to form creative mashups, whereas a business might be motivated to protect their brand, because brand recognition is directly linked to profitability. The Internet can be conceptualized as a communication medium, marketplace, repository, creative space, playground, social meeting place or community, tool for surveillance, and in other ways as well. These conceptual categories may overlap in any particular application, and each is further broken down into thousands of subcategories, which are shaped by users, the tools employed, and the rules of engagement in that corner of the Internet.

Even within a narrow category of internet users, individuals may have different primary reasons for why they engage in using that application of the Internet. I draw on a personal example here. As a blogger who sees the Internet as a communication medium, I have engaged in the blogosphere, specifically in the subgroup of English language blogs focused on retirement, and more narrowly on retirement blogs that have a lifestyle focus.

I am a blogger who uses my lifestyle blog to write myself into understanding about this new retirement stage of life as part of my regular writing practice. Yet, I see the purpose of my blog differently than another member of my blogging community who also writes a retirement lifestyle blog. She says that she mainly sees her blog as way to reach out and build a community of likeminded friends. Yet another blogger that I follow has a primary aim of teaching life strategies that she has learned through her research to other retirees, and another monetizes his blog, as one of his purposes for blogging is to add another income stream during retirement. Even within the same narrow branch of the blogosphere, and within the same blogging community, we each have a different purpose for writing a blog.

That the Internet is heterogeneous, diffuse, and in flux poses a challenge for establishing any specific top-down list of ethical or procedural guidelines [23]. One of the implications for research is that strategies for engagement, ethical considerations, and procedural guidelines might be quite different depending on the purpose of the research and the conceptual model the researcher subscribes to, as well as being shaped by the nature of the web application and the users.

Hine [14], an ethnographer who studies the embedded, embodied, everyday Internet through immersion in an online setting and by interacting with participants in that setting, nevertheless argues that unobtrusive methods of ethnographic observation can be preferable for certain purposes. Traces left by participants can orient a researcher to concerns, values, and forms of expression in that online community. As well, the Internet can function "as a mirror of the everyday" [14], bringing social patterns to notice that otherwise seem mundane and pass by unnoticed. Research ethics requirements and research procedures will differ when the data that are used are archived textual traces as compared with contexts where the researcher actively engages with participants, such as by interviewing them.

Kozinets [15, 16], whose field is social media marketing and research, has coined the term "netnography" as a label for online ethnographic research. He cautions researchers that "the fact that people know that their postings are public does not automatically lead to the conclusion that academics and other types of researchers can use the data in any way that they please" [15]. Netnographic research conducted without careful, ongoing consideration of ethical implications can be invasive of others' privacy, exploitative, and even do harm, he points out [15]. Some of the elements that a researcher must consider include the boundaries between what is public and what is private; issues particular to working with participants' digital doubles (often under a pseudonym) as contrasted with their IRL self; obtaining consent; striving to do no harm; disclosure; citing Internet sources and giving credit; and, where necessary, using strategies to conceal and protect participants' identities [15, 16].

Here I pose the question of whether the purpose of the research matters. For example, Kozinets' [15, 16] field is marketing research, where presumably findings will be used for commercial marketing via social media or will use data gleaned from social media to more precisely target marketing efforts. Failure to disclose one's presence and purpose as a researcher, to obtain consent, to protect participants' identities, or to evaluate whether the participant is a member of an underage or vulnerable population would seem to carry greater risks of harm than some other types of research. Hence, for this field of study, a checklist of ethical principles might be an important starting place for engaging in ethical research practice.

As a contrasting example, consider Hine's [14] work on participants' engagement in the community surrounding The Antiques Roadshow, where her primary purpose in doing the ethnographic research was theoretical, to better understand how the Internet functions as a social forum. She has argued that "rather than the ethical stance preceding the research, it therefore emerges from the engagement with the field which teaches what need to be treated as relevant aspects of context in taking ethical decisions" [14]. From a risk-of-harm perspective, Hine's approach, given her purpose and the type of online community she was studying, may be more well served by a flexible, emergent approach as compared to Kozinets' [15, 16] social media marketing ethnographic research, which requires more checks and balances to avoid potential harm.

As another example, consider research that uses automated processes to collect large social media datasets, or "Big Data" approaches. Webb et al. [30] describe some of the ethical challenges they faced in their research that had the purpose of making recommendations for how to establish responsible governance of online social spaces. Specifically, they used Twitter data to examine the way digital content – "cyber hate" – rapidly spreads across a broad domain. The authors, an interdisciplinary mix of social science researchers, computer scientists, and ethicists, discovered that there is a lack of consensus and guidance about how to proceed ethically, with respect to criteria such as "informed consent, minimizing harm and anonymisation… [as well as] tensions between academic, commercial and regulatory practice… behaving ethically and upholding academic integrity" [30].

For instance, if the researchers [30] anonymized a tweet by covering user names and handles when reproducing it in a research report, it did not truly anonymize the tweet as the original could be retrieved by pasting the text into a search engine, and given that the tweets reflected hate speech, there was a risk of harm to users should they be identified. In addition, anonymization breaches Twitter's policies, which permit only full reproductions of tweets. Webb and her colleagues [30] concluded that some of the conflicts they encountered between different ethical criteria and as well as between the policies of the overseeing bodies are intractable. If all of the various criteria are applied assiduously it could forestall the conduct of many types of important and meaningful research, over time perhaps resulting in a greater social harm than if a responsible, selective approach to research ethics was adopted.

In developing ethical guidelines for digital research, the Association of Internet Researchers (AoIR) [1] has formed an AoIR Ethics Working Committee, charged with developing and updating recommendations. To address the challenge of the heterogeneity of the Internet as a research field, this committee has cross-disciplinary representation from several countries, thus provides a broader lens on Internet research ethics than would be represented by the perspective of a single discipline or political jurisdiction. In the most recent revision of the document, Markham and Buchanan state that "no set of rules or guidelines is static" [23], and that the AoIR guidelines are process-oriented principles, intended "to support and inform those responsible for making decisions about the ethics of internet research" [23].

The AoIR document accepts general ethical guidelines as fundamental to any research endeavour, and in addition, identifies six guiding principles: (a) consider vulnerability of the participant; (b) evaluate the risk of harm within the context; (c) consider whether human subject research principles apply; (d) "balance the rights of the subjects… with the social benefits of the research and researchers' rights to conduct research" (p. 4); (e) address ethical issues as they arise throughout inquiry; and, (f) utilize a consultative, deliberative process to make ethical decisions [23]. Markham and Buchanan recommend an adaptive, inductive approach to ethics, stating: "we advocate guidelines rather than a code of practice so that ethical research can remain flexible, be responsive to diverse contexts, and be adaptable to continually changing technologies" [23].

The document [23] goes on to identify major persistent ethical tensions, and to outline the kinds of dilemmas specific to Internet-based research that a researcher might encounter, along with questions that can help guide the decision-making process. A valuable quality of this document is its flexible application to a vast and changing domain. That said, the responsibility for making appropriate ethical decisions rests squarely with the individual researcher.

3.3 An Example from a Blogging Study

I have been conducting research on personal finance blogs written by young to mid-life adults about their aim to achieve Financial Independence and Retire Early (FIRE) [20]. Although bloggers who describe themselves as FIRE can be quite diverse, and the community includes people in progress towards this goal as well as ones who have already retired at a young age, almost all of them provide an explanation of why they want(ed) to FIRE. They discuss strategies that they are using/used to reach their goal, talk about personal challenges and insights, give details about their personal financial circumstances, and provide pointers to help others progress towards FIRE. There is a strong sense among these bloggers of being members of an online FIRE community.

A longstanding blogger myself, I discovered this subset of the personal finance blogging community at a stage of my academic career when I was trying to decide whether I should retire. I had turned to the blogosphere to find out how other people like me had made the decision to retire, and what their retirement experiences were like. Although I could not be defined as a FIRE blogger myself, being beyond the age for early retirement, I found that these young people described many useful emotional, social, and financial strategies for preparing for and transitioning to retirement. But I also felt considerable curiosity about why they wished to retire at such young ages, which led me to want to study this community.

In order to understand the ethical expectations and constraints for the study I planned to conduct, I read a number of articles and chapters on Internet research ethics, including most of the ones I have cited here, and others. I also read articles on research ethics in qualitative research more broadly, and I reviewed the Tri-Council Policy Statement: Ethical Conduct for Research Involving Humans [3], the policy document provided by Canada's national granting councils.

In my study, I planned to use archived, publicly available blog texts where there is no expectation of privacy (i.e., no account, membership, or login required to view the blog, and no statements on the blog home page restricting use of the material or expressing the expectation of privacy). My study data was to be gathered solely from the blog texts (pages and posts on the blog, and readers' comments on the posts) upon which I would conduct a discourse analysis. I did not plan to interact with the blog writers to gather data, as I would only be looking at archived textual material from the past. I believed that, because of the blog topic area, it was highly likely that these blogs are written by adults who are not members of a vulnerable population. Therefore, I judged that the study posed minimal risk to the blog writers, commenters, and community. Finally, I planned to cite blog authors (by name or pseudonym) in my work, following ethical practices of acknowledgement of authorship and blog etiquette.

Because of the characteristics of the research, I requested and was granted exemption from human ethics review by my university's Research Ethics Board.

One ethical dilemma that I faced as I planned the study was that, prior to deciding to research this blogging community, I had interacted to a small degree with some FIRE bloggers by commenting on their posts. Moreover, my engagement had been via my blogging alias, or pseudonym. I reflected on whether I needed to disclose my presence as a researcher. I also wondered whether I had unintentionally violated an ethical principle. Kozinets states strongly: "Netnographers should never, under any circumstances, engage in identity deception" [15].

The way that I resolved this dilemma was by limiting the blog texts that I studied to archived texts from the past. I identified the date that I had first become aware of FIRE blogs, and only studied blog texts that had been posted in years prior to the time when I had first entered that corner of the blogosphere.

4 Qualitative Inquiry Has Changed

To this point, I have made the claim that most qualitative researchers have been slow to embrace Internet research, and I have suggested that characteristics of the Internet, such its size and complexity, the rapidity with which it evolves technologically and in terms of social usage, and the fact that it crosses jurisdictional boundaries are to blame, in part. I have also suggested that older theoretical paradigms and methodologies often have been a poor fit with the new digital social world, and that new conceptual models and procedural guidelines have been relatively slow to emerge. As an example, I have explored the issue of ethics guidelines as applied to qualitative research on the Internet.

However, another factor is that the enterprise of qualitative inquiry itself has changed over the last several decades. Ethnographers have turned away from realist approaches to ethnographic research, in which, in the most exploitative versions, objective researchers studied the "Other" from a position of power and authority, constructed accounts that spoke for the subjects of the research, published results without considering possible future consequences for participants, and when the researcher's aims were achieved, they exited the field abandoning the relationships they had cultivated.

Although problems inherent in such past research approaches have been addressed over the years through methodological innovations such as researcher reflexivity, participant transcript review, process and relational ethics, co-participatory models of research, action research, and critical and emancipatory approaches, many qualitative researchers have instead simply abandoned doing ethnographic research on or with others. We have seen a turn to inward-looking autoethnography to replace ethnography, and a swing to extreme interpretivism, where empirical data is eschewed. Rather than placing research participants in a position of vulnerability, now researchers display their own vulnerability by revealing and reflecting on intimate details from their personal lives.

It is an irony that qualitive researchers are turning away, en masse, from studying external social phenomena at this time in history. These are troubled times. There is a need for scholars to come together and use our skills as qualitative researchers to make a difference in the pressing issues facing humanity. We have this huge, unprecedented resource of the Internet available as a data source and means of communicating, collaborating, and building community. The Internet is democratically available to most people in the world – not just to elite gate-keepers, or those who currently hold political and economic power. And yet, we are held back by real and perceived constraints, including theories, procedures and practices that arose out of contexts in the past, or in reaction to perceived past mistakes. But, just as mistakes and problems can result in avoidance and closing our eyes to opportunities, they can also be a fruitful catalyst for our most creative breakthroughs.

5 A Way Forward

As qualitative researchers, we can see a way forward to engaging more completely with Internet research by looking at the example of research ethics as a model of how change occurs. When biomedical and psychology researchers conducted research that was ethically egregious, researchers and decision-makers of the past responded by creating codes of research ethics to be administered by REBs/IRBs or professional disciplinary bodies. Once naturalistic qualitative research came under the umbrella of those research ethics codes and ethical review procedures which were found to be ill-fitting, then qualitative researchers developed and advocated for new, elaborated approaches to ethics that better fit the needs of qualitative theoretical paradigms and qualitative methodologies. Similarly, with respect to developing ethical approaches for the Internet, Markham and Buchanan [23], in the AoIR Ethics document, recommend a case-based, process approach.

The limitations of current procedural guidelines are not a reason to turn away from doing Internet research. With thoughtful, careful attention to best practice, new contextualized strategies and guidelines will emerge as we wrestle with the dilemmas we encounter. I do worry, however, about the turn away from researching social phenomena external to the self, and I can only hope that it represents a temporary swing of the pendulum.

6 Conclusion

In the absence of clear ethical guidelines for conducting Internet research in certain contexts, and in fact recognizing that there are some intractable conflicts that cannot be resolved given our current conception of research ethics, Webb et al. [30] say that we still have the obligation to pursue good practice in our research. They ask: "How can we balance the risks of conducting and publishing research against the potential risks of not conducting research" [30]?

While qualitative researchers are struggling with the challenges of conducting online research, big corporations like Google, Facebook, and Amazon are mining our social data from the Internet. They are not doing it primarily because of a love of knowledge or a benevolent desire to help humanity, but in order to increase their profits and corporate power. In a very short time frame, in terms of social evolution, the

Internet and the corporations who employ its affordances to further their own ends have profoundly reshaped how people spend their time and interact with each other socially across the whole planet. To the extent that qualitative researchers choose to opt out of conducting Internet research and turn away from developing principles and guidelines for appropriate online engagement and use of data, there will continue to be limited awareness about how all of our personal data are being used and little regulation of the commercial drivers that shape our digital and off-line lives.

As qualitative researchers in the social sciences, we have a moral obligation to engage, and to use our knowledge and research skills to help humanity understand and manage the profound implications of the Internet on our lives. If we do otherwise and defer to commercial interests, we remove an important source of checks and balances from society, and ultimately risk making ourselves irrelevant.

References

1. Association of Internet Researchers Homepage. https://aoir.org. Accessed 01 Aug 2018
2. Boylorn, R.M.: Blackgirl blogs, auto/ethnography, and crunk feminism. Liminalities: J. Perform. Stud. **9**(2), 73–82 (2013)
3. Canadian Institutes of Health Research, Natural Sciences and Engineering Research Council of Canada, and Social Sciences and Humanities Research Council of Canada: Tri-Council Policy Statement: Ethical Conduct for Research Involving Humans (2014). http://www.pre.ethics.gc.ca/pdf/eng/tcps2-2014/TCPS_2_FINAL_Web.pdf
4. Cannella, G.S., Lincoln, Y.S.: Ethics, research regulations, and critical social science. In: Denzin, N.K., Lincoln, Y.S. (eds.) The Sage Handbook of Qualitative Research, 4th edn, pp. 81–89. Sage, Thousand Oaks (2011)
5. Christians, C.G.: Ethics and politics in qualitative research. In: Denzin, N.K., Lincoln, Y.S. (eds.) The Sage Handbook of Qualitative Research, 4th edn, pp. 61–80. Sage, Thousand Oaks (2011)
6. Denshire, S.: On auto-ethnography. Curr. Sociol. Rev. **62**(6), 831–850 (2014). https://doi.org/10.1177/0011392114533339
7. Ellis, C.: With mother/with child: a true story. Qual. Inq. **7**(5), 598–616 (2001). https://doi.org/10.1177/107780040100700505
8. Ellis, C.: The Ethnographic I. Altamira Press, Walnut Creek (2004)
9. Ellis, C.: "I just want to tell my story": mentoring students about relational ethics in writing about intimate others. In: Denzin, N.K., Giardina, M.D. (eds.) Ethical Futures in Qualitative Research: Decolonizing the Politics of Knowledge, pp. 209–227. Left Coast Press, Walnut Creek (2007)
10. Ellis, C.: Telling secrets, revealing lives: relational ethics in research with intimate others. Qual. Inq. **13**(1), 3–29 (2007). https://doi.org/10.1177/1077800406294947
11. Ellis, C., Kiesinger, C., Tillmann-Healy, L.: Interactive interviewing: talking about emotional experience. In: Hertz, R. (ed.) Reflexivity and Voice, pp. 119–149. Sage, Thousand Oaks (1997)
12. Gee, J.P.: What video games have to teach us about learning and literacy. Palgrave MacMillan, New York (2003)
13. Herring, S.C.: Questioning the generational divide: technological exoticism and adult constructions of online youth identity. In: Buckingham, D. (ed.) Youth, Identity, and Digital Media, pp. 71–92. The MIT Press, Cambridge (2008)

14. Hine, C.: Ethnography for the Internet: Embedded, Embodied, and Everyday. Bloomsbury, London (2015)
15. Kozinets, R.V.: Netnography: Doing Ethnographic Research Online. Sage, Thousand Oaks (2010)
16. Kozinets, R.V.: Netnography: Redefined, 2nd edn. Sage, Thousand Oaks (2015)
17. Lapadat, J.: Technologically mediated delivery in higher education: the margin as a site for change. In: Bastiaens, T., Ebner, M. (eds.) Proceedings of EdMedia: World Conference on Educational Media and Technology 2011, pp. 796–804. Association for the Advancement of Computing in Education (AACE) (2011). https://www.learntechlib.org/p/37958
18. Lapadat, J.C.: Ethics in autoethnography and collaborative autoethnography. Qual. Inq. **23** (8), 589–603 (2017). https://doi.org/10.1177/1077800417704462
19. Lapadat, J.C.: Collaborative autoethnography: an ethical approach to inquiry that makes a difference. In: Denzin, N.K., Giardina, M.D. (eds.) Qualitative Inquiry in the Public Sphere, pp. 156–170. Routledge, New York (2018)
20. Lapadat, J.C.: Bloggers on FIRE performing identity and building community: considerations for cyber-autoethnography (2018, under review)
21. Lapadat, J.C., Atkinson, M.L., Brown, W.I.: What we do online everyday: constructing electronic biographies, constructing ourselves. In: Shedletsky, L., Aitken, J.E. (eds.) Cases on Online Discussion and Interaction: Experiences and Outcomes, pp. 282–301. IGI Global, Hershey (2010)
22. Legal Line website. https://www.legalline.ca/legal-answers/using-copyright-symbol/. Accessed 01 Aug 2018
23. Markham, A., Buchanan, E.: Ethical decision-making and internet research: recommendations from the AoIR ethics working committee, version 2.0. Association of Internet Researchers (2012). https://aoir.org/reports/ethics2.pdf
24. McIntosh, M.J., Morse, J.M.: Institutional review boards and the ethics of emotion. In: Denzin, N.K., Giardina, M.D. (eds.) Qualitative Inquiry and Social Justice, pp. 81–107. Left Coast Press, Walnut Creek (2009)
25. Thomas, A.: Youth Online: Identity and Literacy in the Digital Age. Peter Lang, New York (2007)
26. Tolich, M.: A narrative account of ethics committees and their codes. N. Z. Sociol. **31**(4), 43–55 (2016)
27. Tracy, S.J.: Qualitative quality: eight "big-tent" criteria for excellent qualitative research. Qual. Inq. **16**(10), 837–851 (2010). https://doi.org/10.1177/1077800410383121
28. Tullis, J.A.: Self and others: ethics in autoethnographic research. In: Holman Jones, S., Adams, T.E., Ellis, C. (eds.) Handbook of Autoethnography, pp. 244–261. Left Coast Press, Walnut Creek (2013)
29. Vollmer, T.: Act now to stop the EU's plan to censor the web, Creative Commons. https://creativecommons.org/2018/06/08/act-now-to-stop-the-eus-plan-to-censor-the-web/. Accessed 08 June 2018
30. Webb, H., Jirotka, M., Stahl, B.C., Housley, W., Edwards, A., Williams, M., et al.: The ethical challenges of publishing Twitter data for research dissemination. In: Proceedings of the 2017 ACM on Web Science Conference, WebSci 2017, New York, pp. 339–348 (2017). https://doi.org/10.1145/3091478.3091489

Qualitative Data Analysis Software Packages: An Integrative Review

Luiz Rafael Andrade[1](✉), António Pedro Costa[2], Ronaldo Nunes Linhares[1], Carla Azevedo de Almeida[3], and Luís Paulo Reis[4]

[1] University Tiradentes, Aracaju, Brazil
andrade.luizrafael@gmail.com,
ronaldo_linhares@unit.br
[2] Research Centre in Didactics and Technology in the Training of Trainers (CIDTFF), University of Aveiro, Aveiro, Portugal
pcosta@ludomedia.pt
[3] Faculty of Law, University of Porto, Porto, Portugal
carlazevedoalmeida@gmail.com
[4] LIACC – Artificial Intelligence and Computer Science Lab., FEUP - Faculty of Engineering of the University of Porto, Porto, Portugal
lpreis@fe.up.pt

Abstract. The Computer Assisted Qualitative Data Analysis Software Packages (CAQDAS) can be understood as a means of assisting the researcher in the treatment, organization and analysis of qualitative data for developing research or projects that deal with qualitative data. This type of software stands out and are sought after because they allow the qualitative analysis to be done in diverse formats (audio, video, image, text), levels of collaborative work (individual, two people working at the same time or more than two), and with distinct possibilities of organizing the data surveyed. This article aims to present an integrative review about the most relevant CAQDAS in Ibero-American, highlighting the potential of (i) data organization, (ii) variety of supports, (iii) mobility, and (iv) interaction with social networks.

Keywords: CAQDAS · Integrative review · Qualitative research Ibero-American · NVivo · ATLAS.ti · Dedoose · webQDA · MAXQDA

1 Introduction

The analysis of qualitative data has increasingly become a research method of interest to researchers and professionals in the world. Among this interest, the area of human and social sciences stands out as responsible for the growth of this method, which, among other techniques, has been evidenced in content analysis and ethnographic studies.

In the 1960s, in the United States, there was a need to develop programs that had the initial and even limited capacity of cutting chips, coding texts or even perform data collection in a given master document [1]. In the last two decades, the search for users

A. P. Costa et al. (Eds.): WCQR 2018, AISC 861, pp. 279–290, 2019.
https://doi.org/10.1007/978-3-030-01406-3_24

that access the Internet through tools that can help in the analysis of qualitative data has increased considerably, the number of available solutions/options has grown, as well as the expansion of their functionalities to better meet the specifics of this type of research.

It is possible to identify several Qualitative Data Analysis Software Packages (CAQDAS) available on the Internet. Among these packages, NVivo, ATLAS.ti, Dedoose, MAXQDA, and webQDA [2]. It must be considered that the operations carried out in those said CAQDAS, in relation to analysis and codification, are not made exclusively by the computer, and it is always the responsibility of the researcher to select, indicate and conduct - among the data - the directions of his/her qualitative analysis.

In 2011, about 40 CAQDAS were offered that allow the analysis of qualitative data [3]. Although there are many tools available to the researcher, little has been discussed about their potential, nor has there been a comparison of this scenario. The comparative work proposed here aims to review the characteristics and objectives that are specific to each software, so that the researcher can work with a CAQDAS according to the objectives of his/her research.

This paper presents an integrative review of the qualitative analysis software packages, highlighting their own potentialities in relation to: (i) greater organization of data, (ii) variety of media, (iii) collaboration and (iv) interaction with digital social networks, in the software packages NVivo, ATLAS.ti, Dedoose, webQDA and MAXQDA.

The sections that follow after (1) introduction are presented in an explanation about (2) the CAQDAS and the Ibero-American researcher and later the methodological proposal of the article in the section (3) methodology takes place, followed by the presentation of the (4) integrative review of the selected CAQDAS and, finally, the (5) conclusions.

2 The CAQDAS and the Ibero-American Researcher

During the qualitative data analysis process, the use of the designated CAQDAS tend to facilitate the organization of the data, decrease the time allocated to the analyzes and better triangulate the methods and techniques [4]. Beyond this type of analysis help, the software has allowed the researcher to work collaboratively and, in some cases, simultaneously through the Internet.

Despite the possibilities of supporting the qualitative analysis that CAQDAS can offer the researcher, in Brazil, they have been little used. If we observe the little use by area of study, it is noticed that the Communication Sciences are the most representative [5], followed recently by the Social Sciences [4].

It must be considered that the facilities that CAQDAS can offer to the work of qualitative data analysis, as well as in their organizational capacity, directly imply the need for the researcher to have technical skills geared towards the use of the software, which has specific commands depending on the analysis package. For this reality to take place, the researcher needs to leave his/her comfort zone and devote the necessary time to acquire new skills on a new way of dealing with qualitative data.

In the current scenario, the ways of analyzing qualitative data have increasingly been through the tools of qualitative analysis mediated by digital technologies. In this digital environment, they stand out in two categories: in need of installation (ATLAS.ti, Coding Analysis Toolkit (CAT), ConnectedText, Dedoose, HyperRESEARCH, MAXQDA, NVivo, QDA Miner, Qiqqa, Quirkos, Saturate, XSight) and those based on the cloud, on the Internet, without the need for installation (Computer Aided Textual Markup & Analysis (CATMA), Dedoose, LibreQDA, QCAmap and webQDA).

In the case of an Ibero-American reality, only the qualitative analysis packages webQDA, MAXQDA and NVivo have focused on ease of use in Portuguese and Spanish, the main languages in these regions. Among all the CAQDAS, we selected for an integrative review those with the most market prominence and use between the two categories presented above, therefore: NVivo, ATLAS.ti, Dedoose, webQDA and MAXQDA

3 Methodology

The present research has its proposal to review, in an integrative way, the qualitative analysis software packages available in the Ibero-American scenario. For this, an integrative review has been conceived with six steps, according to Table 1:

Table 1. Steps of an integrative review[a]

Stage of integrative review					
Phase 1	**Phase 2**	**Phase 3**	**Phase 4**	**Phase 5**	**Phase 6**
Identification of the theme and selection of the research question	Establishment of inclusion and exclusion criteria	Identification of pre-selected and selected studies	Categorization of selected studies	Analysis and interpretation of results	Presentation of knowledge review/synthesis
Stages of the first phase	**Stages of the second phase**	**Stages of the third phase**	**Stages of the fourth phase**	**Stages of the fifth phase**	**Stages of the sixth phase**
- Definition of the problem - Formulation of a research question - Definition of the search strategy - Definition of descriptors - Definition of databases	- Use of databases - Search for studies based on inclusion and exclusion criteria	- Reading summary, keywords and titles of publications - Organization of pre-selected studies - Identification of selected studies	- Development and use of the synthesis matrix - Categorization and analysis of information - Formation of an individual library - Critical analysis of selected studies	- Discussion of results	- Creation of a document detailing the review - Proposals for future studies

[a]Integrative review phases originally translated from Botelho et al. [16].

Based on these steps, (i) the research theme was defined with CAQDAS, (ii) between exclusions and inclusions, market demand and Ibero-American use were established as categories, (iii) the selection resulted in five packages to be reviewed

(iv) the analysis of the review will be done by means of a table with established categories, as well as (v) analysis and interpretation of the results, which will finally (vi) present the synthesis of this entire study.

The methodology of the integrative review was used in this article with the purpose of providing us with a synthesis about the CAQDAS available to the Ibero-American researcher, with a focus on highlighting their significant applicability to the practice of this professional in supporting the analysis of qualitative data. For this, we defined as exclusion and inclusion criteria dimensions that would be possible to be identified in most of the revised packages, which are presented in Table 2:

Table 2. Dimensions defined for the integrative review of CAQDAS

Dimension/Software	Objective when selecting dimension
WebPage	Identify address referenced in review
Last version	Identify the version used in the review
Systems on the PC	Identify PC compatibility
Systems on the smartphone	Identify possible compatibility with smartphone
Ibero-American languages	Identify the presence of Ibero-American languages
License	Identify the type of license offered to the researcher
Importing digital social networking data	Identify the possible import of data from digital social networks
Import of data	Identify types of data import
Import of data from other tools/applications	Identify the possible data import from other smartphone tools or applications
Export of data	Identify data export types
Collaborative Work	Identify the possible forms of collaborative work offered to the researcher
Simultaneous use by different users	Identificar as possíveis formas de uso simultâneo por diferentes usuários
Training for users	Identify the possible forms of simultaneous use by different users
Free Trial	Identify if it's offered to the user a free trial

In this study, we sought to identify, in the selected CAQDAS, the dimensions established in Table 2 as a data collection criterion, so that, from this stage, there is the possibility of elaborating a synthesis on what these software packages can offer to the Ibero - The analysis of qualitative data. Before these phases it is important to understand the main characteristics of each of the software selected for this review.

3.1 NVivo

Created by Tom Richards in 1999 and originally named NUDIST, it establishes its current nomenclature from 2002. As a computer software, NVivo allows analysis of qualitative data and treatment with mixed methods. It is produced by QSR International and is primarily intended to contribute to the organization and analysis of data. Among its characteristics, the capacity to store and organize, categorize and analyze, view and discover the data [6].

Based on the ease of accessing digital data through the Internet, NVivo, in its Pro version, considers this important when analyzing and organizing them in the formats of audio, video, digital photo, Word, PDF, spreadsheet, rich text, simple text and social media data and web data. Users of this type of version can also transfer data from programs, such as Microsoft Excel, Microsoft Word, IBM SPSS Statistics, EndNote, Microsoft OneNote, SurveyMonkey, Evernote and TranscribeMe [6]. All this proposal of NVivo in its version 12 Pro is currently available in seven languages: English, French, German, Portuguese, Spanish, Japanese and Chinese, still in a simplified way [7].

3.2 ATLAS.ti

ATLAS.ti [8] can be used as software on the computer, or as an application on smartphones. Therefore, whether in the more conventional media or in increasingly mobile media, this qualitative analysis package aims to help the user to qualitatively analyze data in media varieties: text (.txt, .doc, .docx, .odt, .pdf), audio (.wav, .mp3, . wma, etc.), video (.avi, .mp4, .wmv, etc.), digital social networks (Twitter). In order for the user to work with this media variety, ATLAS.ti [8] offers data processing coding options, interactive margin area, quotation level, network visualization and collaborative work for data processing in six languages: Japanese, Chinese, Spanish, French, English and German.

3.3 Dedoose

Dedoose is essentially a web application for research into qualitative methods and mixed methods developed by UCLA - University of California at Los Angeles. Dedoose is a very good alternative to other qualitative data analysis software packages as it is well-aimed to facilitate rigorous research using mixed methods and including a wide range of quantitative methods. The software is heavily used by researchers funded by the William T. Grant Foundation. Dedoose has essentially gained wide recognition for its integration of qualitative and quantitative data analysis methods in combination with interactive data visualizations [2].

The main focus of the creation and development of Dedoose is to support researchers and evaluators who seek, on the platform, a better approach to their mixed and qualitative methods. Among its characteristics, what stands out is it being totally online, a fact that contributes to a better accessibility of the user from different computers, and to be collaborative, allowing, according to the manufacturers, to work with any number of colleagues in real time and online [9]. About media support, it is divided into five: text (word, text, pdf, htm or html), image (jpg, gif, etc.), audio/video (streaming files) and spreadsheets (xls, xlsx or csv) [9].

3.4 MAXQDA

MAXQDA is a software to support the analysis of qualitative and quantitative data, which was developed by VERBI Software, a German company, with its first version released in 1989, still with the nomenclature "MAX (DOS)". From the date of its launch until 2018 there were more than 10 versions. These include the recent release of smartphone versions for Android and iOS systems [11]. The software offers the user the ability to import, organize and analyze data in pdf, table, doc, photos, videos, web pages, and integration with Twitter. Also among the CAQDAS, MAXQDA offers languages of the Ibero-American region: Portuguese and Spanish [12].

3.5 webQDA

webQDA (Web Qualitative Data Analysis) is a qualitative data analysis software in a collaborative and distributed environment (www.webqda.net). The platform had its first version released in 2010 and can be considered the most recent among those treated in this study. Currently in its version 3.0, webQDA has its main focus on researchers, allowing the treatment with data from diverse types of files (pdf, docx, xlsx, mp3, mp4, png and jpeg), of which we emphasize the possibility of integration with the digital social network YouTube [10]. Among the CAQDAS, it is one of the few to fully contemplate the languages spoken in the Ibero-American region (Portuguese and Spanish).

4 Integrative Review of Selected CAQDAS

In accordance with the characteristics of each of the CAQDAS, the web pages, the latest version of the software, compatibility with the computer and also on smartphones, Ibero-American languages available, the type of license, ability to import digital social network data into qualitative analysis, a tool/application data distribution capability, a data export capability, a collaborative working possibility, a possibility for simultaneous use by different users, a free trial.

Below, in Table 3, we present a synthesis of the results based on the criteria of exclusion and inclusion of data.

With this revision, it has been a tendency among the selected CAQDAS to improve their functionalities based on (i) data organization, (ii) variety of support, (iii) mobility, and (iv) interaction with digital social networks. Among this trend, we highlighted that, increasingly, the companies that manage the CAQDAS have invested in the field of mobility, since with the frequent use of smartphones in everyday social relations and better availability of Internet access in several countries, users have also gradually found the possibility of organizing and analyzing qualitative data through these devices.

Table 3. Integral review of CAQDAS

Dimension/ Software	NVIVO	ATLAS.ti	DEDOOSE	MAXQDA	WEBQDA
WebPage	qsrinternational.com	atlasti.com	dedoose.com	maxqda.de	webqda.net
Last version	NVivo 12	ATLAS.ti 8.1	Dedoose 8.0	MAXQDA 2018	webQDA 3.0
Systems on the PC	Windows/Mac (the Mac version is less developed)	Windows/Mac	Web	Windows/Mac	Web
Systems on the smartphone	–	Android/iOS (para iPads)	Photon Browser in iPads and Android Tablets	Android/iOS	Web
Ibero-American languages	Portuguese and Spanish	Spanish	Spanish	Portuguese and Spanish	Portuguese and Spanish
License	Commercial/Educational	Commercial/Educational/ student/Non Commercial & Governmental	Commercial	Commercial/Educational/ Non-Governmental & Governmental	Commercial/ Educational
Importing digital social networking data	Twitter	Twitter	Twitter	Twitter	YouTube
Import of data	Text, Excel, audio, video, image and HTML	Text, Excel, audio, video, image and HTML	Text, Excel, audio, video, image and HTML	Text, Excel, audio, video, image and HTML	Text, Excel, audio, video, image and HTML
Import of data from other tools/applications	SurveyMonkey Refworks, reference managers for literature review, SPSS	Evernote, reference managers for literature review, Geo-Data	SurveyMonkey, ATLAS.ti, Nvivo, reference managers for literature review,	ASCII (Unicode), Geolink, SPSS, SurveyMonkey, reference managers for literature	–
Export of data	Text, Excel, conceptual maps, images (ex.: word cloud, graphics) and charts (in excel format)	Text, Excel, conceptual maps, images (ex.: word cloud, graphics) and charts (in excel format)	Text, Excel, conceptual maps, images (ex.: word cloud, graphics) and charts (in excel format)	Text, Excel, HTML, XML, conceptual maps, images (ex.: word cloud, graphics) and charts (in excel format)	Text, Excel, images (ex: word cloud), charts (in excel format)
Collaborative Work	Does not allow the user to share is Project file	Allows the joining of projects, users management, documents shared between projects	Allows collaborative work and the control of the level of access of each user to the project	Allows collaborative work and the control of the level of access of each user to the project. Allows the junction between projects	Allows collaborative work and the control of the level of access of each user to the project
Simultaneous use by different users	Only if the user is using windows and have extra tools for Nvivo Teams and Nvivo Server	In a limited way. Allows a validation process through the reports that it makes available	Allows a validation process of the codified data	–	Allows a validation process through the reports that it makes available
Training for users	Yes	Yes	Yes	Yes	Yes
Free Trial	14 days	Unlimited but with large feature restrictions	30 days	14 days	15 days

4.1 Organization of Data

With regards to their functionalities related to the greater organizational capacity that CAQDAS offered during the period of this study, 2018, it could be identified that the organization of data was done through any machine/device with access to the Internet in the webQDA and Dedoose packages. In addition, there are software that allow the user to organize and analyze qualitative data in an offline manner (NVivo, ATLAS.ti and MAXQDA), however, they need to be installed on a particular machine/device.

The companies that provide the CAQDAS have committed to objectify the organization of the qualitative data beginning with the choice of the type of service, whether commercial or educational. Among the packages reviewed, we detected NVivo, MAXQDA and webQDA with these features. Depending on the service purchased, the company can also offer training to users, some training aimed at the organization of data for commercial purposes and others for the organization of data for educational purposes.

4.2 Variety of Supports

Currently academic research has been aimed at understanding and analyzing their realities of study, through diverse supports. Whether it is a video, image, textual or audio analysis, the researcher has the possibility to build data and knowledge. Especially when the focus of analysis is qualitative, then the qualitative perspective becomes fundamental to understand the universe of meanings, motives, beliefs, values and attitudes, corresponding to a deeper space of relationships, processes and phenomena present in the reality [13]. This reality can be understood from different supports.

We detected from the integrative review (see Table 2) which are the ones that the CAQDAS have allowed the researcher/user to have a greater capacity for the organization of qualitative data constructed/collected for the elaboration of analyzes in the formats of text, audio, video and image.

In addition to these compatibilities of diverse media, which allow the organization and analysis of data of different types, it can be considered that the software's ATLAS.ti (for Android and iOS), MAXQDA (for Android and iOS) and webQDA (web) allow researchers/users to access and manipulate their data in different situations: basic for access (computer) and mobile (smartphone). Does the possibility of access to some CAQDAS through mobile applications have consequently generated different mobility of the computer?

4.3 Mobility

The possibility of mobility was detected in most of the revised CAQDAS (webQDA, MAXQDA and ATLAS.ti). We therefore consider mobility as an important element for researchers to work, organize and analyze data in a joint or group way in different everyday situations.

In the review proposed here, we understand the meaning of "mobile" in software to support qualitative analysis, from a society of digital technology, where cult of micro, mobility, light and instantaneous are determining elements [14]. In this reality, with more and less volume, weight, mobility and more capacities, "The lightness of the connected object now surpasses the question of its weight: it refers to the multitude of functionalities that it performs in relation to its ultralight weight" [14]. The software that has invested in providing a form of collaboration, in a mobile and instantaneous manner, is webQDA, MAXQDA with MAXApp and ALTAS.ti (Fig. 1).

Among these packages that allow greater mobility to the user, so that researchers can access and work collaboratively in a fully online way, in the cloud, webQDA was detected through the review for allowing mobile access through the smartphone,

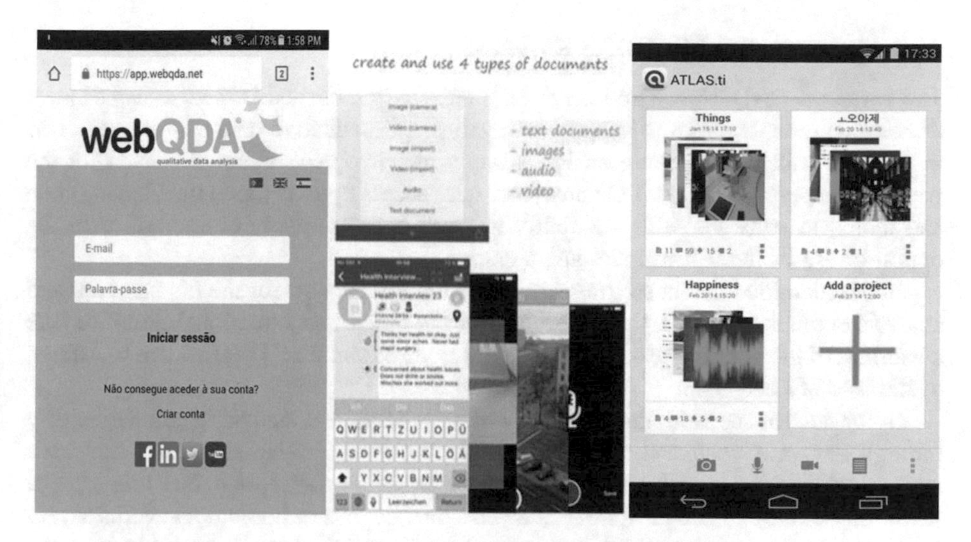

Fig. 1. CAQDAS access interfaces through the smartphone

without the need for installation of applications, through the web browser. But is the user prepared for this mobile and online reality?

In the countries of the Ibero-American region, specifically Brazil, the number of users that have the possibility of accessing the Internet has increased year after year. One fact has drawn attention in this reality of increase, since, according to the Internet Management Committee (IMC) [15], access to the mobile Internet, in turn, has stood out in the country. In addition, the IMC survey in 2016 revealed that "the proportion of households with Internet access, but without a computer, doubled in two years, from 7% in 2014 to 14% in 2016," pointing out that daily access to the Internet has increasingly migrated to mobility through of smartphone.

Access to the smartphone can be done for a variety of reasons, including study, entertainment, work, etc. Regarding the professional reasons, we have drawn attention to the possibility that the applications via smartphone have offered the researcher who seeks tools to support the analysis of qualitative data in digital support. Appears in this scenario a new kind of mobility differs from that offered by the computer.

4.4 Interaction with Digital Social Networks

Besides the work of organization and analysis of the qualitative data being in some cases fully online and in the cloud, we consider that, increasingly, the packages have tried to offer the researcher/user the possibility to export data (videos, image, text) also from digital social networks for analysis. They are NVivo, ATLAS.ti, Deedose and MAXQDA with possibility to export Twitter data, and webQDA with the same possibility through YouTube. Both processes are dependent on Internet access.

4.5 Synthesis of the CAQDAS Reviewed

The reviewed CAQDAS, based on the dimensions established in this research, have directed its functionalities to support the analysis of qualitative data of the researcher, in order to collaborate, facing an increasingly technological digital society, with the variety of supports to support the analyzes, with access options (computer, smartphone and tablet) to software, with the ability to work collaboratively online and with the option of use in Ibero-American languages.

It was identified that in seeking software a type of support for the organization and analysis of qualitative data, the researcher has the opportunity to allow himself to take advantage of the possible "extensions or extensions of abilities" [17] that can be offered to the user of CAQDAS.

In the current scenario, the possibility of organizing and analyzing qualitative data through software, mediated by digital technologies and Internet access, has been increasingly available to the researcher. Here arises the perception that this type of possibility offered by the CAQDAS is a consequence of a technological evolution, for the events of the last sixty years are probably only an initial spark or, if we may, the prehistory of a technological society digital [18]. In this scenario, the use of software starts to contemplate and (re) mix also ways of organizing and analyzing qualitative data.

5 Conclusions

We sought to present in this article an integrative review of the most relevant CAQDAS in Ibero-American. We established categories for this review that made it possible to know better the software packages studied, as well as what each one can offer to the user in terms of data organization, variety of media, mobility, and interaction with digital social networks.

Through the integrative review, we detected that NVivo, MAXQDA and webQDA provide, in addition to English, the Ibero-American Spanish and Portuguese languages. In addition, they offer package licenses for commercial and educational purposes, depending on the user's interest. As a commercial or educational support, the selected CAQDAS provide users with workshops, on-line and face-to-face courses, tutorials and help menus, as well as wide compatibility with different types of available data (text, audio, video, image and digital social network in specific cases).

On what the CAQDAS can offer to users in terms of the organization of qualitative data, we have identified that the CAQDAS were divided into two groups: the first group consists of NVivo, ATLAS.ti and MAXQDA, which allow the user to organize, store and analyze data offline, requiring a software installation on a machine/device. The second group consists of Dedoose and webQDA, which have chosen to provide the user with online qualitative data analysis and organization services. Therefore, one group opts for the ability to work without the need to access the Internet, offline, and the other group has developed its web-centric proposal, thus promoting greater mobility of data access, provided that there is access to the Internet.

The interactions with digital social networks were perceived as service options to support qualitative analysis, which has directed the look also to promote, consequently, an organization and qualitative analysis of the phenomena produced and communicated in digital environments, also allowing the user, for through the ATLAS.ti (Twitter), MAXQDA (Twitter) and webQDA (YouTube) software packages, an analysis interaction with digital social networks. With this, the user has the possibility to qualitatively analyze videos directly from the YouTube platform in webQDA or to qualitatively analyze contents of the Twitter platform through ALTASL.ti and MAXQDA.

On the mobility side, CAQDAS webQDA, MAXQDA and ATLAS.ti have tried to know this field, one of the measures for this improvement allows access to work with qualitative data in increasingly mobile supports, such as smartphones. With this research, we realized that, in the Android and iOS platforms (MAXQDA and ATLAS.ti) or through the Internet browser (webQDA), CAQDAS have achieved the flexibility of everyday relationships allowed by today's smartphone mobility, making, to a certain extent, important protagonists of the current digital technology of the user.

With this review, we consider that currently some CAQDAS allow themselves to live and experience moments of change in the access platform and in the supports, that can be analyzed qualitatively, consequently making this access closer to users' mobility and daily life.

References

1. Orozco Gómez, G., Gonzáles, R.: Una coartada meto-dológica: abordajes cualitativos en la investigación en comuni-cación, medios y audiencias. México, Tintable (2012). 211 p
2. Reis, L.P., Costa, A.P., de Souza, F.N.: A survey on computer assisted qualitative data analysis software, pp. 1–6. IEEE (2016). https://doi.org/10.1109/CISTI.2016.7521502
3. Saillard, E.K.: Systematic versus interpretive analysis with two CAQDAS packages: NVivo and Maxqda. Forum Qual. Soc. Res. **12**(1), art. 34 (2011)
4. Jacks, N., Toaldo, M., Schmitz, D., Mazer, D., Miranda, F.C., Gonçalves, F., Coruja, P.: Uso de softwares na abordagem qualitativa: a experiência da pesquisa "Jovem e Consumo Midiático em Tempos de Convergência", 4, 9 (2016)
5. Teixeira, A.N., Becker, F.: Novas possibilidades da pesquisa qualitativa via sistemas CAQDAS. Sociologias **3**(5), 94–114 (2001)
6. NVivo Website. http://www.qsrinternational.com/. Accessed May 2018
7. Wikipedia –NVivo. https://en.wikipedia.org/wiki/NVivo. Accessed May 2018
8. ATLAS.ti Website. http://atlasti.com/. Accessed May 2018
9. Dedoose WebSite: Great Research. Made Easy! http://www.dedoose.com/. Accessed May 2018
10. webQDA Website: webQDA - Software de Apoio à Análise Qualitativa. https://www.webqda.com/. Accessed May 2018
11. VERBI GmbH: MAXQDA – The Art of Data Analysis. http://www.maxqda.com. Accessed May 2018
12. Wikipedia – MAXQDA. http://en.wikipedia.org/wiki/MAXQDA. Accessed May 2018
13. Minayo, M.C.S., et al.: Investigación social: Teoria, método e criatividade. 1o ed. 1o reimp. Lugar, Buenos Aires (2004)

14. Lipovetsky, G.: Da leveza para uma civilização do ligeiro. Extra colecção (2016)
15. Pesquisa sobre o uso das tecnologias de informação e comunicação nos domicílios brasileiros: TIC domicílios 2016/Survey on the use of information and communication technologies in brazilian households: ICT households 2016 [livro eletrônico]. Núcleo de Informação e Coordenação do Ponto BR [editor]. Comitê Gestor da Internet no Brasil, São Paulo 2017
16. Botelho, L.L.R., Cunha, C.C.A., Macedo, M.: El Método de la revisión integrativa en los estudios organizacionales. Gestão e Sociedade. Belo Horizonte **5**(11), 121–136 (2011)
17. Santaella, L.: Da cultura das mídias à cibercultura: o advento do pós-humano. Revista FAMECOS. Porto Alegre, nº 22 (2003)
18. Lemos, A., Lévy, P.: O futuro da internet: em direção a uma ciberdemocracia planetária. Paulus, São Paulo (2010)
19. Freitas, F., Ribeiro, J., Brandão, C., de Souza, F.N., Costa, A.P., Reis, L.P.: In case of doubt see the manual: a comparative analysis of (self)learning packages qualitative research software. In: Costa, A.P., Reis, L.P., de Souza, F.N., Moreira, A. (eds.) Computer Supported Qualitative Research: Second International Symposium on Qualitative Research (ISQR 2017), pp. 176–192. Springer International Publishing, Cham (2017). https://doi.org/10.1007/978-3-319-61121-1_16
20. MAXQDA: Tutorial. https://www.maxqda.com/learn-maxqda/maxqda-2018-video-tutorials#&id=k02Zlr5hbhk
21. ATLAS.ti: Tutorial. https://atlasti.cleverbridge.com/74/purl-order?_ga=2.58025896.328682098.1532614244-1098966855.1531734517
22. DEDOOSE: Tutorial. https://www.dedoose.com/resources/videos

Qualitative Methodology Helping Police Sciences: Building a Model for Prevention of Road Fatalities in São Tomé and Principe

Sónia M. A. Morgado[1](✉) and Odair Anjos[2]

[1] Higher Institute of Police Sciences and Internal Security, Lisbon, Portugal
smmorgado@psp.pt
[2] Police Officer in Democratic Republic de São Tomé e Principe, São Tomé, São Tomé and Príncipe

Abstract. Road safety has been widely studied in the influence that has on road accidents fatalities. On a global scale, road traffic crashes are the eight cause of death, presenting a growing tendency. Nevertheless, research related with the establishment of intervention models is rare. In this study, the Democratic Republic of São Tomé e Principe's National Polices key role in saving lives and minimizing injury on the road was subject, based on a qualitative approach in order to design a road safety intervention model. The qualitative analysis was undertaken by applying interviews to police officers whose competence placed them in the middle of the road accident prevention field. Studies with empirical data were non-existent in São Tomé e Principe. Therefore, the results allowed us to develop a model so that a national road safety program can be designed, for a pro-active prevention practice.

Keywords: Intervention model · National Police of São Tomé and Príncipe Prevention · Qualitative analysis · Road safety

1 Introduction

The word accident derives etymologically from the Latin word *accidens*, that means "fall in", "gather". It's a fortuity link between elements, as an event of unforeseen nature with negative outcomes.

Road accidents are a civilization phenomenon, resulting from the constant mobility of vehicles and other road users. Hoekstra and Wegman [1], considered that "road traffic today is inherently dangerous" (p. 80). The causes are the result of the dynamics of intertwined factors: human, vehicle, and environment/road. Regardless of the country level of development, road accidents have an impact on daily life, not only by direct effect but also by the collateral ones. Overall, the difference between hazard and safety derives of the human factor.

Alongside with the loss of lives, negatives externalities are due to the phenomenon, either in terms of social, economic, or environmental effects [2, 3].

A. P. Costa et al. (Eds.): WCQR 2018, AISC 861, pp. 291–304, 2019.
https://doi.org/10.1007/978-3-030-01406-3_25

Throughout history, mobility was always a human ambition. The need for rapid transactions of goods and services, for a better quality of life, congregates the rapid growth of mobility aspects. From the XX century, road mobility has been more fluid and had a higher expansion. The development of transport, allowed the reduction of the temporal gap between places, converging to the globalization of economic transactions of goods, services, financial and human capital.

For every intervenient on the road safety or road traffic history, the problem is tackled in three traditional dimensions: (i) human behavior; (ii) infrastructure; and, (iii) vehicles. Wegman [4] points out that the behavior in speeding, alcohol, safety inside the vehicle (seat belts) can be faced through legislation, enforcement, and campaigns. Regarding infrastructures, the planning and the design help to prevent unsafely roads. Last, but not least, active and passive vehicle safety, inspections and better crash worthiness contribute to an effective road policy.

In this study, we want to mention some shortcomings of São Tomé and Principe National Police Force (NPSTP) approach to prevent road accidents. Hence, it bears on the analysis set out in the present study. Our discussion aims to clarify and also to qualify a form of intervention for the police in this subject matter.

The article follows a traditional structure with an introduction, reflecting on the why, how, and to whom, clarifying the problem and the main question in the study. The second part reveals the state of art of the subject, mentioning some relevant references to the present approach. As empirical research, the methodology forms the third part of the paper. It involves the definition of the method and the demonstration of the appropriateness and robustness concerning the objectives. In the method, data collection and interviews, defining a corpus. Afterward, the main results and facts with wider significance are part of the findings in the fourth part. Finally, a conclusion is presented.

2 Road Fatalities

Wegman [4] fundaments the uncertainty of road safety is due to regions characteristics, development, culture, and vehicles age. This vagueness converges to the growth of road death in 66% over the next two decades, sustained by the increase of 92% in low-income countries [5].

This phenomenon is due to a more pro-active approach to the problem in developed countries, that implemented measures in order to improve the safety driving and to overcome the third-person effect, age, and gender aspects. The implementation of regulatory laws, police's intervention, the appliance of technology and road managing, the system and the people from every stakeholder is recurrent and the underlying component.

In fact, driving is a moment of rebellion for some that implement a behavior of driving unsafely [6]. The culture embed in drivers has its tole [7], and that's why awareness is necessary to sustain the undesirable conducts. It is a proven fact, that females respond more favorably, for instance, to campaigns [1].

Due to the different dimensions (road, vehicle, men, environment) involved [6–8], the management, the policy-making and political decisions have a substantial impact on regulatory system, campaigns and control [9].

As a driven-problem for public health, the intervention by training, education, awareness, control, and surveillance are adequate to improve the reduction of numbers in view of a more productive society and safer.

2.1 National Police of São Tomé and Principe and Road Fatalities

Similar to other countries, the quality of driving is asymmetrical. This might be due to unethical behavior, not knowing the rules, unfamiliarity of cultural context, and the risk-lover driving. The context is different from the most known, characterized by the non-respect face to the regulatory traffic rules or the respect for pedestrian.

The driving culture of São Tome and Principe reflects the way of thinking, feel, act and react in the community. In this sense, the road system is produced according to the culture in which the individual is socialized and integrated.

Social and cultural perspective is tremendously important because, for instance, taxi drivers and bikers are the main transgressors in the country. Adding the high cost of the driving license, and the road system, the lack of appropriate management, results in a penurious road safety system. These are the results of incqualitics on income involving the following elements: (i) Gini Coefficient – 50,8; (ii) HDI – 0,558; (iii) average life expectancy of 66,3 years; (iv) average of education years – 4,7; (v) GDP_{pc} $2.713 (PPC, 2011); (vi) inequality human coefficient (2013) – 30,4; and (vii) HDI 142 position. This makes STP one of the poorest countries in the world [10].

Alongside the religion and the deeply-rooted belief overcomes the rationality of fighting the road fatalities or increasing by behavior the road safety. The road fatalities are a non-problem because they believe that the cause §is: (i) "*at that time and hour*", (ii) "*God commands life and death*", and (iii) "*it's written in the stars*". The success of road safety is not as much the action of the stakeholders but the intervention of God ("*it is faith*").

Using the vehicle is a necessity even in the poorest countries. For that reason, the road code is a necessity and has been developing in the country over the years. According to Special Unit of Traffic in STP (SUT), in the last ten years, its observed that road fatalities have increased. In this time frame, 5.948 accidents were registered that represented 12.115 victims. From those, 268 were death causalities, 1.808 seriously injured, 3.372 were slightly injured and the 6.667 material losses not concerning the vehicle (SUT, 2015).

The fluctuation is stronger in the outcome slightly injured. Notwithstanding, in 2009 the highest levels are recorded and in 2013 the lowest ones (see Fig. 1).

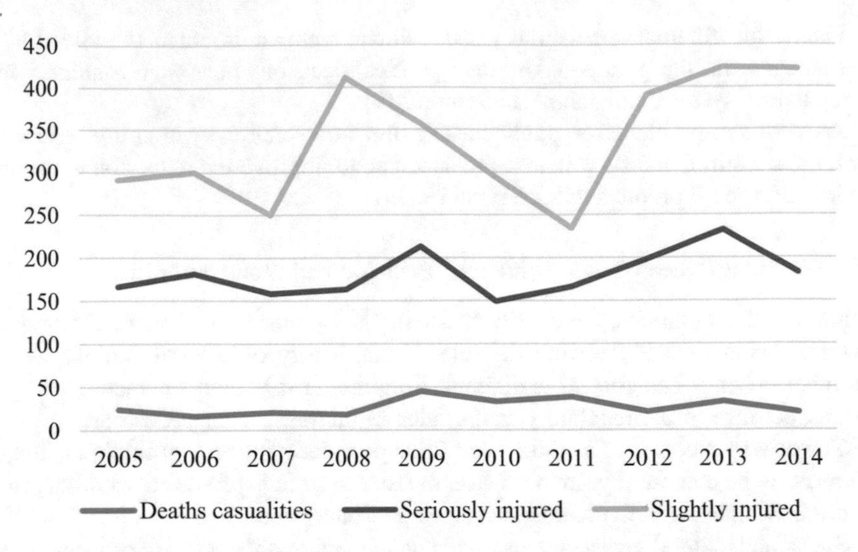

Fig. 1. Human outcomes of road fatalities/ *Source* (SUT, 2015)

The data might be biased because in road code (article 88º) there is no mention of the compulsory requirement to advert the police when an accident with material losses occur. However, its mandatory that police recollect all information about traffic accidents. It's a dubious plight in which code traffic and police legislation aren't' in compliance ("don't see eye to eye").

The taxonomy of the accidents in STP goes from hits, collision and isolated cases. The main issue in STP vis-à-vis traffic road analyses is collisions (see Fig. 2).

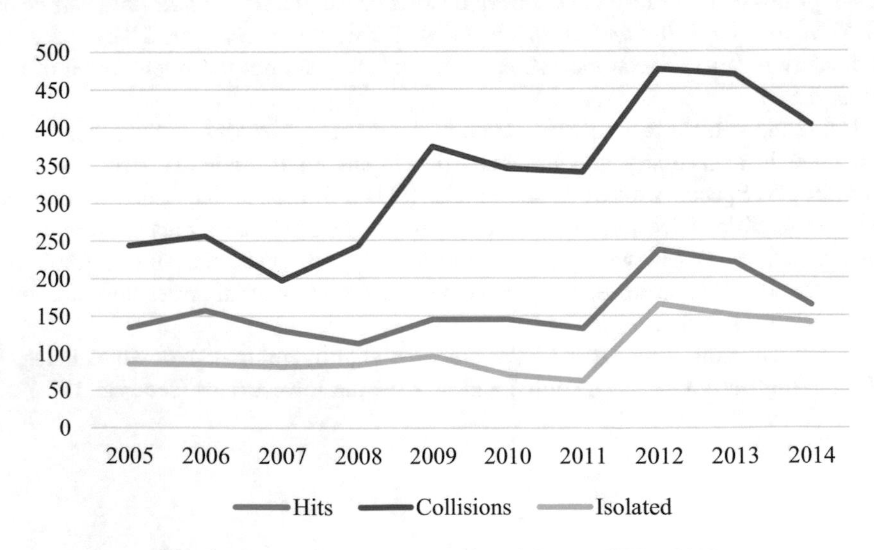

Fig. 2. Taxonomy of road accidents/ *Source* (SUT, 2015)

At a district level, the ones with higher concentrations of accidents are Água Grande and Mé-Zóchi. In the last 10 years, Água Grande, the most populous district, produced 3.437 cases. The other districts, such as Cantagalo, Lobata, Lembá, Caué, Pagué, present residual numbers (SUT, 2015).

The cause of accidents determined by the SUT (2015) are: (i) lack of precaution; (ii) speeding; (iii) distance between vehicles; (iv) traffics overtaking; (v) priority in passing; (vi) absentmindedness of the pedestrian, and (vii) others less common. From the lesser common ones, it was determined the abusive parking in sidewalks, and the lack of zebra cross. These rudiments contribute to committing the crime of manslaughter, the traditional hit-and-run and circulation of pedestrians in traffic lane putting at risk their own physical integrity and also of the drivers.

The role of women in STP society is confined to the home and the care of children and family. As so, the numbers for the road fatalities present a residual contribution from the female sex. For instance, just 2 accidents caused by women were registered in 2012 (SUT, 2015). Although the culture plays a crucial role in the numbers, Fontaine and Saint-Saens [11] mentioned that the use of vehicles for women is made in short journeys, and the mean speed is lower than the men. Also, active individuals drive more frequently at higher speed. The influence is also tested. The findings of the author's state that superior income means longer itineraries.

Might be noticed that at the present moment the fatalities of seriously injured occurred in the hospitals are not considered in the analysis of the data made by SUT and his Commanding officer.

The STP context is set allowing us to continue the path of the present study in which its outcome is of value for the strategic and operational intervention of the police.

3 Methodology

After the presentation of the state of art, the emphasis is on the method and on results. Method and results are the foremost apparatuses for improvement of the operational and strategic policies.

The new problem concerning road accidents is for all of us: prototypes and productivity depletion, loss of lives, same positioning in advanced or developing nations, and segmented interests among them.

3.1 Participants and Data Collection

Eligible participants were experiential experts were selected within the NP-STP, and were willing to participate.

Experiential experts were selected according to the category, association with the problem, and experience on the unit or in police management. The final sample included 9 participants, taking into account the pertinence, validity, and reliability of the information.

The 9 experts were officers of National Police of São Tomé e Príncipe: (i) the previous Commander and Chief of NP; (ii) the District Commanders of Águia Grande,

Cantagal, and Lobata, (iii) the Commander of Traffic Special Unit (TSU), (iv) chiefs of TSU. These were the elements considered relevant for the investigation.

The participants were informed about the procedure and signed a consent form approved by the police ethics. Interviewing is a flexible technique allowing the change in the direction of discussion for capitalizing ideas. This is the reason why they were performed without time limit.

The interviews were undertaken between December and February.

3.2 Conceptual Model and Hypothesis

As noticed in the introduction a society as a whole, reflecting in its cultural behaviors or values shared by a great majority of its citizens, policies must mobilize its resources, intelligence, and energy to undertake to resolve the problems of road fatalities.

For these problems, something more seems required than a descriptive knowledge of the situation. The presentation of a structure adapted to the development of the prevention, mechanisms for articulating strategies and for mobilizing resources to overcome the war to road fatalities.

Hypothesis and investigation questions include the choice of variables and the objective description of the population in the study. These scientifically aspects define the course of the investigation. For perceiving a logical relation between propositions, we formulate hypothesis and investigation questions. In other words, it also can be considered propositions for which we research and try to find out responses.

A conceptual model is presented (see Fig. 3), considering the aim of the study, the literature review, the major factors that might help to mitigate the road fatalities problem and elaborating a prevention model.

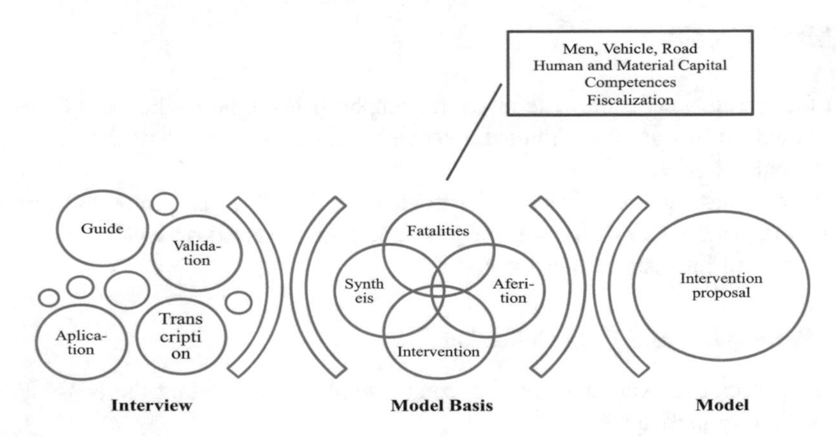

Fig. 3. Conceptual model

With the objective to define the model and its guidelines of intervention, some specific goals were determined: (i) understand the domains of NPSTP action in the prevention; (ii) contribute to a convergence between road fatalities and positive results.

In this view, the questioning is set according to the literature.

Christoforou et al. [2] suggested, that the main elements for road fatalities are the man, the vehicle, the road and the environment. This is also true for Chengye and Ranjitkar [12]. According to Vlassenroot et al. [6] people are motivated to drive fast and sometimes with insecurity. Other authors also consider gender, third-person effect and age as a complement of men in road fatalities [7, 13]. Thus, the following question was tested:

Q_1: Men, vehicle, road, and environment are the main elements of road fatalities in São Tomé and Princípe.

The approach to technology in road safety has many dimensions, from the road itself, vehicle and police. Hence, considering Gichaga [8], Lewis et al. [7], Vlassenroot et al. [6], and others, we can build the second question:

Q_2: Human and material resources in PN-UET are enough to prevent road fatalities.

As an influential issue, road safety is a part of public health agenda. As it is measured is real actions, a more efficacy in leadership, political priority, funding, expertise is needed to face the problem. On the other hand, to build up to the point of pro-active approach, a Safe System Approach [4], Gichaga [8] and Milenkovic and Glavić [14] also proposes that the design of the road is also a predictor to consider. In light of this, the third question is suggested.

Q_3: Efficacy in road fatalities combat is due to the competencies in management and domain of the road system stakeholders.

Elder, Shults, Sleet, Nichols, Thompson, Rajab, and Task Force on Community Preventive Services [15], consider that there were social benefits of mass media campaigns in road safety. They also pointed out that high visibility enforcement also has a major role in road fatalities' prevention. Therefore, we determine the fourth question:

Q_4: Increasing the campaigns and monitoring actions is a reduction factor of the phenomenon.

The presented questions is the mainframe of the road accident as a fact object to investigation and able to be improved or combated.

3.3 Method

There are four prerequisites to the method. The researcher must have a problem that causes some restlessness. After determining it, the researcher must make rational choices about the ways in which might address the problem. Subsequently, it must be able to define the appropriate method in view to the literature review (concepts and methodologies) applied in different cases. Finally, applying the method to be able to present the responses needed to bring the desired model for intervention.

The systematization of methods enables us to present an exploratory study with a qualitative approach. The qualitative method allows a higher quality of data recollection and the development of concepts as a result of a squabble between the data obtained from the interviews.

Focusing on the way interviewers give sense to their experiences is the core of qualitative investigation [16].

As a recollection method the interviews consists in oral conversation, individual or in groups, to gather description and describe the phenomena [17], enabling a "in-depth information" [18]. On the other hand, the appliance allows to undertake a comprehensive analysis of the content [19–21].

The interviewed presented the correct knowledge of the topic, allowing the analysis of the specific domain, and information about the research question. The individual discourse embodies a richer and profound inner view of the subject, and according to Hastie and Hay [22], it allows accessing on how the participants observe, feel and think about a subject.

The interview was composed by 12 questions, that presented a transversal approach to every domain of road fatalities, without a compliance of answering them in the order proposed.

For obtaining the results, the path involves different stages. First the elaboration of the interview guide. Second, the revision of it, for validation in terms of form, content, and writing by 4 persons (Researchers, Commander and Chief from NPSTP and Commander of Traffic unit in National Police of Portugal. Third, the process of interviewing the designated elements. Fourth, the transcription of the interviews. The analysis of it constitutes the fifth step, followed the consequent the presentation of the results. Results presentation followed the stages proposed by some authors, for instance, Erlingsson and Brysiewicz [23]. The starting point was the determination of the meaning unit. Afterward, the condensation of the meaning unit, followed by coding the units according to the research question. The categorization precedes the code and allows the establishment of a theme for each element of the model. All this process contains, in itself, an array of institutional and formal procedures. Using the taxonomy proposed by Resende [21], these five stages, can be grouped in one acronym ETCI (Interview, Transcription, Categorization and Interpretation).

For providing the results, and ensure the correct analysis of the data obtained, we employ the content analysis. For Krippendorf [24] the content analysis is the technique that allows to replicate the valid inferences of the matter to the context. It also helps to determine, in frequencies, the number of occurrences. This knowledge confines in establishing the categorization of the more noticeable elements. This introduces us into Bardin [25] concept which states that content analysis in the treatment of the results and its interpretation. At this moment a critical reflection about all the information is done. Charmaz [26, 27] claims that it's the moment the researcher has intimate familiarity with the content, which means an in-depth knowledge.

For Campenhoudt and Quivy [28] there is a strict association between content analysis and interviews. In fact, it should provide the maximum of information and reflection that will withstand the response to the intersubjectivity of the process.

In this sense, the erstwhile results analysis promotes the compilation and ordination of the data that permits responding the problem and present a solution. It's the final conclusion of the HOMER process (Hypothesize, Operationalize, Measure, Evaluate, Replicate, revise and report) [29].

4 Presentation and Discussion of Results

One of the oldest and most important uses of qualitative data is its application to solving problems in empirical and exploratory themes.

The output that measured the causes of road fatalities of STP determined that alcohol and road maintenance are the main issues. Other aspects mentioned are the speeding and disrespect of the code (see Fig. 4).

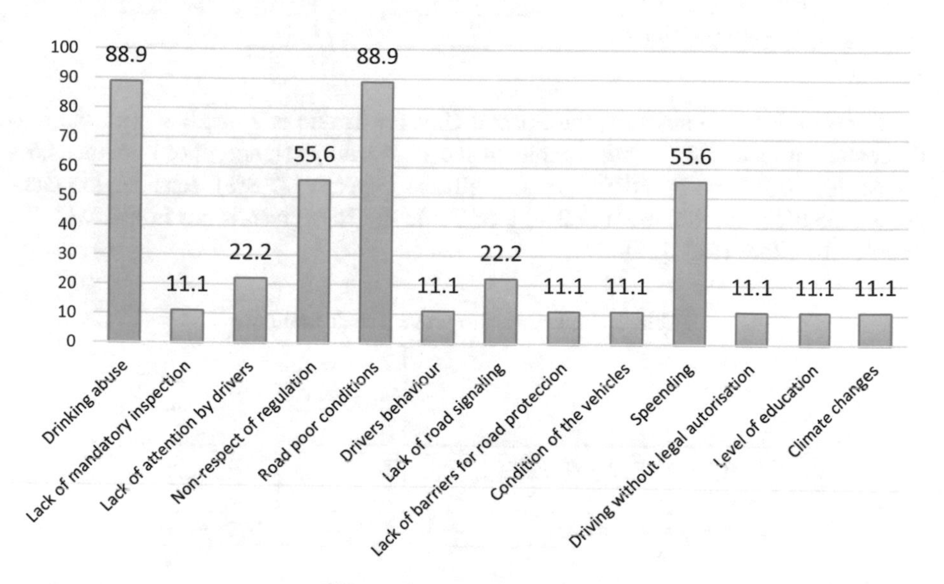

Fig. 4. Road fatalities causes

The inputs from the interviewers mentioned that the best predictors how to deal with road safety are: (i) increasing monitoring (44,4%); (ii) sensitization programs (33,3%); (iii) modern technology for police intervention (22,2%); (iv) compliance with the road regulations and (22,2%), and (v) police formation.

The results pronouncing road fatalities predictors were the vehicle driver (88,9% of responses), vehicle, road and weather conditions (22,2% of responses, each). Some other aspects were mentioned, viz, regulative and technical measures (11,1%). This compiles to the fact that that road user behavior has negative effects on road fatalities [8] (Table 1).

Considering the binomial - driver and road - the main aspects designated referring to the question "higher safety in roads embodies more safety for citizens" were that enhanced safety in roads enables more security for people and goods (55,6%), linked with a higher number of policies.

Upon regulatory code of road, the elements interviewed considered that is fundamental to form a civil, and responsible driver in the use of the elements because it allows the organization of the sector (55,6%). Though, some mentioned the lack of regulatory code is specific matters and the need for actualization (33,3%). A minority stated that it difficult police intervention (11,6%).

Table 1. Road fatalities predictors.

	Predictors	
	N	%
Ensure compliance with traffic legislation	2	22,2
Appliance of modern technology	2	22,2
Police education and training	1	11,1
Awareness progress	3	33,3
Increase surveillance	4	44,4

For 88,9% of the answers, the Special Unit for Traffic is embodied with the fundamental competencies to intervening in the road safety. Hence, these competences reveal the work done by NPSTP in surveillance increase (77,8%), sensitization campaigns (55,6%), and training (22,2%) (see Table 2). These results are in line with the exposed by Elder et al. [15].

Table 2. Techniques to prevent road fatalities.

	Road security management	
	N	%
Awareness-raising campaign	5	55,6
Training	7	77,8
Surveillance operations	7	77,8

Even though the work that was done by NPSTP, with positives results, the fact is that for the majority of the interviewers the human and material resources aren't the most appropriate to the prevention of road fatalities (66,7%). From this fact, it is undeniable the need to keep up with the society evolution, by adopting measures and up-date methods for the operational intervention. This follows the path proposed by Gichaga [8].

Going beyond what is done by police, some other reflections were determined has a complement and a more effective intervention in road security. Improving road security starts with sensitization actions (55,6%), followed by surveillance actions and equipment acquisitions (33,3%).

It's a fact that awareness-raising actions plus surveillance based on modern technology influence behavior [1, 30]. Being that the effectiveness increases when there is a combination of other factors, namely enforcement and education [8, 31], it's natural that the increasing of police presence (22,2%), a firm and clear police intervention (11,1%) were elements pointed out. Alongside, with training in control in road security (11,1%).

Reviewing the elements exposed by the interviewed, and the ones considered by the authors Elvik, Vaa, Hoye, Erke, and Sorensen [31], the enforcement (surveillance actions; police presence, and firm and clear police intervention) represent 66,6%, on the other hand, the sensitization programs 55,6%, and the education, 11,1% (see Table 3).

Table 3. NPSTP and road safety intervention

	Road safety improvements	
	N	%
Awareness programs	5	55,6
Surveillance actions	3	33,3
Police presence in roads	2	22,2
Clear and firm police intervention	1	11,1
Equipment's acquisition	3	33,3
Education in road safety	1	11,1

Questions about NPSTP were proposed considering road safety. Some questions concern the "black spots" of roads (meaning the points where more accidents occur), improving road behaviors and decreasing road fatalities.

Most of the experts, 88,9% considered that the identification of "black spots" is the core element for defining an intervention plan, followed by better police intercession (33,3%).

Taking into account previous questions it embodies the simplification of building road safety measures in terms of awareness (100%); surveillance actions (44,4%), control of vehicles (11,1%) and more police elements in the road (11,1%).

The improvement of behavior and decreasing road fatalities is a journey that prevails with NPST makes a better prevention plan (55,6%), works with objective driven for road safety good results (22,2%), and prevention of future predictors (22,2%) (see Fig. 5).

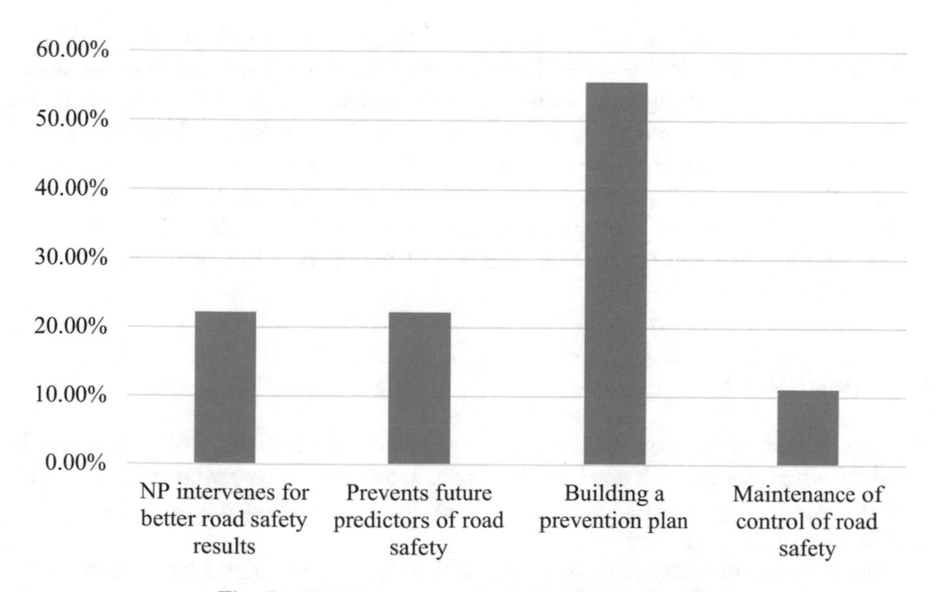

Fig. 5. NPSTP measures for improving road safety

In setting the appropriate inclusion from appraising the common elements between interviewers, plus categorization and adjusting margins a conception of the model PPSM was achieved with the fundamental aspects recollected from content analysis (see Fig. 6).

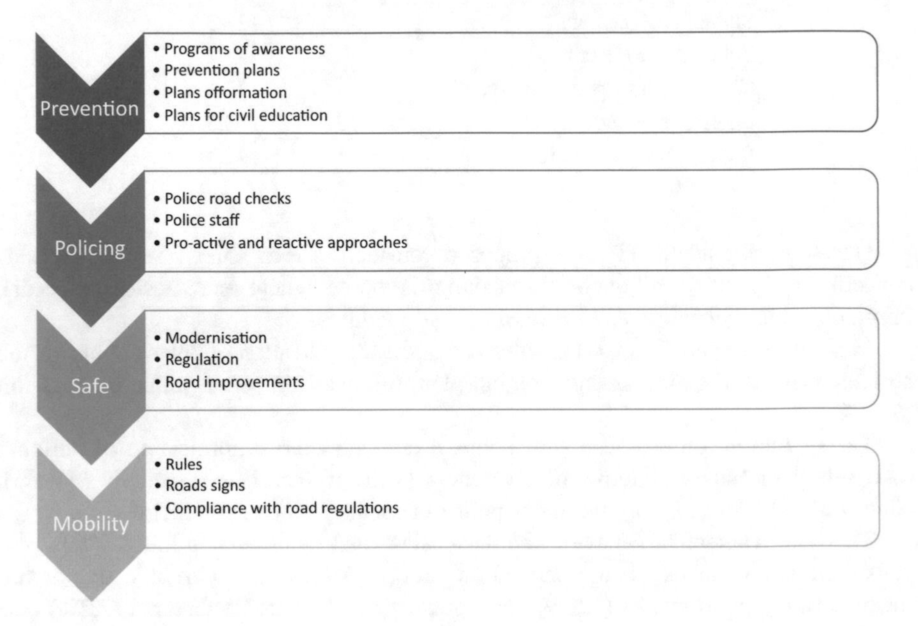

Fig. 6. PPSM model

In fact, an applicable critical appraisal of all the content analysis for this study was made, and therefore, a tool was developed by the authors to present the fore mentioned model for NPSTP. Assigning words for each categorization and determining the strength of its quality might be biased based on the authors' understanding, and experience on the subject. However, this may be, the experience of working in the organization, the use of qualitative strategies, and assessing the quality of the interviewed, helped to form the conceptual footing for intervention. It also helps to illustrate the relationship between the various elements of the intervention, our predictors, and the potential outcomes.

5 Conclusions

It is conceivable that upon reflection a majority of predictors are not a new lens to consider when designing a road safety model intervention. Nevertheless, its consideration sustains a pro-active intervention in establishing a good setting in road safety and ineluctably in road fatalities.

For a fact road safety and road fatalities' metaphor, we might embrace the economic cycle for this purpose. The expansion of road fatalities in STP occurred in 2011, having a peak in 2012. Since then until 2014 (last year of data available) there is a decrease (recession), and we can anticipate, that with the development of the country, with the increase of vehicles in the road a recovery is due. And a new cycle of road fatalities begins.

The findings support the literature review done, and the need for the use of a model to be intentional explicitly identified and clearly outlined on how to do the intervention, the means needed and at the NPSTP disposable to achieve the outcome of improving road safety, by reducing road fatalities. Furthermore, this study highlights the likeness between Western and African countries concerning road safety. It also helps to expand the knowledge and the understanding of the subject, and the best way to fit international experiences in the conceptual context of STP.

This pioneer empirical analyses of road fatalities led, in fact, to the development of a model for NPSTP.

These findings, even if pioneer they represent, are limited to NPSTP context. As a result, the listed elements in the model, may not be exhaustive or applied to another police force. Further, the authors of this paper identified the categorization according to the matrix provided by the answers, which can be somewhat subjective.

References

1. Hoekstra, T., Wegman, F.: Improving the effectiveness of road safety campaigns: current and new practices. ITSS Res. **34**(2), 80–86 (2011)
2. Christoforou, Z., Cohen, S., Karlaftis, M.G.: Integrating real-time data in road safety analysis. Procedia – Soc. Behav. Sci. **48**, 2454–2463 (2012)
3. Oliveira, P. Os factores potenciadores da sinistralidade rodoviária: Análise dos factores que estão na base da sinistralidade. http://www.aca.org/w/images/3/3d/Factores_potenciadores_sinistralidade_rodoviaria.pdf. Accessed 20 Dec 2017
4. Wegman, F.: The future of road safety: a worldwide perspective. ITSS Res. **40**(2), 66–71 (2017)
5. Kopits, E., Cropper, M.: Traffic fatalities and economic growth. Accid. Anal. Prev. **37**(1), 169–178 (2005)
6. Vlassenroot, S., Broekx, S., De Mol, J., Panis, L.I., Brijs, T., Wets, G.: Driving with intelligent speed adaptation: final results of the Belgian ISA-trial. Transp. Res. Part A Policy Pract. **41**(3), 267–279 (2007)
7. Lewis, I., Watson, B., Tay, R.: Examining the effectiveness of physical threats in road safety advertising: the role of the third-person effect, gender, and age. Transp. Res. Part F Traffic Psychol. Behav. **10**(1), 48–60 (2007)
8. Gichaga, F.J.: The impact of road improvements on road safety and related characteristics. IATSS Res. **40**(2), 72–75 (2017)
9. Wegman, F., Berg, H.-Y., Cameron, I., Thompson, C., Siegrist, S., Weijermars, W.: Evidence-based and data-driven road safety management. IATSS Res. **39**(1), 19–25 (2015)
10. UNDP. Human development report-sustaining human progress: reducing vulnerabilities and building resilience. http://hdr.undp.org/sites/default/files/hdr2014_pt_web.pdf. Accessed 20 June 2018
11. Fontaine, H., Saint-Saens, I.: L' exposition au risque des conducteurs de vehicules legers. Institut de Recherche des Transports Paris (1988)

12. Chengye, P., Ranjitkar, P.: Modelling freeway accidents using negative binomial regression. J. East. Asia Soc. Transp. Stud. **10**, 1946–1963 (2013)
13. Ferreira, S., Amorim, M., Couto, A.: Risk factors affecting injury severity determined by the MAIS score. Traffic Inj. Prev. **18**(5), 515–520 (2017)
14. Milenković, M., Glavić, D.: Analysis of relations between freeway geometry and traffic characteristics on traffic accidents. In: Hadžikadić, M., Avdaković, S. (eds.) Advanced Technologies, Systems, and Applications II, IAT 2017, pp. 539–548. Springer, Cham (2018)
15. Elder, R.W., Shults, R.A., Sleet, D.A., Nichols, J.L., Thompson, R.S., Rajab, W., Task force on community preventive services: Effectiveness of mass media campaigns for reducing drinking and driving and alcohol-involved crashes: a systematic review. Am. J. Prev. Med. **27**(1), 57–65 (2004)
16. Sparkes, A.C., Smith, B.: Qualitative Research Methods in Sport, Exercise and Health: From Process to Product. Routledge, London (2014)
17. Kvale, S.: Interviews: An Introduction to Qualitative Research Interviewing. Sage, Thousand Oaks, CA (1996)
18. Schostak, J.: Interviewing and Representation in Qualitative Research. McGraw-Hill, London (2006)
19. Culver, D., Gilbert, W., Trudel, P.: A decade of qualitative research in sport psychology journals: 1990–1999. Sport Psychol. **17**, 1–15 (2003)
20. Culver, D.M., Gilbert, W., Sparkes, A.: Qualitative research in sport psychology journals: the next decade 2000–2009 and beyond. Sport Psychol. **26**(2), 261–281 (2012)
21. Resende, R.: Técnicas de investigação qualitativa: ETCI. J. Sport. Pedagog. Res. **2**(1), 50–57 (2016)
22. Hastie, P., Hay, P.: Qualitative approaches. In: Armour, K., Macdonald, D. (eds.) Research Methods in Physical Education and Youth Sport, pp. 79–94. Routledge, London (2012)
23. Erlingsson, C., Brysiewicz, P.: A hands-on guide to doing content analysis. Afr. J. Emerg. Med. **7**(3), 93–99 (2017)
24. Krippendorff, K.: Content Analysis: An Introduction to Its Methodology, 3rd edn. Sage, Thousand Oaks (2013)
25. Bardin, L.: L'analyse de contenu. PUF, Paris (2013)
26. Charmaz, K.: Premises, principles, and practices in qualitative research: revisiting the foundations. Qual. Health Res. **14**(7), 976–993 (2004)
27. Charmaz, K.: A future for symbolic interactionism. In: Denzin, N.K., Giardina, M.D. (eds.) Studies in Symbolic Interaction, pp. 51–59. Emerald Group Publishing Limited, London (2008)
28. Campenhoudt, L.V., Quivy, R.: Manuel de recherche en science sociales, 4th edn. Dunod, Paris (2011)
29. Lakin, J.L., Giesler, R.B., Morris, K.A., Vosmik, J.R.: HOMER as an acronym for the scientific method. Teach. Psychol. **34**(2), 94–96 (2007)
30. WHO Homepage. http://www.who.int/gho/road_safety/en/. Accessed 18 June 2018
31. Elvik, R., Vaa, T., Hoye, A., Erke, A., Sorensen, M.: The Handbook of Road Safety Measures, 2nd edn. Elsevier, Amsterdam (2009)

Fostering Geogaming Pedagogical Integration: A Case Study Within a Portuguese School

Vânia Carlos(✉), António Moreira, and Cecília Guerra

Research Centre "Didactics and Technology in the Education of Trainers", Department of Education and Psychology, University of Aveiro, Aveiro, Portugal
{vania.carlos,moreira,cguerra}@ua.pt

Abstract. This paper presents results emerged from the ENAbLE project. The project was focused on the development of a Geogaming app - the OriGami tool – with the intention of contributing to students' spatial literacy. The app was developed by a multidisciplinary team (educational researchers, experts in technology, experts in Geography; and teachers and students from a Portuguese school. Following a case study, data was collected (2015 to 2017) in order to evaluate the usability of the OriGami tool and to identify the impact of the project (e.g., development of spatial literacy of students). Following a research-based-design approach, teachers and students were involved in the evaluation process of the app. Potentials, constrains and suggestions for improvement of the OriGami tool emerged through the monitoring activities (e.g., usability tests, virtual meetings and teacher-training workshops with students and teachers). Results showed that the coordinator' leadership, and the active involvement of project's team, allowed to sustain the outcomes of the project (e.g., the involvement of teachers and students of this school in upcoming projects focused on the development of spatial literacy).

Keywords: Geogaming · Monitoring · Spatial literacy Pedagogical integration

1 Introduction

Despite young people being eager to use emerging geospatial technologies, these are still insufficiently integrated into teaching. One of the reasons could be related with the Teachers' resistance to integrate technologies in educational contexts that could difficult the innovation in their professional practices [22]. Due to the students' enthusiasm to use of innovative technologies (and particularly geospatial technologies), their involvement as co-promoters of change in teaching and learning practices is essential. This was a starting point for the implementation of ENABLE, aiming the pedagogical integration of an interdisciplinary mobile Geogame in school context.

The "Educational Advancement of ICT-Based Spatial Literacy in Europe" - ENAbLE project (2014-2017), funded by the Erasmus+ program, results from a strategic partnership between three European Universities (Munster, Germany; Castellón, Spain; Aveiro, Portugal), with comprehensive expertise in ICT-based spatial literacy education

A. P. Costa et al. (Eds.): WCQR 2018, AISC 861, pp. 305–315, 2019.
https://doi.org/10.1007/978-3-030-01406-3_26

and complementary competences in education and technology development. Spatial literacy, learning about and improving interaction with one's surroundings, is a transdisciplinary competence overarching Science, Technology, Engineering and Mathematics (STEM), Social Sciences and Arts, where Geogamification can play a significant role.

The project developed a game-based orientation app for browsers and tablets (OriGami), which fosters spatial orientation and competencies such as map understanding/navigation [20]. It consists of a simple basemap and displayed route instructions, allowing users to add waypoints to the map and describe verbal instructions for reaching each waypoint. Feedback, hints and game elements allow students to orientate and find reference points in the map/real world. Smilies provide feedback on the current position of players, changing color/friendliness (smile/scowl), and thus giving intuitive hints if the player is moving in the right direction towards the next waypoint [20]. Thematic tasks (answering questions and georeferencing pictures) may be assigned to each waypoint, constituting the interdisciplinary potential of OriGami (see Fig. 1).

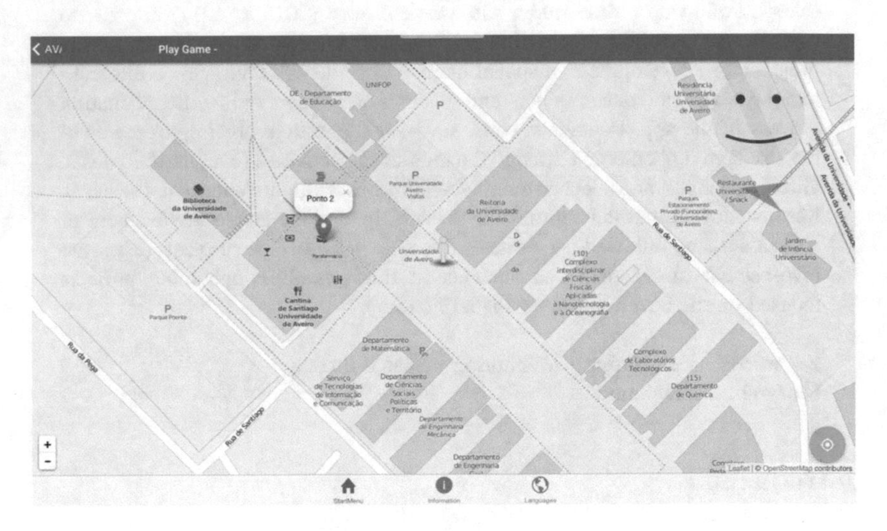

Fig. 1. OriGami.

Therefore, besides developing a teacher-training concept, the project focused on giving students voice and leadership in the curricular integration of the proposed Mobile Geogame (OriGami) by assisting their teachers. Three main actions were undertaken:

(i) an online Community of Learning and Practice developed in a virtual environment (CoL&P), engaging students and teachers, led by students;
(ii) the development of OriGami, based on user centred design (UCD) principles since students regularly conduct usability tests and develop didactic activities;
(iii) students develop and will conduct teacher-training workshops.

It was necessary to understand how to evaluate the impact of the ENAbLE project, as well as to recognize ways to enhance it. This study aims, therefore, to answer the following questions:

- How to use digital technologies to foster student-centred learning on Geogaming?
- How to monitor the impact of the ENAbLE project on the CoL&P?

Thus, this study also aims to develop a theoretical-practical framework with the purpose of monitoring the impact of ENAbLE. This framework is based on the literature review, particularly centred on project monitoring (i.e., project design) and research sustainability (i.e., community involvement), and results emerged from the first phase of ENAbLE.

Particularly in what regards to the educational potential of geotechnologies (e.g., GIS, remote sensing, GPS, virtual globes), Baker and colleagues [2] state that knowledge in this field remains scarce and inconsistent, requiring well-structured, systematic, replicable studies, using multidisciplinary approaches. In this sense, the authors propose a research agenda around four pillars: relations between geotechnologies and spatial thinking; geo-technologies learning; curriculum and student learning through geo-technologies; and geo-technologies professional development of teachers [2].

2 Theoretical Framework

2.1 Online Gamification

Spatial literacy is an inherently transdisciplinary competency transcending from Science, Technology, Engineering, and Mathematics (STEM) to social sciences and Arts that is essential in many professional and daily activities. Spatial literacy is the competency that allow any citizen see the value of Geography as a basis for organizing and discovering information, and comprehend such basic concepts as scale and spatial resolution [5]. However, despite young people being very eager to use emerging geospatial technologies (e.g. GPS, tagging technologies, and other sensors on smartphones), these are still insufficiently integrated into current teaching practices [4].

Meanwhile, as the ease of students' access to mobile devices increases in many educational settings, the debate around concepts like Bring Your Own Device - BYOD [1] and Mobile Learning [17], and their educational potential, gain acuity. Naismith et al. [17] stress the greater impact on learning that mobile technologies will have in the near future, arguing that learning will move outside the classroom into both real and virtual learning environments. Among other potentials mentioned by the authors, context-aware applications are stated as relevant to enable learners to capture and record their daily-life events, allowing recalling and sharing experiences for collaborative reflection [17].

Moreover, digital games are one of the emerging educational resources which allow students to develop social competences such as teamwork and, simultaneously, gain experience in the use of digital technologies [18]. Furthermore, gamifying elements of a given technology should allow it to become more user enjoyable and engaging [9, 11].

The importance of promoting spatial literacy in the curriculum lies not only in the individual development of the student, but the use of spatial information in social terms [7, 13]. The development of spatial orientation competences (e.g., the ability to describe locations and routes) are fundamental in the curricula within Geographic Education [4, 20]. Students' spatial literacy development can, thus, be supported by Geogaming [6]. For instance, the Foursquare® location service uses game elements to motivate and encourage its use. Given its engagement potential, gamification can be a way to encourage the use of geo-technologies in educational contexts, since Geogaming can be fun and support the development of spatial literacy of its users [15].

[3] emphasize several mobile Geogames based on the movement of the player in real environments, allowing a greater impact on the perception of the surrounding environment and the development of spatial abilities, in particular: Ingress®, Actionbound®, MapAttack®, GeoTicTacToe®, City Poker®, Feeding Yoshi® and Neocartographer®. Regarding the main requirements for the design of mobile educational Geogames for the development of spatial literacy, the same authors reported that:

(i) in the field of spatial literacy, competences and guidance of reading and interpreting maps are central to the curriculum;
(ii) in the field of gamification, an educational game has to motivate students while developing spatial competences, particularly fostering teamwork, competition, an objective, customization;
(iii) in the field of technology, mobile Geogames allow to develop tasks in real time and contexts, and the geotechnologies enable us to locate the student and keep the path followed and the time taken to accomplish a given task.

3 Methodology

The study focuses on the Portuguese context. Since October 2015, a group of sixteen students at the Secondary School of Gafanha da Nazaré, supported by two teachers (ICT and Geography) and one researcher (University of Aveiro), were given the challenge of leading the ENAbLE implementation in Portugal, aiming to: improve teachers' practices for teaching and learning innovation; promote students' social and spatial awareness through Mobile Geogames (OriGami); and develop students' research and leadership competences.

All members became co-participants during the development process of the mobile learning games, such the Origami app. The User Centred Design (UCD) approach for carrying out rigorous research consisting of a set of techniques, methods, procedures and processes that places the user at the centre of the development process [8, 19]. The goal of applying UCD during the development process of artefacts is to attempt to satisfy users via producing usable and understandable products that meet their needs and interests [10]. Prensky [18] states that using strategies to actively involve students during the development of digital games could foster their learning processes. UCD suggests active participation of users in the conception and/or evaluation of prototypes through the development of educational resources.

3.1 How to Use Digital Technologies to Foster Student-Centred Learning in Geogaming?

Microsoft® OneNote™ was elected to encourage virtual collaborative work among students and their teachers (by negotiation with students and under access convenience, being available for students and teachers' mobile devices), on the following activities (see Fig. 2):

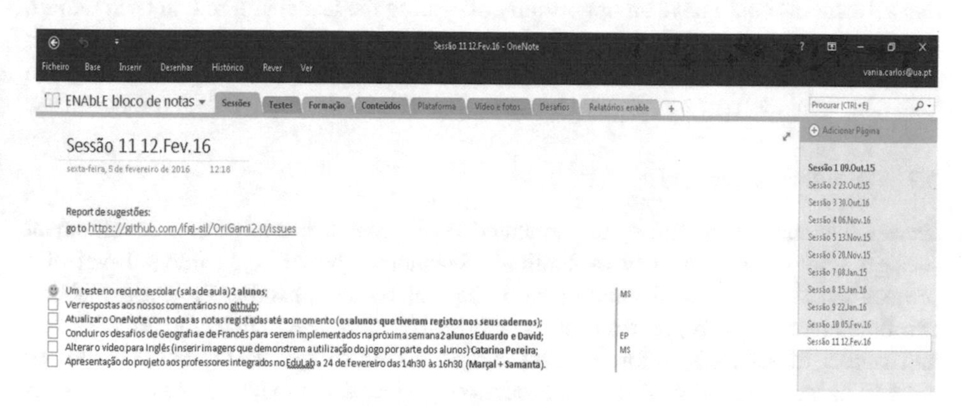

Fig. 2. OneNote® to support collaborative work within the CoL&P.

A student-centred CoL&P was promoted and sustained using this online tool. Figure 3 shows students co-organized in four groups:

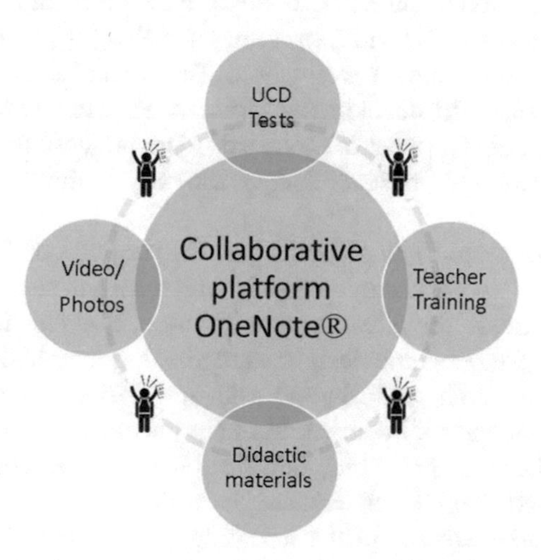

Fig. 3. How to use digital technologies to foster student-centred learning

Each group had a specific task, allowing students to play an active role on the development of the following activities, which correspond to one folder of OneNote®. Group 1 – UCD tests - was asked to weekly promote OriGami usability tests. This testing was centred around three criteria - usability, response time and satisfaction. Group 2 – didactic materials - developed educational activities. Group 3 – teacher-training - conducted teacher-training activities. Group 4 – Video/photos - created an OriGami promotional video, made available on the ENAbLE website. To guarantee that all students could have an opportunity of sharing the leadership role in each Group, a strategy was designed for promoting an inter-flowing dynamics between groups in each session, as shown in Fig. 3. Each student has an active 'voice' on choosing the task s/he wants to perform in each session, that was made possible using OneNote®.

3.2 Evaluation Process

Research evaluation should be implemented in different stages of project development (design, implementation, final evaluation). Evaluation should be an integral part of a project because it can contribute to the impact of research results. Project evaluation can be understood as a process seeking to identify relevance, efficiency, effectiveness and impact of research, before, during and after completion [2, 12, 14]. Monitoring aims to understand and describe the process of project development in order to deliver timely progress indicators, or lack of results to policymakers and key stakeholders. However, monitoring in itself is not sufficient to identify reasons why certain changes occur during project development. Project evaluation must be carried out whether the project is still underway or has already finished [2, 14]. Research impact involves the extent to which research evaluation influenced the development of individuals, the group and the organization/community to which they are bound.

Authors like Stern et al. [21] and Morra Imas and Rist [16] argue that the Theory of Change (ToC) may contribute to evaluation. ToC includes five project evaluation components: inputs (used to develop the project); activities (what is being done to develop the project); outputs (what is produced in the project); results/outcomes (what the project aims to achieve); impacts (changes the project aims at as a consequence of results).

Project evaluation should use different quantitative and qualitative methodologies in order to collect and analyse observed data in with regard to several aspects, including understanding the process for effective monitoring and evaluation; identifying gaps and/or constraints in the geogame design; strengthening the validity of the theoretical construction of the research, contributing with qualitative indicators provided by a complex and multi-dimensional context and help understand the meaning of the collected statistical indicators; providing information about the context in which each project was developed, helping understand the differences in results; and delivering regular and timely assessment feedback during project implementation, providing indications of these results. These aims help monitor and evaluate the impact of ENAbLE on the educational community.

3.3 How to Evaluate the Impact of the ENAbLE Project on the Educational Community?

The dynamics developed in the CoL&P using OneNote®, the weekly OriGami usability tests, the recordings transcriptions of videoconferencing between students and OriGami programmers and the focus group interviews with students and teachers about the potential pedagogical integration of OriGami will allow to evaluate the impact of ENAbLE on the educational community design (see Fig. 4).

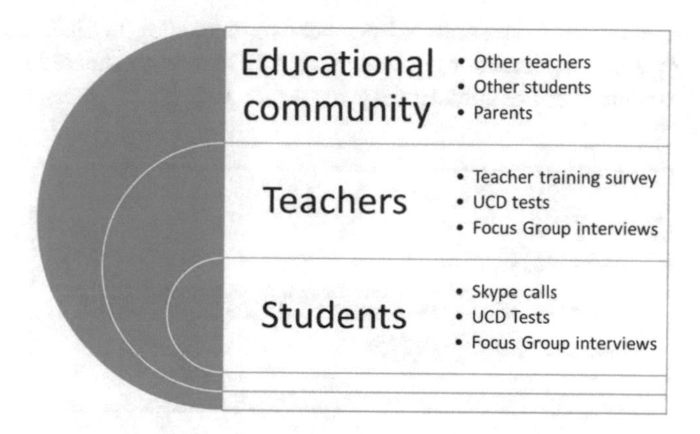

Fig. 4. How to evaluate the impact of ENAbLE on the educational community?

Students and teachers become, therefore, OriGami development evaluators and contributors, as well as of its didactic and pedagogical materials.

3.4 Data Collection

Students involved were supporting the development of Mobile Geogames, by testing the OriGami prototype and developing didactic activities (questions/answers for thematic tasks and design of OriGami games, with routes and destination points inside the school ground). Through a weekly UCD report (online questionnaire), students provide an in-depth analysis on OriGami and its didactic integration potential. Students' views on OriGami are being taken into account in its design, since the UCD reports and didactic activities are sent weekly to the German research team (software developers and educational researchers). Moreover, pre-scheduled videoconferencing takes place between students and developers, to facilitate the feedback process on the OriGami testing and communication of suggestions for improvement, at technological and pedagogical levels. Besides usability tests and didactic activities, students also started planning and implementing a Teacher-Training Workshop for the OriGami curricular integration.

Data was collected through: interaction on the CoL&P (students, teachers and researchers); OriGami game usability test reports (students, teachers and researchers); videoconferences between students and research team developers; and group interviews with students along the year. Further on, content analysis will be applied in order to: monitor and evaluate the impact of the ENAbLE project (i.e., how students engage in project tasks – OriGami usability test and didactic activities conception); validate the conceived framework; and sustain research outputs of the project (i.e., workshop replicated in the partners' contexts).

4 Results and Discussion

Presenting some results, from students' perspective, according to UCD usability tests reports (paper pencil questionnaire, online, etc.), OriGami is considered to be easy for 11 out of 16 students, and extremely easy for 4 (see Fig. 5).

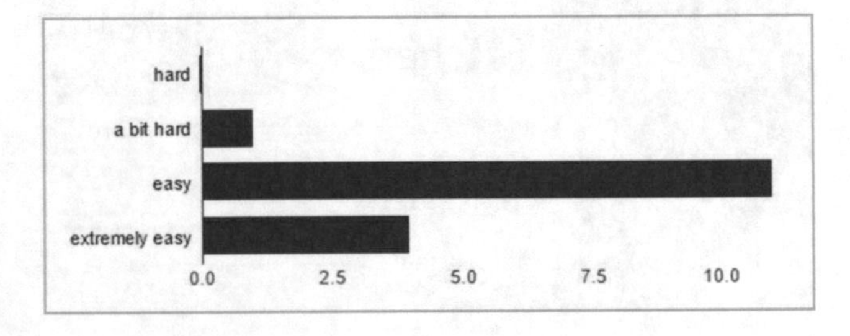

Fig. 5. Is OriGami user-friendly (easy to use)?

Regarding the second criteria of the UCD tests – response time, 7 students considered it good, contrasting with other 7 who considered it a bit long (see Fig. 6).

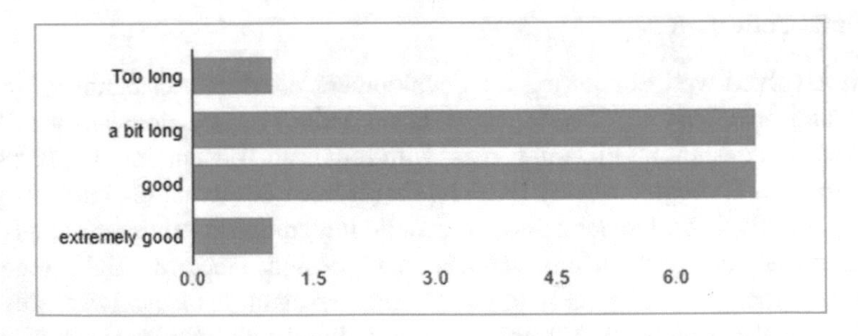

Fig. 6. Does OriGami offer adequate response time to perform tasks?

Finally, for the third UCD criteria – level of satisfaction, students presented two main opinions: 8 considered it fun, and 7 a bit fun (see Fig. 7).

To complement the data presented, some students' quotations reinforce that the level of satisfaction is conditioned by the stage of development of OriGami: "As for satisfaction, it's being fun and a creative way to learn"; "It's fun to use in our activities"; "It isn't fully developed, but it is going to be interesting"; "It's not a lot of fun

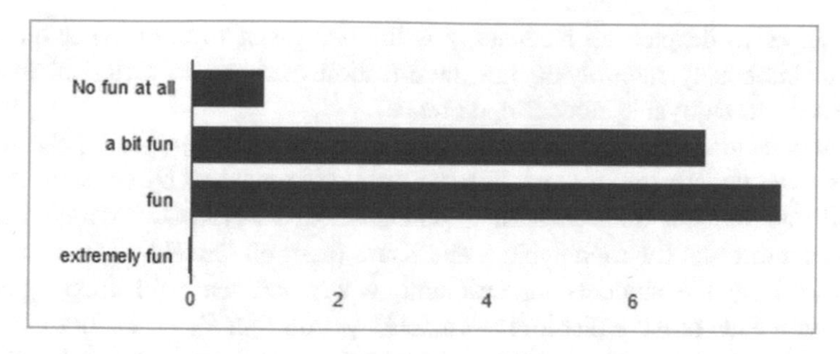

Fig. 7. What was your level of satisfaction while playing with OriGami?

because we can't play the game properly, since it doesn't track our location and the app doesn't let us save the game"; "Although there are some functionalities still in a bit of a problem, we are enjoying the project"; "Georeferencing isn't working right".

Aiming at sharing the students' evaluation of OriGami, a virtual meeting was conducted between students and researchers/developers, where students reported difficulties encountered when testing the game, suggested improvements for some functionalities and asked for information to prepare the next tasks.

Within the difficulties reported, students stressed that: "games are not always saved (only on android smartphones, not on windows tablets)"; "in some devices OriGami doesn't recognize location, even with internet connection and location definitions enabled"; "the game doesn't allow to change the basemap and we need the satellite basemap to test activities inside the school grounds"; "the basemap symbol shows up overlapping another symbol"; "the thematic tasks are not working yet and we need them to properly test the game, work further on the teacher-education concept, and conduct the teacher-training workshop". Students also suggested that OriGami should have the ability to define user profiles, with locations of registration and routes per user, record routes taken by students available to teachers, access real-time visualization of the routes other competing colleagues are following, and score points based on time taken carrying out tasks or route followed. Finally, students asked for additional information, such as a list of hardware/software specifications for OriGami to work properly and the next OriGami' developing stages.

5 Conclusions

A good leadership is the key for sustaining a CoL&P: creating proper conditions for all members to be involved in activities, leading momentarily if they so wish. From different levels of participation, depending on whether the elements belong to the 'active group' or to the remaining members, different roles emerge as promoting activities, events, etc.

As noted by Thomas Bartoschek et al. [4], although there are several Geogames available, these were not specifically developed for education. OriGami provides some developments arising from UCD process undertaken by students in the Portuguese ENAbLE project case. Developments based on students UCD reports and virtual meeting allow to deepen the feedback possibilities given to the teacher on learning, while simultaneously multiplying the gamification elements of OriGami in order to promote a more enjoyable student experience.

To involve teachers and the educational community in this study, the following activities were undertaken: a workshop for teachers organized by students, to present ENAbLE and explain the educational potential of OneNote®, inviting teachers to create thematic tasks for each subject and share them on OneNote®; a workshop for teachers lead by the students on OriGami, where teachers will create games and compete with each other; a dissemination activity with four Classes (5th to 9th grades) to make OriGami and its educational potential known; an international project meeting where results of the Portuguese case will be announced, with the presence of two teachers and two students.

After 2015/2016, to foster the pedagogical integration of the game, validate and test the pedagogical integration capabilities of OriGami, the pedagogical methodology developed was replicated in the German and Spanish contexts, based on the evaluation of OriGami usability, students' collaborative work methodology, teacher-training, co-constructed didactic materials (including students and teachers), where students, teachers and the educational community become evaluators/contributors of the OriGami pedagogical integration.

Acknowledgments. This work was financially supported by the European Commission within the Erasmus+ program (2014–2017, VGSPSNW140007143) and by National Funds through FCT – Fundação para a Ciência e a Tecnologia, I.P., under the project UID/CED/00194/2013.

References

1. Attewell, J.: BYOD bring your own device: a guide for school leaders. Brussels Eur. Sch. (EUN Partnersh. AISBL) (2015)
2. Baker, T.R., et al.: A research agenda for geospatial technologies and learning. J. Geog. **114**, 3 (2015)
3. Bartoschek, T., et al.: Mobile geogames for spatial literacy. In: 18th AGILE International Conference on Geographic Information Science, Lisbon (2015)
4. Bartoschek, T., et al.: OriGami: a mobile geogame for spatial literacy. In: Geogames and Geoplay, pp. 37–62. Springer (2018)
5. Bednarz, S.W., Kemp, K.: Understanding and nurturing spatial literacy. Procedia Soc. Behav. Sci. **21**, 18–23 (2011)
6. Carlos, V., Moreira, A.: Aprendizagem situada e jogos digitais significativos: uma proposta de referencial para a conceção de geojogos (projeto ENAbLE). Indagatio Didact., vol. 9, no. 4 (2017)
7. Council, N.R., Committee, G.S.: Learning to Think Spatially. National Academies Press (2005)
8. Dell'Era, C., Landoni, P.: Living lab: a methodology between user-centred design and participatory design. Creat. Innov. Manag. **23**(2), 137–154 (2014)
9. Deterding, S., et al.: From game design elements to gamefulness: defining gamification. In: Proceedings of the 15th International Academic MindTrek Conference: Envisioning Future Media Environments, pp. 9–15. ACM (2011)

10. Detweiler, M.: Managing UCD within agile projects. Interactions **14**(3), 40–42 (2007)
11. Fudenberg, D., Levine, D.K.: The Theory of Learning in Games. MIT Press, Cambridge (1998)
12. Gertler, P.J., et al.: Impact Evaluation in Practice. The World Bank (2016)
13. Goodchild, M.F., Janelle, D.G.: Toward critical spatial thinking in the social sciences and humanities (2010)
14. Leeuw, F.L., Vaessen, J.: Impact evaluations and development: NONIE guidance on impact evaluation. Network of networks on impact evaluation (2009)
15. Levandoski, J.J., et al.: Lars: a location-aware recommender system. In: 2012 IEEE 28th International Conference on Data Engineering (ICDE), pp. 450–461. IEEE (2012)
16. Morra Imas, L.G., Rist, R.: The Road to Results: Designing and Conducting Effective Development Evaluations. The World Bank (2009)
17. Naismith, L., et al.: Literature Review in Mobile Technologies and Learning (2004)
18. Prensky, M.: Digital Game-Based Learning. Paragon House, St. Paul (2007)
19. Santos, O.C., Boticario, J.G.: User-centred design and educational data mining support during the recommendations elicitation process in social online learning environments. Expert Syst. **32**(2), 293–311 (2015)
20. Schwering, A., et al.: Gamification for spatial literacy: the use of a desktop application to foster map-based competencies. In: 17th AGILE International Conference on Geographic Information Science-Workshop Games, Castellón (2014)
21. Stern, G.E., et al.: Broadening the range of designs and methods for impact evaluations
22. Tondeur, J., et al.: Understanding the relationship between teachers' pedagogical beliefs and technology use in education: a systematic review of qualitative evidence. Educ. Technol. Res. Dev. **65**(3), 555–575 (2017)

Author Index

A. P. Costa et al. (Eds.): WCQR 2018, AISC 861, pp. 317–318, 2019.
https://doi.org/10.1007/978-3-030-01406-3

Printed in the United States
By Bookmasters